Editorial Board Members' Collection Series: "Biodiversity and Ecosystem Function under Global Change"

Editorial Board Members' Collection Series: "Biodiversity and Ecosystem Function under Global Change"

Guest Editors

Shucun Sun
Panayiotis G. Dimitrakopoulos

MDPI

Basel • Beijing • Wuhan • Barcelona • Belgrade • Novi Sad • Cluj • Manchester

Guest Editors
Shucun Sun
School of Life Sciences
Nanjing University
Nanjing
China

Panayiotis G. Dimitrakopoulos
Department of Environment
University of the Aegean
Mytilene
Greece

Editorial Office
MDPI AG
Grosspeteranlage 5
4052 Basel, Switzerland

This is a reprint of the Special Issue, published open access by the journal *Biology* (ISSN 2079-7737), freely accessible at: https://www.mdpi.com/journal/biology/special_issues/V4288LIAF8.

For citation purposes, cite each article independently as indicated on the article page online and as indicated below:

Lastname, A.A.; Lastname, B.B. Article Title. *Journal Name* **Year**, *Volume Number*, Page Range.

ISBN 978-3-7258-5961-0 (Hbk)
ISBN 978-3-7258-5962-7 (PDF)
https://doi.org/10.3390/books978-3-7258-5962-7

Contents

About the Editors

Shucun Sun

Shucun Sun is a Professor of Ecology at the Department of Ecology, School of Life Sciences, Nanjing University, Jiangsu Province, China. His research is centered on trophic interaction and food web ecology. In particular, along with his colleagues and students, he has carried out field investigations, field manipulative experiments, and laboratory studies in alpine meadows in eastern Tibet Plateau for about 20 years. Under the support of National Natural Science Foundation of China, he has successfully addressed the brown-world role of predators in detrital food webs, the relationship between network architecture and species coexistence across plant–herbivorous insect–parasitoid food chains, and food web response to global warming. He has currently authored >100 papers in peer-reviewed journals, with H-index = 33.

Panayiotis G. Dimitrakopoulos

Panayiotis G. Dimitrakopoulos is Professor of Functional Ecology at the Department of Environment of the University of the Aegean, Greece. He obtained his Ph.D. degree from the University of the Aegean in 2001. His postdoctoral research was carried out in the University of Zurich, Switzerland, and was funded by the European Science Foundation (LINKECOL program). His research focuses on functional plant ecology, community ecology, biodiversity conservation, and conservation policy. He has served as President of the Hellenic Ecological Society (2012–2014), member of the "Natura 2000" National Committee (2010–2020), and Honorary Visiting Senior Fellow at Anglia Ruskin University, Cambridge (2019–2022). He serves as Section Editor-in-Chief of Ecology in *Biology* (Basel) and is a member of the editorial board of eight journals and referee for more than 70 journals. He has published more than 100 papers in peer-reviewed journals.

Preface

This Special Issue aims to advance our understanding of the mechanisms governing the responses of community biodiversity and ecosystem functioning to global change drivers. It compiles studies on the direct and indirect effects of climate change (warming, altered precipitation regimes), nitrogen deposition, and biological invasions on species interactions, biological communities, and ecosystem processes, such as plant productivity, nutrient cycling, and decomposition, across a wide range of ecosystems. This Special Issue is intended for researchers in the field of ecology, policymakers and practitioners interested in the mechanistic basis of biodiversity–ecosystem function relationships. By linking scientific knowledge with applied relevance, we hope this Special Issue will contribute to better predicting, managing, and conserving biodiversity and ecosystem processes and services on a rapidly changing planet.

Shucun Sun and Panayiotis G. Dimitrakopoulos
Guest Editors

biology

MDPI

Editorial

Biodiversity and Ecosystem Function Under Global Change: An Editorial for the Special Issue

Shucun Sun [1,*] and Panayiotis G. Dimitrakopoulos [2,*]

[1] Department of Biology, School of Life Sciences, Nanjing University, Nanjing 210093, China
[2] Biodiversity Conservation Laboratory, Department of Environment, University of the Aegean, 81132 Mytilene, Greece
* Correspondence: shcs@nju.edu.cn (S.S.); pdimi@aegean.gr (P.G.D.)

Climate change, land use changes, pollution (e.g., nitrogen deposition), and biological invasions are features of global environmental change, affecting biodiversity and ecosystem functioning worldwide [1]. In recent decades, many studies have been conducted to investigate how biodiversity loss affects ecosystem functioning under mean or variable environmental conditions [2–5]. However, the biological and ecological mechanisms through which global changes affect biodiversity and ecosystem functioning remain incompletely understood. This Special Issue, through its ten articles that cover a wide range of ecosystems—from alpine meadows and temperate steppes to semi-arid grasslands and polar marine environments— aims to address this gap by offering new knowledge about the mechanisms linking global change drivers to ecosystem structure and function.

Several research studies focus on understanding climate and nutrient interactions on carbon sequestration plant productivity, soil processes, and ecosystem dynamics in terrestrial ecosystems [6,7]. Bebaba et al. [8] showed that species richness and functional diversity of plants decreased with increasing nitrogen deposition rate and that nutrient enrichment affects the relationship between plant functional diversity and productivity but not that with taxonomic or phylogenetic diversity. Lou et al. [9] revealed that soil nitrogen supply levels affect variability in seed production and enhance reproductive success across years in a perennial herb in the eastern Tibetan plateau. Feng et al. [10] showed that long-term increases in daytime temperatures, more so than increases in nighttime temperatures, cause changes in the composition of soil microbial communities, reducing the ratio of fungi to bacteria, which enhances soil respiration in semi-arid grasslands.

Species interactions and ecological responses under changing environmental conditions is an important issue shaping the response of biological communities to global change [11]. An and Sun [12] found a clear impact of introduced Western honeybees on native bee species, especially rare species that present greater niche overlaps defined based on their food sources, underscoring that the development of beekeeping practices in natural ecosystems must consider the diversity and rarity of wild pollinators and the availability of food resources to them. Sun et al. [13] noted the role of rainfall timing but not rainfall intensity in the structure of dung beetle communities and the rate of dung decomposition in an alpine meadow in Tibet, explaining how changing precipitation patterns shape ecosystem function. Futai and Ishiguro [14] demonstrate that sexually mature *Monochamus* beetles are key drivers of pine wilt disease spread by transmitting nematodes to nearby healthy trees, leading to the formation of asymptomatic carriers that sustain and expand infections in pine forests.

On broader biogeographical scales, Abedin et al. [15] modeled the future loss and fragmentation of the habitat of the vulnerable Red Goral (*Naemorhedus baileyi*) due to cli-

 https://doi.org/10.3390/biology14111503

mate change in the temperate mountains of South Asia, highlighting the need for effective conservation strategies. Focusing on Antarctic marine ectotherms, Morley et al. [16] investigated their thermal limits and physiological plasticity, providing critical information on the mechanisms of species-specific thermal tolerance in polar ecosystems. Bei et al. [17] present an example of adaptation of *Elaeagnus angustifolia* to arid environments, which has specialized umbrella-shaped trichomes on its leaves that allow it to collect water from the atmosphere while reflecting solar radiation, reducing water loss.

Finally, Pang et al. [18] present a bibliometric analysis of global research efforts (2002–2023) on carbon sources and sinks in farmland ecosystems, identifying emerging trends and research gaps in carbon management under global change.

Taken together, the studies presented in this Special Issue highlight the complex interconnections between biodiversity, ecosystem functioning, and environmental variability. Further research is needed to clarify how global drivers alter species interactions and ecosystem processes to guide effective conservation and policy responses to the alarming environmental changes influencing species and ecosystems worldwide.

Conflicts of Interest: The authors declare no conflicts of interest.

References

1. Keck, F.; Peller, T.; Alther, R.; Barouillet, C.; Blackman, R.; Capo, E.; Chonova, T.; Couton, M.; Fehlinger, L.; Kirschner, D.; et al. The Global Human Impact on Biodiversity. *Nature* **2025**, *641*, 395–400. [CrossRef] [PubMed]
2. Balvanera, P.; Pfisterer, A.B.; Buchmann, N.; He, J.S.; Nakashizuka, T.; Raffaelli, D.; Schmid, B. Quantifying the Evidence for Biodiversity Effects on Ecosystem Functioning and Services. *Ecol. Lett.* **2006**, *9*, 1146–1156. [CrossRef] [PubMed]
3. Tilman, D.; Isbell, F.; Cowles, J.M. Biodiversity and Ecosystem Functioning. *Annu. Rev. Ecol. Evol. Syst.* **2014**, *45*, 471–493. [CrossRef]
4. van der Plas, F. Biodiversity and Ecosystem Functioning in Naturally Assembled Communities. *Biol. Rev.* **2019**, *94*, 1220–1245. [CrossRef] [PubMed]
5. Puppe, D.; Dimitrakopoulos, P.G.; Lu, B. Biodiversity and Ecosystem Functioning in Naturally and Experimentally Assembled Communities. *Biology* **2023**, *12*, 835. [CrossRef] [PubMed]
6. Ostertag, R.; DiManno, N.M. Detecting Terrestrial Nutrient Limitation: A Global Meta-Analysis of Foliar Nutrient Concentrations after Fertilization. *Front. Earth Sci.* **2016**, *4*, 23. [CrossRef]
7. Stocker, B.D.; Dong, N.; Perkowski, E.A.; Schneider, P.D.; Xu, H.; de Boer, H.J.; Rebel, K.T.; Smith, N.G.; Van Sundert, K.; Wang, H.; et al. Empirical Evidence and Theoretical Understanding of Ecosystem Carbon and Nitrogen Cycle Interactions. *New Phytol.* **2025**, *245*, 49–68. [CrossRef] [PubMed]
8. Debaba, G.H.; Li, K.; Wang, X.; Wang, Y.; Bai, W.; Li, G. Effect of Nitrogen Application Rate on the Relationships between Multidimensional Plant Diversity and Ecosystem Production in a Temperate Steppe. *Biology* **2024**, *13*, 554. [CrossRef] [PubMed]
9. Lou, Y.; Wang, R.; Che, P.; Zhao, C.; Chen, Y.; Yang, Y.; Mu, J. Nitrogen Addition Affects Interannual Variation in Seed Production in a Tibetan Perennial Herb. *Biology* **2023**, *12*, 1132. [CrossRef] [PubMed]
10. Feng, J.; Ru, J.; Song, J.; Qiu, X.; Wan, S. Long-Term Daytime Warming Rather Than Nighttime Warming Alters Soil Microbial Composition in a Semi-Arid Grassland. *Biology* **2023**, *12*, 699. [CrossRef] [PubMed]
11. Åkesson, A.; Curtsdotter, A.; Eklöf, A.; Ebenman, B.; Norberg, J.; Barabás, G. The Importance of Species Interactions in Eco-evolutionary Community Dynamics under Climate Change. *Nat. Commun.* **2021**, *12*, 4759. [CrossRef] [PubMed]
12. An, R.; Sun, S. Introduced Western Honeybees Dramatically Reduce the Abundance of Wild Bees in Alpine Meadows, Eastern Tibet Plateau. *Biology* **2025**, *14*, 1186. [CrossRef]
13. Sun, W.; Tang, W.; Wu, Y.; He, S.; Wu, X. The Influences of Rainfall Intensity and Timing on the Assemblage of Dung Beetles and the Rate of Dung Removal in an Alpine Meadow. *Biology* **2023**, *12*, 1496. [CrossRef]
14. Futai, K.; Ishiguro, H. Hidden Threats: The Unnoticed Epidemic System of Pine Wilt Disease Driven by Sexually Mature *Monochamus* Beetles and Asymptomatic Trees. *Biology* **2025**, *14*, 485. [CrossRef] [PubMed]
15. Abedin, I.; Mukherjee, T.; Abedin, J.; Kim, H.-W.; Kundu, S. Habitat Loss in the IUCN Extent: Climate Change-Induced Threat on the Red Goral (*Naemorhedus baileyi*) in the Temperate Mountains of South Asia. *Biology* **2024**, *13*, 667. [CrossRef] [PubMed]
16. Morley, S.A.; Bates, A.E.; Clark, M.S.; Fitzcharles, E.; Smith, R.; Stainthorp, R.E.; Peck, L.S. Testing the Resilience, Physiological Plasticity and Mechanisms Underlying Upper Temperature Limits of Antarctic Marine Ectotherms. *Biology* **2024**, *13*, 224. [CrossRef] [PubMed]

17. Bei, Z.; Zhang, X.; Tian, X. The Mechanism by Which Umbrella-Shaped Ratchet Trichomes on the *Elaeagnus angustifolia* Leaf Surface Collect Water and Reflect Light. *Biology* **2023**, *12*, 1024. [CrossRef] [PubMed]
18. Pang, Y.; Zhang, M.; Zhong, H.; Cevin, T.; Sun, C.; Zhang, S.; Li, X.; Dai, J.; Liu, C.; Zhang, C. Current Progress and Future Trends in Carbon Sources and Sinks in Farmland Ecosystems: A Bibliometric Analysis (2002–2023). *Biology* **2025**, *14*, 365. [CrossRef] [PubMed]

biology

MDPI

Article

Effect of Nitrogen Application Rate on the Relationships between Multidimensional Plant Diversity and Ecosystem Production in a Temperate Steppe

Gossaye Hailu Debaba [1], Kunyu Li [1], Xiaowei Wang [1], Yanan Wang [1], Wenming Bai [2] and Guoyong Li [1,*]

1 International Joint Research Laboratory for Global Change Ecology, School of Life Sciences, Henan University, Kaifeng 475004, China; gossaye@henu.edu.cn (G.H.D.); 104753170857@vip.henu.edu.cn (K.L.); wangxw0331@henu.edu.cn (X.W.); wangyanan@henu.edu.cn (Y.W.)

2 State Key Laboratory of Vegetation and Environmental Change, Institute of Botany, Chinese Academy of Sciences, Beijing 100093, China; bwming@ibcas.ac.cn

* Correspondence: ligy535@henu.edu.cn

Simple Summary: The continuous rise in anthropogenic nitrogen input, as one of the global change drivers, could have significant effect on terrestrial ecosystems, altering plant diversity and production. However, the effect of the nitrogen deposition rate on the multidimensional plant diversity–production relationship is poorly understood. Here, we investigated how varying rates of nitrogen deposition affect multidimensional plant diversity, biomass production, and its correlation in a temperate steppe of northern China. Biomass production increased initially and reached the maximum, then decreased with nitrogen deposition rates. Nitrogen deposition reduced species richness and plant functional diversity, while it enhanced plant functional trait identity such as plant height and leaf chlorophyll content. The phylogenetic structure of the plant community shifted from clustering to overdispersion along the nitrogen deposition gradient. However, nitrogen deposition did not change the relationships of species richness and phylogenetic structure with production but affected the functional diversity–production relationships. The results imply the robust species and phylogenetic diversity–production relationships and the varying functional diversity–production correlations under nitrogen deposition. The findings highlight the importance of a trait-based approach in studying the linkage between biodiversity and ecosystem function and facilitate the development of effective management strategies to maintain biodiversity and ecosystem function in the temperate steppe.

Abstract: Nitrogen (N) deposition, as one of the global change drivers, can alter terrestrial plant diversity and ecosystem function. However, the response of the plant diversity–ecosystem function relationship to N deposition remains unclear. On one hand, in the previous studies, taxonomic diversity (i.e., species richness, SR) was solely considered the common metric of plant diversity, compared to other diversity metrics such as phylogenetic and functional diversity. On the other hand, most previous studies simulating N deposition only included two levels of control versus N enrichment. How various N deposition rates affect multidimensional plant diversity–ecosystem function relationships is poorly understood. Here, a field manipulative experiment with a N addition gradient (0, 1, 2, 4, 8, 16, 32, and 64 g N m^{-2} yr^{-1}) was carried out to examine the effects of N addition rates on the relationships between plant diversity metrics (taxonomic, phylogenetic, and functional diversity) and ecosystem production in a temperate steppe. Production initially increased and reached the maximum value at the N addition rate of 47 g m^{-2} yr^{-1}, then decreased along the N-addition gradient in the steppe. SR, functional diversity calculated using plant height (FDis-Height) and leaf chlorophyll content (FDis-Chlorophyll), and phylogenetic diversity (net relatedness index, NRI) were reduced, whereas community-weighted means of plant height (CWM_{Height}) and leaf chlorophyll content ($CWM_{Chlorophyll}$) were enhanced by N addition. N addition did not affect the relationships of SR, NRI, and FDis-Height with production but significantly affected the strength of the correlation between FDis-Chlorophyll, CWM_{Height}, and $CWM_{Chlorophyll}$ with biomass production across the eight levels of N addition. The findings indicate the robust relationships of taxonomic and phylogenetic diversity and production and the varying correlations between functional diversity and production under increased N deposition in the temperate steppe, highlighting the importance of a trait-based approach in studying the plant diversity–ecosystem function under global change scenarios.

Keywords: biodiversity and ecosystem functioning (BEF); species richness; phylogenetic diversity; functional diversity; nitrogen deposition; grassland ecosystem; global change

1. Introduction

The interrelationship between plant diversity and ecosystem productivity is one of the most debated subjects in ecological research [1]. With rapid environmental changes, species loss is recognized as a threat to ecosystem function because species diversity has been documented as a major determinant of ecosystem productivity, stability, invasibility, and nutrient dynamics [2,3]. Recently, there has been a growing body of literature evidence that shows how diversity controls productivity [3,4]. However, the effect of environmental changes such as nitrogen (N) deposition on the diversity–productivity relationship is poorly understood [5].

Nitrogen is a key element controlling plant growth, survival, and many other biological processes, especially in N limited ecosystems [6]. Reactive N deposition from intensive agricultural, industrial, and other anthropogenic sources has been frequently recognized as a threat to global terrestrial biodiversity [6–8]. Global decline in biodiversity in turn will have direct negative impact on ecosystem functioning [2]. For instance, chronic N deposition induces the loss of species [9,10], homogenization of ecosystems, and alters the terrestrial carbon cycle [11]. Therefore, the relationship between plant diversity and productivity may be altered by N deposition in the terrestrial ecosystem. Some studies have found that N enrichment changes the strength or directions of the relationships of species richness (SR) with productivity and its stability [12]. In contrast, other studies have demonstrated that nutrient addition had no effect on the pattern of the SR-productivity relationships in grasslands [4] or an old field [13]. Apart from the studied ecosystem context, the level of N enrichment may be a dominant driver of the abovementioned contrasting results. However, the impact of simulating N deposition on the SR-productivity relationships is examined by comparing one level of N deposition simulation treatment with a control (i.e., N addition vs. without N addition) [8,14,15]. Few studies have been performed to test the SR-productivity relationship with varying levels of N deposition. In addition, most of previous studies about N deposition's effect on the relationship between plant diversity and ecosystem productivity considered SR as the sole taxonomic diversity (TD) metric [5,10], neglecting how the underlying processes are mediated by different dimensions of plant diversity metrics such as phylogenetic (PD) and functional diversity (FD). Exploring phylogenetic community structure along the environmental gradient is vital for predicting ecosystem processes under global change [16]. Thus, studying the evolutionary relationships between plant species in natural communities under nutrient enrichment can reveal how global change drivers may impact different lineages and functional groups differently based on their shared evolutionary history and trait profile [16,17]. The trait-based approach is expected to explain species fitness, niche differences, and the community productivity response to global change [18]. Accordingly, integrating the phylogenetic and functional dimensions of biodiversity in the diversity–productivity relationship study can provide additional insights into biodiversity effects on ecosystem functioning rather than giving much attention to only one biodiversity dimension exclusively [17]. However, there are relatively few instances simultaneously assessing N deposition rates' impacts on the various facets of biodiversity, which limits our comprehensive understanding of the effects of N deposition on the plant diversity–productivity relationships. Therefore, it is essential to examine the effect of N deposition rates on a comprehensive suite of diversity metrics beyond TD, which will enhance our ability to predict and gain a more complete understanding of the consequences of N deposition in the face of global change.

Grasslands cover 40% of the land area in China [19]; however, nitrogen enrichment has negatively affected plant diversity and the ability of grasslands to absorb carbon [11]. The temperate steppe grassland of Inner Mongolia supports plant diversity, but it is highly sensitive to N deposition, overgrazing, and climate change [20]. A very limited number of studies have been undertaken on the impact of N deposition on plant diversity and productivity. Here, a manipulative field experiment with N addition rates was conducted in 2003 to explore how plant community structure and ecosystem function change along the N addition gradient in a temperate steppe of Inner Mongolia, Northern China. During the growing seasons of 2015 and 2016, plant community composition and biomass production were surveyed and plant functional traits such as plant maximum height and leaf chlorophyll content were measured in this study. TD, PD, and FD were calculated based on community investigation and functional traits measurement, and the mixed-effects model was employed to test N addition rates' effects on the three dimensions of plant diversity. An analysis of covariance (ANCOVA) was used to examine the impact of N addition on the relationship between multidimensional plant diversity and production. The objectives of this study were to investigate the responses of TD, PD, and FD to N addition rates and to explore whether N addition rates affect multidimensional plant diversity–production relationships in the temperate steppe.

2. Materials and Methods

2.1. Study Site

The study was conducted in a typical semiarid temperate steppe (116°17′ E, 42°02′ N) of Inner Mongolia in northern China. The study area is characterized by a temperate monsoon climate, having irregular rainfall and drastic changes in temperature. The mean annual temperature was 2.1 °C with the monthly mean temperature ranging from −17.5 °C in January to 18.9 °C in July. The amount of precipitation during the growing season (May–October) was 290.0 and 369.4 mm in 2015 and 2016, respectively. The soil at the study site is categorized as chestnut according to the Chinese soil classification system, and as Calsic-orthic Aridisol according to the US soil Taxonomy classification. The soil texture composition consisted of 62.75 ± 0.04% sand, 20.30 ± 0.01% silt, and 16.95 ± 0.01% clay [21].The soil available N content at a depth 0–10 cm was 0.19 mg g^{-1} and the pH value was 6.84 ± 0.07. The dominant grass species were *Stipa krilovii* and *Leymus chinensis* and the forb species were *Artemisia frigida* and *Potentilla acaulis* in the temperate steppe grassland [22].

2.2. Experimental Design

An extensive experimental research project, involving N addition and mowing treatments which simulated N deposition and grassland utilization, respectively, was launched to explore community structure and ecosystem function in 2003 in the temperate steppe of northern China. Eight blocks were established and each block consisted of eight plots which were randomly assigned to eight levels of N addition (N0, N1, N2, N4, N8, N16, N32, and N64 denote the N addition levels of 0, 1, 2, 4, 8, 16, 32, and 64 g N m^{-2} yr^{-1}). Each plot had dimensions of 15 m × 10 m with a 4 m buffer zone between any two adjacent plots. Eight rates of N addition in the form of urea ($CO(NH_2)_2$) were added manually in each plot during mid-July from 2003. Among the eight blocks, four of them were mowed, and the clipped plants were moved out of the blocks at the end of each growing season since the project initiated. As part of the project, this study was performed to examine the effect of N addition rates on the relationships between multidimensional plant diversity and biomass production during two mid-growing seasons (June, July, and August) in 2015 and 2016 in the four unmowed blocks (each block consisted of 8 N addition plots, for a total of 32 plots).

2.3. Community Survey and Plant Biomass Measurement

The community survey and biomass harvest were conducted during the middle of the local growing season (i.e., June, July, and August) in 2015 and 2016. A 50 cm × 50 cm quadrat was randomly placed in each plot for the vegetation survey, by placing the quadrats a minimum of 50 cm away from the edges of the plot to prevent any edge effects. All the vascular plants in each quadrat were identified to species level and recorded. The total coverage of individual plant species (%) in each sampling quadrat was estimated visually based on their occurrence. Based on cotyledon number and N fixation feature, all plant species were categorized into 3 different functional groups: forbs, grass, and legumes. Plant maximum height was calculated by averaging the height of three tallest individuals of each species in each quadrat, which was measured as height from ground level to the tip of the highest vegetative tissue using tapeline (cm). Leaf chlorophyll content was also measured by averaging the chlorophyll content of the same three individuals of each species using a Konica Minolta SPAD-502 Plus chlorophyll meter before harvesting plant biomass. The SPAD value typically ranges from 0 to 99, with higher values indicating a higher chlorophyll content. All aboveground plant tissues within the quadrat were clipped at the soil surface and sorted based on species identity before being put into different marked paper bags. Subsequently, belowground plant tissues and soil were sampled with a soil auger (20 cm in length, 5 cm in diameter). Two soil cores were collected and mixed for each quadrat. Plant roots and soil samples were separated from the mixed soil using a 2 mm sieve. Plant roots were cleaned with water and placed into marked paper bags. All plant samples were oven-dried at 70 °C for more than 48 h and weighed to determine AGB and BGB for each quadrat. AGB was estimated using the dry mass of all aboveground living plant tissues divided by the area of the quadrat, and BGB was evaluated with the dry mass of belowground plant tissues divided by the area of the two soil cores. The sum of AGB and BGB was used as a surrogate of ecosystem biomass production.

2.4. Plant Diversity Metrics

We have examined three dimensions of biodiversity including TD, PD, and FD. TD included T0, T1, and T2 calculated using species-based Hill's numbers (effective number of species) and represented species richness, the Shannon's taxonomic diversity index, and the inverse Simpson diversity index, respectively, using the *hillR* package in R [23,24]. SR represents the number of species per plot. Community PD including the net relatedness index (NRI) and nearest taxon index (NTI) was estimated and a phylogenetic tree was made using the *Picante* and *ape* packages in R [25,26]. The phylogenetic tree generated from the dated molecular phylogeny of land plants was constructed by Zanne et al. [27]. The phylogenetic tree was extracted by using the *V.PhyloMaker* package in R [28]. The phylogenetic tree counted 40 tips, 1 for each species in the vegetation matrix, and had 39 internal nodes (see Figure S2). The NRI had a standardized effect size of the mean phylogenetic distance (MPD) of all species in the community [29]. The significance of NRI for an individual quadrat was assessed by comparing the observed MPD to a null distribution of MPD measured on 999 null communities. Thus, NRI is calculated as

$$\mathrm{NRI} = -1 \times (\mathrm{MPD}_{\mathrm{randomized}} - \mathrm{mean}\,(\mathrm{MPD}_{\mathrm{null}}))/\mathrm{sd}(\mathrm{MPD}_{\mathrm{randomized}})$$

where $\mathrm{MPD}_{\mathrm{randomized}}$ and $\mathrm{sdMPD}_{\mathrm{randomized}}$ represent the mean and standard deviation of MPD estimated using 999 randomly sampled communities with given species diversity, respectively. Positive NRI values indicate that coexisting taxa are more related to each other than expected by chance (clustered dispersion) and negative NRI values indicate that coexisting taxa are more distantly related to each other than expected by chance (overdispersion) [29]. Similarly, NTI was calculated for each quadrat as alternate indicators of the plant community evolutionary relationships, by comparing the observed mean nearest taxon distance (MNTD) to a null distribution of MNTD measured on 999 null communities to see if they deviated significantly from random chance. NTI is calculated as

$$\mathrm{NTI} = -1 \times (\mathrm{MNTD}_{\mathrm{obs}} - \mathrm{mean}\,(\mathrm{MNTD}_{\mathrm{randomized}}))\mathrm{sd}(\mathrm{MNTD}_{\mathrm{randomized}})$$

Two key functional traits were chosen based on the plant ecology strategy scheme because they are known to be functionally important and closely related to ecosystem functions (i.e., plant height and leaf chlorophyll content). These traits are also known to influence plants' response to N addition [30]. The functional dispersion (FDis) of traits (the distribution of traits within a community) refers to the variation in functional trait values among species, which describes the mean distance in the trait space of individual species to the centroid of all species [31] calculated using both plant height and leaf chlorophyll content (i.e., FDis-Height and FDis-Chlorophyll content). In order to determine how plant functional traits responded to the experimental treatments at the community level, the community-weighted mean (CWM) value of each functional trait (functional composition of the community) was calculated using the plot level mean value of the trait weighted by the abundance of each species within the community. FD was computed using the *dbFD* function in *FD* package [32]. To avoid the analysis of highly correlated variables, we reduced the set of diversity metrics to SR, NRI, FDis-Height, FDis-Chlorophyll, CWM_{Height}, and $CWM_{Chlorophyll}$, which were strongly correlated to other diversity metrics (T1, T2, NTI, MPD, and MNTD) (Figure S2). Moreover, phylogenetic signals in functional traits (height and chlorophyll content) were tested following Pagel's λ statistics [33] using the *phylosig* function in the *phytools* package to assess whether traits were phylogenetically conserved (i.e., to assess whether more closely related species have more similar functional traits) because the evolutionary relatedness of species may have an influence on the values of their traits. Thus, phylogenetic signals could indicate a tendency for closely related species to display similar trait values as a consequence of their phylogenetic proximity.

2.5. Statistical Analysis

The Shapiro–Wilk normality test was employed to examine the normality of the data before statistical modeling. Biomass production was log transformed because the data showed strong positive skewness and to meet the assumptions of normality and heteroscedasticity. Correlations between diversity metrics and biomass production were assessed with Pearson's correlation coefficients before modeling (Figure S1; Table S2) to test the relationship between plant diversity and biomass production. We found strong correlations among the set of predictor variables (between SR and T1, T2, and PD) (Figure S1). Therefore, T1, T2, and PD were eliminated from further analysis. To explore the effect of N addition with time on the diversity metrics and biomass production, a linear mixed-effects model was used to assess the significance of the fixed effects using Type I analysis of variance (ANOVA) with N addition and time as fixed factors and plots set as the random factor. In the mixed-effects model, diversity metrics and biomass production were the response variables to N fertilizer over time and showed the interactive effect of N addition and time. The mixed-effect models were fitted using the *lmer* function in the *lme4* package [34]. Data were fitted to the models using the maximum likelihood to generate unbiased estimates of the model parameters. The homogeneity of variance among the means of response variables across treatment levels was visually examined using quantile–quantile plots of residuals for all models' normality to assess whether model assumptions were met. A post hoc test (Tukey multiple comparison of treatment means across measuring times at a significance level of $p < 0.05$) was performed using the *emmeans* function in the *emmeans* package, to test the effect of various rates of N addition on plant diversity indices and biomass production. Moreover, regression analysis was employed to test the effect of the N addition rate on plant biomass production.

To test the effect of N addition on the pattern of the relationship between diversity metrics and biomass production, an analysis of covariance (ANCOVA) with the generalized additive model (GAM), generalized linear model (GLM), and generalized linear mixed model (GLMM) was implemented using the gam function in the *mgcv* package [35] and the *stats* and *glmer* function in the *lme4* package in R, respectively. In the GAM, GLM, and GLMM, N addition was treated as categorical factor and diversity indices were treated as covariates, while plot was included as a random factor. Both linear and quadratic forms of the covariates were included in the models. Diversity metrics in the GAM model were represented as a penalized regression spline (smoother) to detect nonlinear relationships, while N addition and measuring time were categorical factors. Optimal model selection was performed based on the lowest Akaike Information Criterion value (AIC: the lower the AIC value, the better the model). The final model was selected by simplifying the full GLM, GAM, and GLMM models in a stepwise deletion procedure using ΔAIC > 3 as the criterion for term deletion [36], following the procedure of Crawley [37]. GLM, GAM, and GLMM models were fitted with a Gaussian error distribution as the response was a continuous variable. We then used locally weighted regression smoothing with the *lowess* (scatterplot smoothing) function following model fitting to examine multidimensional plant diversity–biomass production relationships. All statistical analyses were performed using R 4.3.1.

3. Results

3.1. Biomass Production along the N Addition Gradient

The range of production across all plots and measuring times was from 301.76 to 5985.30 g m^{-2} along the N addition gradient (Figure 1). The lowest production was observed in the N0 plot in June 2015, while the highest production was observed at the rate of 64 g N m^{-2} yr^{-1} in August 2016. When averaged for each treatment, production exhibited a nonlinear response pattern along the experimental N gradient (Figure 2). That is, biomass production initially increased and then decreased, with saturation thresholds appearing at approximately 47 g N m^{-2} yr^{-1}. Maximum production (1909.91 g m−^{2}) was calculated from the nonlinear regression at the N addition rate of 47 g N m^{-2} yr^{-1} in the temperate steppe (Figure 2). In this study, both the N addition and measuring time had a significant effect on production (both $p < 0.001$; Table 1). Compared with that in the N0 plots, production was increased by 74.9, 104.8, and 102.4% in the N16, N32, and N64 plots, respectively (all $p < 0.01$; Figure 1).

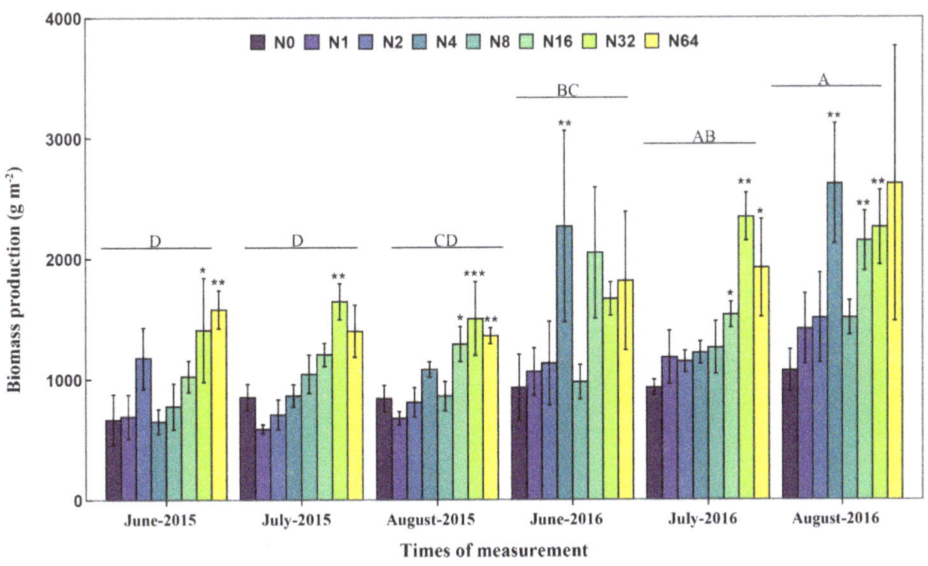

Figure 1. Effect of N addition rate on plant biomass production (mean ± SE) during the growing seasons of 2015 and 2016 in a temperate steppe of Inner Mongolia, northern China (n = 192; Error bars represent the standard error of the mean). N0, N1, N2, N4, N8, N16, N32, and N64 represent nitrogen fertilizer rates of 0, 1, 2, 4, 8, 16, 32, and 64 g N m^{-2} yr^{-1}, respectively. Asterisks indicate significant differences between the treatment and control groups * $p < 0.05$, ** $p < 0.01$, *** $p < 0.001$. Different letters indicate significant differences between measuring times ($p < 0.05$).

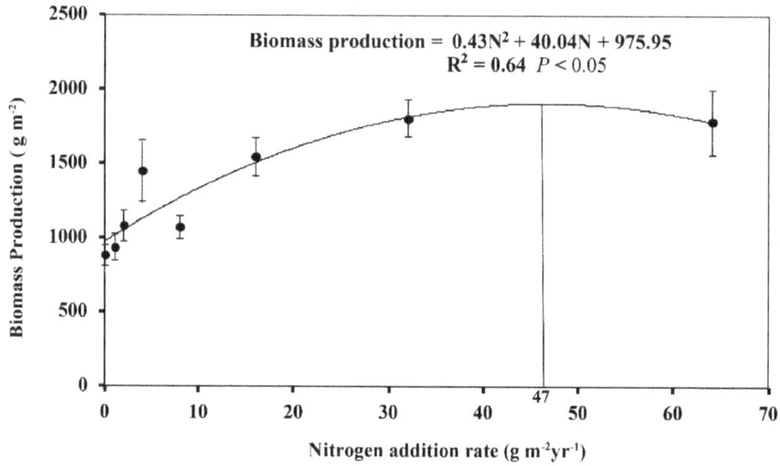

Figure 2. Response of biomass production to N addition rate in a temperate steppe of Inner Mongolia, northern China. Maximum biomass production (i.e., 1909.91 g m^{-2}) was calculated from the second-degree quadratic regression at the rate of 47 g N m^{-2} yr^{-1}. Each data point represents the mean (±SE) biomass production across all measuring times. The vertical line represents the point where maximum biomass production was calculated from second-degree quadratic regression. Error bars show the mean ± 95% confidence intervals.

Table 1. Summary (F-values) of mixed−effects model for the effects of nitrogen (N), time, and their interaction on plant diversity indices (species richness, SR; net relatedness index, NRI; functional diversity of plant height, FDis−Height; functional diversity of chlorophyll content, FDis−Chlorophyll; community−weighted mean of plant height, CWM_{Height}; and community−weighted mean of chlorophyll content, $CWM_{Chlorophyll}$) and biomass production in a temperate steppe of Inner Mongolia, northern China.

Source	df	SR	NRI	CWM_{Height}	$CWM_{Chlorophyll}$	FDis-Height	FDis-Chlorophyll	Production
N	7	32.79 ***	4.08 **	11.66 ***	15.38 ***	4.49 **	5.36 ***	8.66 ***
Time	5	1.45	3.04 *	44.23 ***	24.17 ***	4.18 **	7.70 ***	15.56 ***
N × Time	35	1.06	1.07	2.11 **	1.74 *	0.87	1.78 *	0.97

Note: Statistical significance is indicated as * $p < 0.05$, ** $p < 0.01$, *** $p < 0.001$, df, degree of freedom.

3.2. *Multiple Plant Diversity along the N Addition Gradient*

Species richness across all plots and measuring times varied from 1 to 17 along the N addition gradient in this study. The lowest SR occurred at the N64 plot in June 2015 and the highest SR was recorded at the N2 plot in August 2016. The significant proportional loss of SR relative to the control was 31.1, 43.3, 55.2, 64.1, and 73.7% under N4, N8, N16, N32, and N64, respectively (all $p < 0.01$; Figure 3a). NRI changed from positive (phylogenetic clustering) to negative (phylogenetic overdispersion) along the N addition gradient (Figure 3b). The relative change in NRI was −3.2 and −2.30 under N32 and N64 compared with that under N0, respectively (both $p < 0.05$ Figure 3b). The lowest FDis-Height and FDis-Chlorophyll (both 0.01) were observed in the N64 plots in August 2016 while the highest FDis-Height (1.95) and FDis-Chlorophyll (2.56) were observed in the N2 and N1 plots in July 2016, respectively (Figure 3c,d). FDis-Height was significantly decreased by 62.11 and 72.6% under N8 and N16, respectively (both $p < 0.05$; Figure 3c). FDis-Chlorophyll was reduced by 60.2 and 68.7% under N16 and N64, respectively (both $p < 0.05$; Figure 3d). The highest (63.30) and lowest (2.12) CWM_{Height} were observed under N64 in August 2016 and N0 in June 2015, respectively. The highest (55.13) and lowest (0.53) $CWM_{Chlorophyll}$ occurred in the N64 plot in June 2015 and N0 plot in July 2016, respectively.

CWM_{Height} and $CWM_{Chlorophyll}$ were significantly affected by N addition and measuring time as well as the interaction between N addition and measuring time (all $p < 0.05$; Table 1). Compared with N0, CWM_{Height} was enhanced by 75.20 and 137.04% under N32 and N64, respectively (both $p < 0.05$; Figure 3e). $CWM_{Chlorophyll}$ was stimulated by 28.1 and 111.6% under N16 and N64, respectively (both $p < 0.05$; Figure 3f). Phylogenetic signal Pagel's λ values were 0.63 and 0.41 for height and chlorophyll content, respectively, suggesting that closely related species did not show similar trait values and the phylogenetic niche related to these traits was not conserved during evolution (Figure S3). Furthermore, based on the plants' relative percent cover, *Artemisia_frigida*, *Potentilla acaulis*, and *Potentilla bifurca* were dominant forbs in the low N plots (i.e., in the control and 1 g N yr^{-1}), whereas *Leymus chinensis* and *Stipa krylovii* were dominant grass species in the 32 and 64 g N yr^{-1} plots (Figure S4; Table S1).

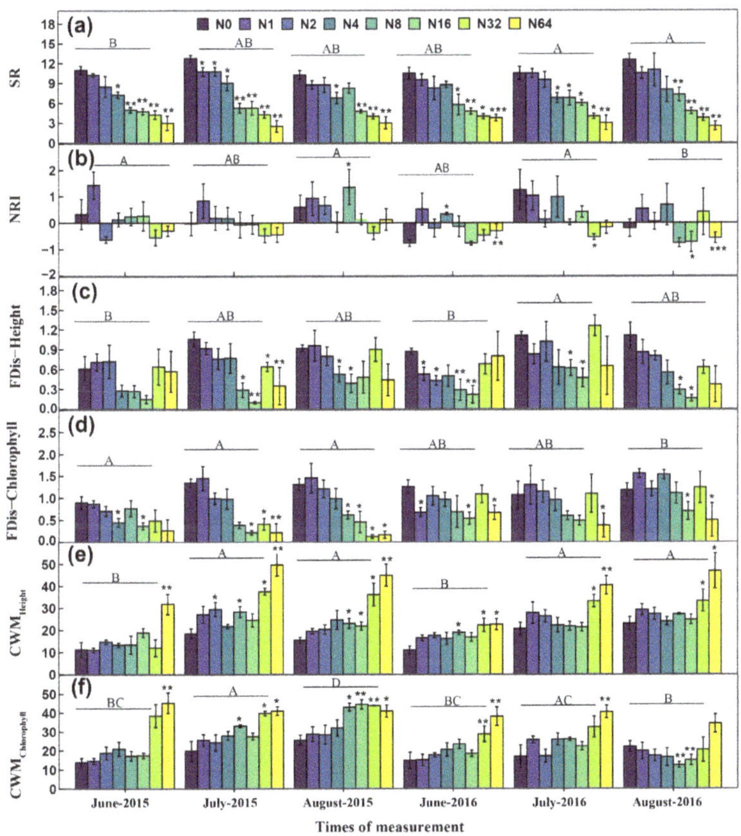

Figure 3. Effect of N addition gradient on multidimensional plant diversity metrics (mean ± SE) (Tukey pairwise tests on the difference of plant diversity indices) including species richness (SR; (**a**)), net relatedness index (NRI; (**b**)), functional diversity of plant height (FDis−Height; (**c**), functional diversity of chlorophyll content (FDis−Chlorophyll; (**d**)), community−weighted mean of plant height (CWM_{Height} (cm); (**e**)), and community−weighted mean of chlorophyll content ($CWM_{chlorophyll}$; (**f**)) in a temperate steppe of Inner Mongolia, northern China. N0, N1, N2, N4, N8, N16, N32, and N64 represent nitrogen fertilizer rates of 0, 1, 2, 4, 8, 16, 32, and 64 g N m^{-2} yr^{-1}, respectively. Error bars represent standard error of the mean (n = 192). Asterisks indicate significant differences between the treatment and control groups * $p < 0.05$, ** $p < 0.01$, *** $p < 0.001$. Different letters indicate significant differences between measuring times ($p < 0.05$).

3.3. Relationship between Plant Diversity and Biomass Production along the N Addition Gradient

The generalized additive model was retained for describing the impact of N addition on the relationships between plant diversity indices and biomass production based on the lowest AIC (Table S5). The GAM model retained for interpretation, while the GLM and GLMM models were not retained for interpretation due to high model AIC (Tables S3 and S4). The GAM model revealed that N addition had no significant effect on the relationships of SR and NRI with production (Table 2). Both SR and NRI were negatively related to production across N addition rates in this study (Figure 4a,b). N addition did not affect the correlation between FDis−Height and production, but it significantly altered the relationships of FDis−Chlorophyll, CWM_{Height}, and $CWM_{Chlorophyll}$ with production (all $p < 0.05$, adjusted $R^2 = 0.53, 0.49, 0.64$, respectively; Table 2). FDis−Chloropyll was negatively associated with production ($p < 0.01$; Figure 4d; Table 2), whereas CWM_{Height} and $CWM_{Chlorophyll}$ exhibited positive effects on production (Figure 4e,f).

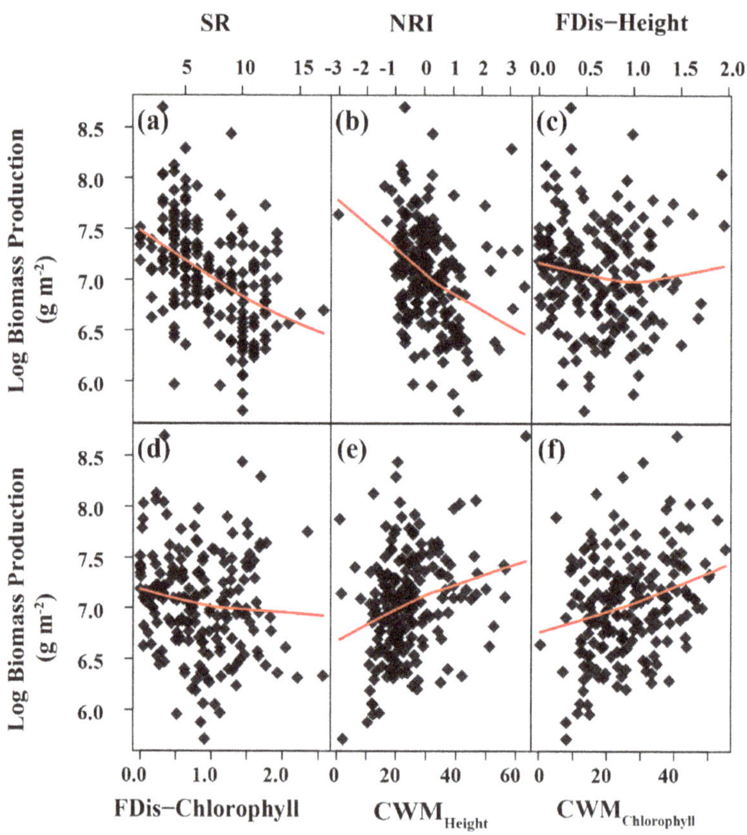

Figure 4. Relationships between biomass production and plant diversity indices including species richness (SR; (**a**)), net relatedness index (NRI; (**b**)), functional diversity of plant height (FDis−Height; (**c**)), functional diversity of chlorophyll content (FDis-Chlorophyll; (**d**)), community−weighted mean of plant height (CWM_{Height}; (**e**)), and community−weighted mean of leaf chlorophyll content ($CWM_{Chlorophyll}$; (**f**)) across eight levels of nitrogen (N) addition during the growing season in a temperate steppe of Inner Mongolia, northern China. Solid lines are fitted Lowess-smoothed curves. Each data point represents the measured value of individual plots for each measuring time (n = 192).

Table 2. Analysis of deviance results for the generalized additive model (GAM) testing the responses of biomass production to species richness (SR), net relatedness index (NRI), functional diversity of plant height (FDis−Height), functional diversity of chlorophyll content (FDis−Chlorophyll), community−weighted mean of plant height (CWM_{Height}), and community−weighted mean of chlorophyll content ($CWM_{Chlorophyll}$) across eight levels of nitrogen (N) addition in a temperate steppe of Inner Mongolia, northern China. $CWM_{Height} \times$ N: interactive effects of CWM_{Height} and N addition; qFDis−Height and $qCWM_{Height}$: second degree of the quadratic term of FDis−Height and CWM_{Height}, respectively; $CWM_{Height} \times$ N: interactive effects of CWM_{Height} and nitrogen; $CWM_{Chlorophyll} \times$ Time: interactive effects of $CWM_{Chlorophyll}$ and measuring times; N: various rates of nitrogen fertilizer addition; Time: measuring time (i.e., repeated measures in June, July, and August of 2015 and 2016); N × Time: interactive effects of N and time retained in the model.

Response	**Term**	**Statistics**			
Biomass Production	Taxonomic diversity	df	F	*p*	AIC
	s(SR)	1	0.95	0.32	177.22
	s(qSR)	1	2.27	0.13	
	N	7	2.49	**<0.05**	
	Time	5	16.27	**<0.001**	
	Phylogenetic pattern	df	F	*p*	AIC
	s(NRI)	1	0.27	0.61	179.17
	N	7	7.74	**<0.001**	
	Time	5	15.15	**<0.001**	
	Functional Diversity	df	F	*p*	
	s(FDis-Height)	1	1.68	0.90	178.80
	s(qFDis-Height)	1	1.75	0.18	
	N	7	7.10	**<0.001**	
	Time	5	15.72	**<0.001**	
	s(FDis-Chlorophyll)	1	5.26	**<0.05**	170.20
	FDis-Chlorophyll × N	7	3.28	**<0.01**	
	N	7	4.72	**<0.001**	
	Time	5	15.00	**<0.001**	
	Functional composition	df	F	*p*	AIC
	s(CWM_{Height})	1	5.17	**0.05**	183.57
	s($qCWM_{Height}$)	1	4.84	**<0.05**	
	$CWM_{Height} \times$ N	7	2.50	**<0.05**	
	$qCWM_{Height} \times$ N	7	2.27	**<0.05**	
	N	7	3.21	**<0.01**	
	Time	5	14.63	**<0.001**	
	s($CWM_{Chlorophyll}$)	1	32.65	**<0.001**	141.64
	$CWM_{Chlorophyll} \times$ N	5	8.35	**<0.001**	
	N	7	2.22	**<0.05**	
	Time	5	9.32	**<0.001**	
	N × Time	35	2.33	**<0.001**	

Significant effects ($p < 0.05$) are indicated in bold.

4. Discussion

4.1. N Deposition Affects Biomass Production

The current study shows that biomass production initially increases along the N addition gradient until the optimum is reached and then slowly decreases with the N addition rate, indicating that the positive effect of N deposition on biomass production may diminish with a higher rate of N addition when N loading surpasses the N demand of plants. Previous studies have documented that after long-term chronic N inputs, N supply may exceed plant and microbial requirements, resulting in N saturation [9]. Numerous studies which only have two N addition rates (i.e., with N addition at the level of 6 or

10 g m^{-2} yr^{-1} and without N addition) have demonstrated that N addition increases biomass production [8,14,15]. However, in the present study under a low-to-moderate N addition rate (<47 g N m^{-2}), biomass production increases, which might partly be attributed to the enhanced size of individual plants such as plant height and leaf chlorophyll content of dominant species (i.e., CWM plant height and leaf chlorophyll content) in line with previous studies [14,38]. This indicates that plant growth in this ecosystem was limited by nitrogen. Our study site in Inner Mongolia's steppe soil had low N content (0.19 mg g^{-1}), resulting in plant growth being limited by the low soil N content, as reported in a previous study [22]. Plant species that require higher nutrient levels could experience growth inhibition if they face prolonged nutrient deficiencies. This nutrient deficiency over time will limit the overall biomass production of those plant species [15]. Earlier studies showed that leaf chlorophyll content and plant height have a substantial impact on grassland productivity [39], as plant height significantly influences how plant communities respond to the addition of N by increasing light interception [30]. This could be attributed to the increased production of the dominant grasses offsetting the negative effects of species loss on biomass production. Hence, the dominant species, as controllers of ecosystem function, can provide short-term resistance to reductions in biomass production when species loss is nonrandom due to N deposition [6]. This indicates that N addition stimulates biomass production and promotes the turnover of species composition by favoring a few opportunistic species [40], which then neutralize the species loss effect on production. When the N enrichment rate approximately reaches 47 g N m^{-2} yr^{-1}, biomass production reaches its maximum, indicating that N was a limiting factor at rates below 47 g N m^{-2} yr^{-1}. It may be that N addition does lead to increased production but only up to a point where plants no longer efficiently utilize the added N. The reason for this pattern is likely due to the decrease in SR and FD because previous studies have shown that higher SR and variation in plant functional strategies increase biomass production [41,42]. Higher N addition rates promote competitive stress in the plant communities, as species have no similar competitive or stress-tolerant capability, as such, this result is partly in support of the stress gradient hypothesis, which suggests that the positive effect of biodiversity on ecosystem functioning can be enhanced when the environment experiences moderate levels of stress. However, if the stress levels increase beyond a certain point, the relationship between biodiversity and ecosystem functioning tends to diminish [43]. This indicates that traits related to resource acquisition, such as height and chlorophyll content, are no longer capable of offsetting the decline in biomass production resulting from the loss of species following N enrichment.

In addition, changes in community composition as a result of high N supply may lead to the dominance of a small number of species, which may limit biomass production [44]. Meanwhile, under a higher rate of N addition (>47 g m^{-2} yr^{-1}), biomass production declined, likely due to light competition and reduction in vegetation light penetration (the portion of light reaching the ground surface) [45], leading to species loss, specifically the elimination of highly productive forb species such as *Artemisia frigida* and *Potentilla acaulis* [20]. This shows that the negative effect of N addition on SR gradually dominates with increasing biomass production. The results suggest that higher N availability can offset the positive effects of diversity on biomass production. Nitrogen addition to grasslands increases the dominance of nitrogen-demanding grasses, which then suppress other slow-growing plant species, resulting in decreased productivity [6]. The results indicate that higher-rate N addition would be weaken commonly observed positive diversity–productivity relationships. In addition, the N−induced decline in biomass production at higher-rate N addition might be attributed to the loss of plant species due to the direct toxicity of N, depletion of soil nutrients like calcium, potassium, and phosphorus, lower soil pH and soil CEC, and increase in Mn ion toxicity [6,9]. Thus, this result suggests that competition and an increase in litter accumulation can reduce seedling establishment following a higher rate of N addition leading to light and water limitation [9], resulting in a greater reduction in plant diversity and plant biomass [40]. The fact that N enrichment may make the water limitation in temperate steppe grasslands worse [44] may also be a contributing factor to the decreased biomass production following high rates of N addition. The increased biomass production resulting from N addition can lead to higher plant water demand for transpiration and metabolism [42]. Consequently, our findings confirm that a low rate of N addition induces a trade-off between plant diversity and biomass production, but higher rates of N addition decrease both plant diversity and biomass production in a temperate steppe of Inner Mongolia. This finding is consistent with research studies reporting that diversity and productivity are interdependent [3]: resource supply drives diversity, and diversity drives resource usage, which then influence biomass production simultaneously. In addition, the contrasting effect of N deposition on FD and biomass production, with the former decreasing and the latter increasing with an increasing N rate, might be attributable to intraspecific trait variation promoted by N addition, which was observed in previous studies [38,46]. Given the positive effect of N addition on CWM_{Height} and $CWM_{Chlorophyll}$ (community functional trait composition) and the negative effect of N addition on FDis-Height and FDis-Chlorophyll content with biomass production, our results likely reflect that N addition promotes the growth of highly productive species at the expense of short-statured and slow-growing species. Consequently, the biomass production under multiple N additions in a temperate steppe is determined by the functionally dominant species that attained the largest biomass (i.e., dominance effect). This is consistent with the results of the positive links between CWM of plant traits and biomass production under nutrient enrichment, emphasizing that the functional traits related to resource acquisition (plant height and leaf chlorophyll content) may play a significant role in response to N addition, driving the plant diversity—biomass production relationship in the temperate steppe. Thus, we found support for the mass ratio hypothesis [47], which indicates that biomass production is primarily determined by the functional traits of the dominant species instead of variation in ecological strategies, reflecting from the fact that a community composed of species with a high trait value could increase biomass production. That is, the high biomass production under low N addition rates would depend on the functional traits of the dominant grass species [48]. Tatarko et al. [38] also observed that dominant species' traits can play a significant role in driving grassland primary biomass production under N enrichment. In part accordance with our expectations, this may be a result of intense interspecific competition following N addition, indicating that the addition of N shifted few species toward extreme trait values [48]. Therefore, this finding suggests that, to reduce light competition resulting from N enrichment and improve biomass production in the grassland community of inner Mongolia, moderate mowing or grazing that can remove or open up the litter layer is vital to enhance seedling establishment and maintain biodiversity and production, particularly in the plant communities with abundant litter.

4.2. *Effect of N Addition Rate on Multidimensional Plant Diversity*

The present study supports previous findings showing that species richness decreases under N addition [46,49]. The decline in species richness may be attributed to size asymmetry light competition and changes in underground processes, such as soil acidity and the direct toxicity of N, which impairs electron transport in chloroplasts [6,9,45]. Additionally, this decrease may also be partially driven by nutrient imbalance in plants [6]. In general, according to a global meta-analysis [50], species loss induced by elevated N might be explained by two aspects. On one hand, the accumulation of a large amount of standing litter and higher live biomass production increases size asymmetry competition for light [45,51]. Previous nutrient addition tests have demonstrated that under high N supplies, plants' competition shifts from soil N to light [20,52], which may also be a factor in the loss of species along the N gradient by decreasing the proportion of photosynthetically active radiation transmitted through the canopy to the ground surface [40]. On the other hand, N deposition alters the soil chemistry, resulting in the limitation of basic cations such as calcium, potassium, and phosphorus in the soil [30], which can cause soil acidification and ammonium toxicity when the magnitude of N input surpasses the plants' ability to utilize it [7,53], thereby reducing SR [10,49] and altering the composition of the soil microbial community [9]. In the Inner Mongolia temperate steppe, it has been reported that N addition decreased forb species due to the accumulation of inorganic N, a decrease in soil pH, and an increase in toxic metallic ion concentrations such as Mn and Al, resulting in the depletion of basic cations such as Ca, Mg, and K [54]. The increase in CWM_{Height} along the gradient of N addition may have stimulated rising competition for light among coexisting species. Species that are short-statured, leguminous, locally rare, and capable of tolerating low nutrient levels are likely to be especially susceptible to loss due to competitive exclusion following N-induced increases in biomass production [8,55]. Meanwhile, our results demonstrated a considerable variation in the response pattern of plants from different functional groups following varying rates of N addition, which was consistent with previous findings [22,56]. The higher rate of N addition (32 g m^{-2} N yr^{-2} and above) leading to the loss of dominant forbs such as *Artemisia frigida* and *Potentilla acaulis* (on average, made up in the order of 35% and 20% of the relative percent cover, respectively) in the control (N0) and N1 (1 g N yr^{-1}) plots, respectively. In the higher-rate treatments (i.e., over 16 g N yr^{-1}), perennial rhizomatous grasses such as *Leymus chinensis* (a perennial rhizomatous clonal grass; on average, comprises 65% relative percent cover) dominated, which to a large extent supported the previous results [20]. This indicates that there is variability in the responses of different functional groups of plants to nutrient addition. This finding is consistent with previous studies, which have found that forb species in a temperate steppe of Inner Mongolia grassland are more sensitive to N−induced stress, whereas grass such as *Leymus chinensis* is positively associated with N addition [44,54]. Grass could have a higher nitrogen utilization efficiency compared to forbs [56]. As such, grass species outcompete forbs by growing at a faster rate and exploiting light more efficiently, which hinders the growth of forbs [45]. Similarly, Song et al. [20] observed that higher rates of N addition promote the dominance of grasses, which outcompete forbs in a temperate steppe ecosystem. Although, a study conducted by De-Malach et al. [45] demonstrated a decrease in forb diversity following N enrichment in the Mediterranean grassland. This result suggests that N−induced changes in plant diversity are species-specific, possibly due to the difference in the sensitivity of species to nutrient addition and resource utilization strategies [38], and in turn, fast-growing and short-lived forb species are more sensitive to N deposition [19]. This might be related to, at least in part, the relatively large rhizosheath size on grass species providing resistance against N induced stress compared to forbs. More recently, it has been reported that rhizosheath size plays a substantial role in grass relative to forbs in providing resistance against N induced metal stress by releasing protective compounds (such as organic acids and mucilaginous compounds), which in turn regulate the competitive advantage for grass at the cost of forbs by colonizing the belowground habitat [49]. In addition, clonal plants (such as *L. chinensis* in our experiment) have competitive advantages over forbs in acquiring soil nutrients and light, due to clones and greater root biomass [54]. Thus, N regulates the growth of grass to become taller and possess a dominant canopy. This allows the grass to intercept available light and increase resource use efficiency such as light. This result provides support for the light asymmetry hypothesis, which posits that tall individuals receive a higher amount of light per unit biomass than shorter individuals [45], which, in turn, results in the loss of forb species in grasslands [30,56]. Another important process, beside light competition, that can likely lead to species loss related to N addition is that high levels of soil resources reduce the niche dimension [40,44], which in turn favors a few dominant plant species by affecting competitive balance among species owing to a reduction in the number of limiting resources. Our study in a temperate steppe of Inner Mongolia revealed loss of species along the N addition gradient probably resulting from both light competition and N-induced change in soil chemical properties [49].

In the present study, the shifting of phylogenetic diversity from phylogenetic clustering to overdispersion (i.e., high phylogenetic dissimilarity) owing to elevated N fertilization is consistent with previous findings in a semi-natural plant community [57]. However, given that closely related species are more likely to share similar traits (i.e., have a high phylogenetic signal), our findings are in contrast to the prediction that closely related species may exhibit similar sensitivity or resistance to a particular perturbation [58]. However, our findings are congruent with the theory that more closely related species are less likely to coexist, which means that closely related species compete more intensely for resources such as light [29], resulting in phylogenetic overdispersion [16]. A possible reason is that some functionally important and productive clades outcompeted other short-statured groups resulting in species loss under N enrichment, which may lead to phylogenetic dissimilarity among coexisting plants. This suggests that plant species that are resilient to stress caused by N are not limited to a certain plant family and may have evolved separately in different lineages [57]. Thus, competition might be one possible reason and our work provides evidence for this phenomenon. N addition may act as an environmental filter on ecologically similar non-related species, which might cause phylogenetic overdispersion (i.e., convergent evolution) by filtering lineages based on their functional traits [59]. This suggests that light competition among coexisting species under N enrichment and N toxicity might be a potential cause of phylogenetic overdispersion, as it increases the likelihood of extinction of short-statured and rare species while promoting the coexistence of multiple distinct lineages [16]. Consequently, N−induced competition among closely related species leads to favoring the coexistence of distantly related species with distinct ecological strategies [57]. Plant height and chlorophyll content were used in the current investigation but failed to capture a strong phylogenetic signal, indicating that these traits may have experienced phylogenetically convergent evolution throughout the course of evolutionary time [58]. This suggests the possibility that the management strategies to conserve plant diversity in the temperate steppe of Inner Mongolia, such as sowing seeds of phylogenetically distantly related species (particularly grass species), could be crucial to maintain plant diversity in the face of N deposition.

Functional diversity decreased along the N addition gradient in the current study, indicating that the nutrient gradient in grassland affects the functional structure of communities [59]. For instance, Xu et al. [41] observed that biotic processes such as competition can produce trait divergence by excluding similar species (i.e., promoting niche difference). This implies that dominant plants tend to eliminate neighboring short-statured plants that have similar traits. N addition may lead to the emergence of new environmental conditions that favor the selection of new traits, while disadvantaging the common species and accelerating rates of competitive exclusion [45,55]. In turn, nutrient addition can promote the dominance of species with similar functional traits that have less overlap in their use of resources and facilitate the coexistence of species that share similar traits, leading to a more homogeneous plant community. One possible reason for this phenomenon is that N-induced competition for light affects plant functional traits that are related to light acquisition, such as the height of dominant species. This increase in height then leads to a reduction in functional diversity [51], by causing the elimination of short-statured species. Consequently, the overall FD of the plant community becomes low. As such, the potential factor for the loss of species might be linked to the differences in resource acquisition and resource storage strategies among species (grass vs. forbs) [44]. These results are similar to those of other studies [19,57], which indicate that N fertilization enhances competitive response traits such as CWM_{Height}, which may accelerate light competition intensification among coexisting plant species. Plant height is a key functional trait that affects the competitive ability for light. Plants with a short stature and resource storage strategy particularly suffer more from reduced light availability [9]. However, our results contradicted some other previous findings [60] that N addition had no direct and significant effect on FD and PD in alpine meadow and alpine steppe grassland. The discrepancy may be related to many reasons, including the difference in annual mean temperature, plant species composition, different forms of N added, and soil properties between alpine grasslands and temperate steppe [42]. The mean annual temperature is higher in the temperate steppe compared to the alpine meadow and alpine steppe, as higher temperature promotes the uptake of soil N by plants [55]. Moreover, N addition increases $CWM_{Chlorophyll}$, which is an important pigment for photosynthesis [39]. For instance, a previous study found that N addition increases $CWM_{Chlorophyll}$ by 34% (i.e., 111 mg cm^{-2}) [38], compared with the control. Through this understanding, the results further confirmed that the dominating species with a resource-acquisitive strategy was a mediator of the effects of N on plant diversity [53]. These results suggest that nutrient availability increase plant strategy towards resource acquisition [38]. Therefore, future management that focuses on locally susceptible species to N-induced stress, particularly short-statured forbs, will be vital to maintain the highest levels of biodiversity in the temperate steppe.

4.3. *Effect of N Addition Rate on the Relationship between Multidimensional Plant Diversity and Biomass Production*

Our results have demonstrated that N addition did not significantly alter the strength or direction of the relationship of SR, NRI, and FDis-Height with biomass production, despite the fact that N addition significantly decreases the aforementioned diversity indices. It is observed that there is an inverse relationship between the increase in biomass production caused by N addition and the decline in plant diversity. In other words, as the biomass production increases, the plant diversity decreases, and vice versa following N enrichment, which agrees with the results of previous global syntheses [4,52]. The direction and strength of the relationship between SR and biomass production in response to nutrient addition varies between contexts, systems, and sites [40]. The present study revealed that the plant community tends to be phylogenetically overdispersed at higher N addition rates, implying that species are distantly related, but N addition did not change the negative correlation between NRI and biomass production. A possible reason could be that elevated N may exacerbate intense light competition between closely related species, which may have increased the likelihood of the coexistence of distantly related species that are more tolerant of elevated N. The functional dispersion of leaf Chlorophyll content exhibited a negative relationship with biomass production under the N addition gradient. This may result from the N-induced changes in functional diversity, which decrease resource use efficiency such as light [51], presumably reflecting the effect of functional filtering, which in turn reduces biomass production by inhibiting the diversity of traits and decreasing the resource use efficiencies owing to declining diverse nutrient-uptake strategies during the different periods of growing [4,30]. Our findings showed that functional trait diversity in response to N enrichment has a negative relationship with biomass production. This indicates that the deposition of N leads to a decrease in trait spaces, which in turn influences the link between functional diversity and biomass production in the temperate steppe grassland. Plant communities have species which possess distinct traits, utilize diverse resources, and maintain biomass production [53], but the enrichment of varying rates of N promotes the growth of plant species that have similar functional strategies [59]. However, these species may have low efficiency in utilizing resources, which leads to a decrease in biomass production [30]. In accordance, numerous research studies have demonstrated that in long-term grassland experiments, nitrogen enrichment usually increases grassland productivity while decreasing plant diversity [5,15,51], suggesting that the generally observed positive correlation between diversity and biomass production would be diminished with N addition [40,52]. Taken together, our findings in a temperate steppe of Inner Mongolia demonstrate that elevated N alters the composition and structure of the community. Consequently, this could counteract the positive correlation between plant diversity and biomass production (i.e., biomass production was influenced by community structure), consistent with prior investigations [20,44]. Our findings indicate that the addition of nitrogen decreases SR, NRI, and FDis-Height, and biomass production was also negatively associated with the aforementioned facets of diversity metrics. However, the CWMs of plant height and leaf chlorophyll content were significantly positively related to biomass production under N addition. The results suggest the dampening of positive interactions among the grassland species and an increasing contribution of a few dominant species to a higher biomass production in response to increasing N accumulation. Measuring only two functional traits may have been a limitation of our study because the effect of functional diversity on the relationship between plant diversity and biomass production depends on which traits are included in the analysis [53]. Moreover, we only conducted this experiment for two years. For a comprehensive understanding of the impact of N addition on biodiversity–productivity relationships, a longer time series of data is required. The response of the plant diversity–production relationship to N deposition varies with the rate and duration of N addition [7], and these limitations may introduce bias into the results due to the relatively short duration of the study.

5. Conclusions

In this study, N deposition significantly altered multidimensional plant diversity and ecosystem production in the temperate steppe. Forbs exhibited more sensitive responses to N deposition-induced stress than grass, leading to a decline in SR along the N deposition gradient. The phylogenetic structure changed from clustering to overdispersion with N deposition rates by filtering out lineages with similar functional traits. Plant height and leaf chlorophyll content-based functional dispersion reduced, but CWM_{Hight} and $CWM_{Chlorophyll}$ increased with N deposition rates, implying that N deposition−induced stress favored tall species with high leaf chlorophyll content in the temperate steppe. Ecosystem production increased, reached the maximum, and then decreased with N deposition rates. However, N deposition had no impact on the relationships of SR and PD with production, but it affected the relationships of FDis-Chlorophyll, CWM_{Height}, and $CWM_{Chlorophyll}$ with production. The findings highlight the importance of a trait-based approach for understanding the multidimensional biodiversity–ecosystem function in grassland ecosystems under N deposition scenarios. For a better understanding of the N deposition effect on ecosystem structure and function, long-term manipulative experiments incorporating multiple functional traits, environmental factors, and soil microbial analysis are desirable. Based on the findings, to mitigate N−induced species loss and improve ecosystem production in the grassland of Inner Mongolia, moderate mowing or grazing may be desirable to facilitate seedling establishment and maintain plant diversity. Additionally, it is recommended to establish a critical limit for N deposition in order to implement the idea of critical loads. This limit would help to restrict the amount of N load and mitigate its subsequent negative effects on both biodiversity and ecosystem production.

Supplementary Materials: The following supporting information can be downloaded at https://www.mdpi.com/article/10.3390/biology13080554/s1, Detailed see the supplementary files. Figure S1: Result of Pearson's correlation analysis between plant diversity metrics and biomass production; Figure S2: Phylogenetic tree of the plant species recorded in the experimental plots; Figure S3: The distribution of plant height across the phylogeny; Figure S4: Top most abundant plant species based on percent relative cover; Table S1: Most abundant plant functional group selected based on the relative cover; Table S2: Pearson's correlation result showing relationship between productivity and multidimensional diversity metrics; Table S3: A summary of the final generalized linear models (GLM); Table S4: Analysis of deviance table for the final generalized linear mixed model (GLMM); Table S5: The Akaike Information Criterion (AIC) of generalized additive models (GAM), generalized linear models (GLM), and generalized linear mixed models (GLMM).

Author Contributions: Designed Research and Conceptualization, G.L.; Data Analysis, Original Draft Preparation, and Visualization, G.H.D., Review and Editing, G.L., K.L. and W.B.; Investigation, G.L.; Data collection, K.L., X.W. and Y.W.; Funding Acquisition, G.L.; all authors contributed to revisions of the manuscript. All authors have read and agreed to the published version of the manuscript.

Funding: The study was financially supported by the National Natural Science Foundation of China, award number 31770522; 32130066.

Institutional Review Board Statement: Not applicable.

Informed Consent Statement: Not applicable.

Data Availability Statement: The datasets analyzed during the current study are available from the corresponding author on reasonable request.

Acknowledgments: We thank Zhenxing Zhou, Mingxing Zhong, Jingyi Ru, Huanhuan Song, Danyu Sun, and all the staff of Duolun Restoration Ecology Station of the Institute of Botany, Chinese Academy of Sciences, for their help in the field survey.

Conflicts of Interest: The authors have declared that no competing interests exist.

References

1. Grace, J.; Anderson, T.; Seabloom, E.; Borer, E.; Adler, P.; Harpole, W.; Hautier, Y.; Hillebrand, H.; Lind, E.; Pärtel, M.; et al. Integrative modelling reveals mechanisms linking productivity and plant species richness. *Nature* **2016**, *529*, 390–393. [CrossRef]
2. Tilman, D.; Isbell, F.; Cowles, J. Biodiversity and ecosystem functioning. *Annu. Rev. Ecol. Evol. Syst.* **2014**, *45*, 471–493. [CrossRef]
3. Cardinale, B.J.; Gross, K.; Fritschie, K.; Flombaum, P.; Fox, J.W.; Rixen, C.; van Ruijven, J.; Reich, P.B.; Scherer-Lorenzen, M.; Wilsey, B.J. Biodiversity simultaneously enhances the production and stability of community biomass, but the effects are independent. *Ecol. Ecol. Soc. Am.* **2013**, *94*, 1697–1707. [CrossRef] [PubMed]
4. Craven, D.; Isbell, F.; Manning, P.; Connolly, J.; Bruelheide, H.; Ebeling, A.; Roscher, C.; van Ruijven, J.; Weigelt, A.; Wilsey, B.; et al. Plant diversity effects on grassland productivity are robust to both nutrient enrichment and drought. *Philos. T. R. Soc. B* **2016**, *371*, 20150277. [CrossRef] [PubMed]
5. Isbell, F.; Reich, P.B.; Tilman, D.; Hobbie, S.E.; Polasky, S.; Binder, S. Nutrient enrichment, biodiversity loss, and consequent declines in ecosystem productivity. *Proc. Natl. Acad. Sci. USA* **2013**, *110*, 11911–11916. [CrossRef]
6. Stevens, C.J.; David, T.I.; Storkey, J. Atmospheric nitrogen deposition in terrestrial ecosystems: Its impact on plant communities and consequences across trophic levels. *Funct. Ecol.* **2018**, *32*, 1757–1769. [CrossRef]
7. Midolo, G.; Alkemade, R.; Schipper, A.M.; Benítez-López, A.; Perring, M.P.; De Vries, W. Impacts of nitrogen addition on plant species richness and abundance: A global meta-analysis. *Glob. Ecol. Biogeogr.* **2019**, *28*, 398–413. [CrossRef]
8. Chen, Q.; Blowes, S.A.; Ladouceur, E.; Harpole, W.S.; Seabloom, E.W.; Tognetti, P.M.; MacDougall, A.; Daleo, P.; Hautier, Y.; Stevens, C.; et al. Nutrient addition in grasslands worldwide reveals proportional plant diversity decline across spatial scales but little change in beta diversity. *BioRxiv* **2024**. [CrossRef]
9. Ma, F.; Song, B.; Quan, Q.; Zhang, F.; Wang, J.; Zhou, Q.; Niu, S. Light competition and biodiversity loss cause saturation response of aboveground net primary productivity to nitrogen enrichment. *JGR BiogeoSci.* **2020**, *125*, e2019JG005556. [CrossRef]
10. Namuhan; Wang, J.; Yang, G.; Song, Y.; Yu, Y.; Wang, J.; Wang, X.; Shi, Y.; Shen, Y.; Han, X.; et al. Mechanisms of biodiversity loss under nitrogen enrichment: Unveiling a shift from light competition to cation toxicity. *N. Phytol.* **2024**. *early view*. [CrossRef]
11. Lu, C.; Tian, H.; Liu, M.; Ren, W.; Xu, X.; Chen, G.; Zhang, C. Effect of nitrogen deposition on China's terrestrial carbon uptake in the context of multifactor environmental changes. *Ecol. Appl.* **2012**, *22*, 53–75. [CrossRef] [PubMed]
12. Hautier, Y.; Seabloom, E.W.; Borer, E.T.; Adler, P.B.; Harpole, W.S.; Hillebrand, H.; Lind, E.M.; MacDougall, A.S.; Stevens, C.J.; Bakker, J.D.; et al. Eutrophication weakens stabilizing effects of diversity in natural grasslands. *Nature* **2014**, *508*, 521–525. [CrossRef] [PubMed]
13. Li, G.; Barfknecht, D.F.; Gibson, D.J. Disturbance effects on productivity-plant diversity relationships from a 22-year-old successional field. *J. Veg. Sci.* **2021**, *32*, e12970. [CrossRef]
14. Wang, X.; Yan, X.; Huang, K.; Luo, X.; Zhang, Y.; Zhou, L.; Yang, F.; Xu, X.; Zhou, X.; Niu, K.; et al. Nitrogen enrichment and warming shift community functional composition via distinct mechanisms: The role of intraspecific trait variability and species turnover. *Funct. Ecol.* **2022**, *36*, 1230–1242. [CrossRef]
15. Zhao, Y.; Yang, B.; Li, M.; Xiao, R.; Rao, K.; Wang, J.; Zhang, T.; Guo, J. Community composition, structure and productivity in response to nitrogen and phosphorus additions in a temperate meadow. *Sci. Total Environ.* **2019**, *654*, 863–871. [CrossRef]
16. Barber, N.A.; Farrell, A.K.; Blackburn, R.C.; Bauer, J.T.; Groves, A.M.; Brudvig, L.A.; Jones, H.P. Grassland restoration characteristics influence phylogenetic and taxonomic structure of plant communities and suggest assembly mechanisms. *J. Ecol.* **2019**, *107*, 2105–2120. [CrossRef]
17. Le Bagousse-Pinguet, Y.; Soliveres, S.; Gross, N.; Torices, R.; Berdugo, M.; Maestre, F.T. Phylogenetic, functional, and taxonomic richness have both positive and negative effects on ecosystem multifunctionality. *Proc. Natl. Acad. Sci. USA* **2019**, *116*, 8419–8424. [CrossRef]
18. Cadotte, M.W. Functional traits explain ecosystem function through opposing mechanisms. *Ecol. Lett.* **2017**, *20*, 989–996. [CrossRef]
19. Liu, L.; Sun, K.; Sun, R.; Ma, Q.; Wang, Y.; Jia, B.; Zhou, G.; Xu, Z.; Zhang, F. Effects of nitrogen deposition with phosphorus addition on desert steppe plant communities. *Agric. Ecosyst. Environ.* **2024**, *366*, 108954. [CrossRef]
20. Song, L.; Bao, X.; Liu, X.; Zhang, Y.; Christie, P.; Fangmeier, A.; Zhang, F.J.B. Nitrogen enrichment enhances the dominance of grasses over forbs in a temperate steppe ecosystem. *Biogeosciences* **2011**, *8*, 2341–2350. [CrossRef]
21. Niu, S.; Wu, M.; Han, Y.; Xia, J.; Li, L.; Wan, S. Water-mediated responses of ecosystem carbon fluxes to climatic change in a temperate steppe. *N. Phytol.* **2008**, *177*, 209–219. [CrossRef] [PubMed]

22. Tian, Q.; Lu, P.; Ma, P.; Zhou, H.; Yang, M.; Zhai, X.; Chen, M.; Wang, H.; Li, W.; Bai, W.; et al. Processes at the soil–root interface determine the different responses of nutrient limitation and metal toxicity in forbs and grasses to nitrogen enrichment. *J. Ecol.* **2021**, *109*, 927–938. [CrossRef]
23. Li, D. hillR: Taxonomic, functional, and phylogenetic diversity and similarity through Hill Numbers. *J. Open Source Softw.* **2018**, *3*, 1041. [CrossRef]
24. R Development Core Team. *R: A Language and Environment for Statistical Computing*; R Foundation for Statistical Computing: Vienna, Austria, 2024.
25. Kembel, S.W.; Cowan, P.D.; Helmus, M.R.; Cornwell, W.K.; Morlon, H.; Ackerly, D.D.; Blomberg, S.P.; Webb, C.O. Picante: R tools for integrating phylogenies and ecology. *Bioinformatics* **2010**, *26*, 1463–1464. [CrossRef] [PubMed]
26. Paradis, E.; Schliep, K. ape: R package version 5.7.1 an environment for modern phylogenetics and evolutionary analyses in R. *Bioinformatics* **2019**, *35*. [CrossRef] [PubMed]
27. Zanne, A.E.; Tank, D.C.; Cornwell, W.K.; Eastman, J.M.; Smith, S.A.; FitzJohn, R.G.; McGlinn, D.J.; O'Meara, B.C.; Moles, A.T.; Reich, P.B.; et al. Three keys to the radiation of angiosperms into freezing environments. *Nature* **2014**, *506*, 89–92. [CrossRef] [PubMed]
28. Jin, Y.; Qian, H.V. PhyloMaker: An R package that can generate very large phylogenies for vascular plants. *Ecography* **2019**, *42*, 1353–1359. [CrossRef]
29. Webb, C.O. Exploring the phylogenetic structure of ecological communities: An example for rain forest trees. *Am. Nat.* **2000**, *156*, 145. [CrossRef] [PubMed]
30. Zhang, D.; Peng, Y.; Li, F.; Yang, G.; Wang, J.; Yu, J.; Zhou, G.; Yang, Y. Trait identity and functional diversity co-drive response of ecosystem productivity to nitrogen enrichment. *J. Ecol.* **2019**, *107*, 2402–2414. [CrossRef]
31. Laliberté, E.; Legendre, P. A distance-based framework for measuring functional diversity from multiple traits. *Ecology* **2010**, *91*, 299–305. [CrossRef]
32. Laliberté, E.; Legendre, P.; Shipley, B. FD: Measuring Functional Diversity from Multiple Traits, and Other Tools for Functional Ecology. *R Package Version 1.0.12.3*. 2014, Volume 1, 01–12. Available online: https://CRAN.R-project.org/package=FD (accessed on 13 February 2024).
33. Pagel, M. Inferring the historical patterns of biological evolution. *Nature* **1999**, *401*, 877–884. [CrossRef] [PubMed]
34. Bates, D.; Mächler, M.; Bolker, B.; Walker, S. Fitting linear mixed-effects models using lme4. *J. Stat. Softw.* **2014**, *67*, 1–48. [CrossRef]
35. Wood, S.N. Fast stable restricted maximum likelihood and marginal likelihood estimation of semiparametric generalized linear models. *J. R. Stat. Soc.* **2011**, *73*, 3–36. [CrossRef]
36. Zuur, A.F.; Ieno, E.N.; Elphick, C.S. A protocol for data exploration to avoid common statistical problems. *Methods Ecol. Evol.* **2010**, *1*, 3–14. [CrossRef]
37. Crawley, M.J. *The R Book*, 2nd ed.; John Wiley & Sons: San Jose, CA, USA, 2012; pp. 537–554. [CrossRef]
38. Tatarko, A.R.; Knops, J.M.H. Nitrogen addition and ecosystem functioning: Both species abundances and traits alter community structure and function. *Ecosphere* **2018**, *9*, e02087. [CrossRef]
39. Zhang, Y.; He, N.; Li, M.; Yan, P.; Yu, G. Community chlorophyll quantity determines the spatial variation of grassland productivity. *Sci. Total Environ.* **2021**, *801*, 149567. [CrossRef] [PubMed]
40. Harpole, W.S.; Sullivan, L.L.; Lind, E.M.; Firn, J.; Adler, P.B.; Borer, E.T.; Chase, J.; Fay, P.A.; Hautier, Y.; Hillebrand, H.; et al. Addition of multiple limiting resources reduces grassland diversity. *Nature* **2016**, *537*, 93–96. [CrossRef]
41. Xu, Z.; Li, M.-H.; Zimmermann, N.E.; Li, S.-P.; Li, H.; Ren, H.; Sun, H.; Han, X.; Jiang, Y.; Jiang, L. Plant functional diversity modulates global environmental change effects on grassland productivity. *J. Ecol.* **2018**, *106*, 1941–1951. [CrossRef]
42. Tang, Z.; Deng, L.; An, H.; Yan, W.; Shangguan, Z. The effect of nitrogen addition on community structure and productivity in grasslands: A meta-analysis. *Ecol. Eng.* **2017**, *99*, 31–38. [CrossRef]
43. Baert, J.M.; Eisenhauer, N.; Janssen, C.R.; De Laender, F. Biodiversity effects on ecosystem functioning respond unimodally to environmental stress. *Ecol. Lett.* **2018**, *21*, 1191–1199. [CrossRef]
44. He, K.; Huang, Y.; Qi, Y.; Sheng, Z.; Chen, H. Effects of nitrogen addition on vegetation and soil and its linkages to plant diversity and productivity in a semi-arid steppe. *Sci. Total Environ.* **2021**, *778*, 146299. [CrossRef] [PubMed]
45. DeMalach, N.; Zaady, E.; Kadmon, R. Light asymmetry explains the effect of nutrient enrichment on grassland diversity. *Ecol. Lett.* **2017**, *20*, 60–69. [CrossRef] [PubMed]
46. Zheng, S.; Chi, Y.; Yang, X.; Li, W.; Lan, Z.; Bai, Y. Direct and indirect effects of nitrogen enrichment and grazing on grassland productivity through intraspecific trait variability. *J. Appl. Ecol.* **2021**, *59*, 598–610. [CrossRef]
47. Grime, J.P. Benefits of plant diversity to ecosystems: Immediate, filter and founder effects. *J. Ecol.* **1998**, *86*, 902–910. [CrossRef]
48. Sonkoly, J.; Kelemen, A.; Valko, O.; Deak, B.; Kiss, R.; Toth, K.; Miglecz, T.; Tothmeresz, B.; Torok, P. Both mass ratio effects and community diversity drive biomass production in a grassland experiment. *Sci. Rep.* **2019**, *9*, 1848. [CrossRef] [PubMed]
49. Tian, Q.; Lu, P.; Zhai, X.; Zhang, R.; Zheng, Y.; Wang, H.; Nie, B.; Bai, W.; Niu, S.; Shi, P.; et al. An integrated belowground trait-based understanding of nitrogen-driven plant diversity loss. *Glob. Chang. Biol.* **2022**, *28*, 3651–3664. [CrossRef] [PubMed]
50. Band, N.; Kadmon, R.; Mandel, M.; DeMalach, N. Assessing the roles of nitrogen, biomass, and niche dimensionality as drivers of species loss in grassland communities. *Proc. Natl. Acad. Sci. USA* **2022**, *119*, e2112010119. [CrossRef] [PubMed]
51. Hautier, Y.; Niklaus, P.A.; Hector, A. Competition for light causes plant biodiversity loss after eutrophication. *Science* **2009**, *324*, 636–638. [CrossRef]

52. Yue, K.; Jarvie, S.; Senior, A.M.; Van Meerbeek, K.; Peng, Y.; Ni, X.Y.; Wu, F.Z.; Svenning, J.C. Changes in plant diversity and its relationship with productivity in response to nitrogen addition, warming and increased rainfall. *Oikos* **2020**, *129*, 939–952. [CrossRef]
53. Liu, X.; Shi, X.; Zhang, S. Soil abiotic properties and plant functional diversity co-regulate the impacts of nitrogen addition on ecosystem multifunctionality in an alpine meadow. *Sci. Total Environ.* **2021**, *780*, 146476. [CrossRef]
54. Tian, Q.; Yang, L.; Ma, P.; Zhou, H.; Liu, N.; Bai, W.; Wang, H.; Ren, L.; Lu, P.; Han, W.; et al. Below-ground-mediated and phase-dependent processes drive nitrogen-evoked community changes in grasslands. *J. Ecol.* **2020**, *108*, 1874–1887. [CrossRef]
55. Clark, C.M.; Cleland, E.E.; Collins, S.L.; Fargione, J.E.; Gough, L.; Gross, K.L.; Pennings, S.C.; Suding, K.N.; Grace, J.B. Environmental and plant community determinants of species loss following nitrogen enrichment. *Ecol. Lett.* **2007**, *10*, 596–607. [CrossRef] [PubMed]
56. Liu, C.; Liu, J.; Wang, J.; Ding, X. Effects of short-term nitrogen additions on biomass and soil phytochemical cycling in alpine grasslands of Tianshan, China. *Plants* **2024**, *13*, 1103. [CrossRef] [PubMed]
57. Pellissier, L.; Wisz, M.S.; Strandberg, B.; Damgaard, C. Herbicide and fertilizers promote analogous phylogenetic responses but opposite functional responses in plant communities. *Environ. Res. Lett.* **2014**, *9*, 024016. [CrossRef]
58. Blomberg, S.P.; Garland, T., Jr.; Ives, A.R. Testing for phylogenetic signal in comparative data: Behavioral traits are more labile. *Evolution* **2003**, *57*, 717–745. [CrossRef] [PubMed]
59. Li, S.; Dong, S.; Shen, H.; Xu, Y.; Gao, X.; Han, Y.; Zhang, J.; Yang, M.; Li, Y.; Zhao, Z.; et al. Nitrogen addition gradient can regulate the environmental filtering of soil potassium or phosphorus in shaping the community assembly of alpine meadow. *Ecol. Indic.* **2020**, *109*, 105774. [CrossRef]
60. Li, S.; Dong, S.; Shen, H.; Han, Y.; Zhang, J.; Xu, Y.; Gao, X.; Yang, M.; Li, Y.; Zhao, Z.; et al. Different responses of multifaceted plant diversities of alpine meadow and alpine steppe to nitrogen addition gradients on Qinghai-Tibetan Plateau. *Sci. Total Environ.* **2019**, *688*, 1405–1412. [CrossRef]

biology

MDPI

Article

Nitrogen Addition Affects Interannual Variation in Seed Production in a Tibetan Perennial Herb

Yuanxin Lou [1], Ruolan Wang [1], Peiyue Che [1], Chuan Zhao [2], Yali Chen [1], Yangheshan Yang [3] and Junpeng Mu [1,*]

1 Ecological Security and Protection Key Laboratory of Sichuan Province, Mianyang Normal University, Mianyang 621000, China; 18623782240@163.com (Y.L.); wangruolan2023@126.com (R.W.); a28589924@163.com (P.C.); chenyali5549@163.com (Y.C.)

2 Key Laboratory of Mountain Ecological Restoration and Bioresource Utilization, Chengdu Institute of Biology, Chinese Academy of Sciences, Chengdu 610041, China; zhaochuan@cib.ac.cn

3 School of Ecological and Environmental Sciences, East China Normal University, Shanghai 200241, China; yangheshanyang@gmail.com

* Correspondence: gbmujp@163.com

Simple Summary: There are various factors linked to global change that possess the capacity to alter the interannual variability in seed production. Our current knowledge regarding the impact of nitrogen availability on the year-to-year fluctuations in seed production patterns of perennial plants is limited. A multiyear field experiment was undertaken to examine the impact of nitrogen addition on the interannual seed production patterns of perennial plants. The introduction of element N had a significant impact on the preservation of aboveground biomass in plants, as well as the stability of flower traits. Consequently, this factor had an impact on the frequency of pollinator preference and the annual seed output. The findings of our study indicate that increasing the nitrogen content in the soil has the potential to alter the natural fluctuations in seed production that occur over different years. The results of this study possess the capacity to improve our understanding of the effects of nitrogen supplementation on the reproductive success of perennial herbaceous plants, as well as the fundamental mechanisms driving biodiversity in the face of worldwide environmental shifts.

Abstract: The variability observed in the annual seed production of perennial plants can be seen as an indication of changes in the allocation of resources between growth and reproduction, which can be attributed to fluctuations in the environment. However, a significant knowledge gap exists concerning the impacts of nitrogen addition on the interannual seed production patterns of perennial plants. We hypothesized that the addition of nitrogen would impact the annual variations in the seed production of perennial plants, ultimately affecting their overall reproductive efficiency. A multiyear field experiment was conducted to investigate the effects of varying nitrogen supply levels (e.g., 0, 4, and 8 kg N ha^{-1} yr^{-1} of N0, N4, and N8) on vegetative and floral traits, pollinator visitation rates, and seed traits over a period of four consecutive years. The results showed that the N0 treatment exhibited the highest levels of seed production and reproductive efficiency within the initial two years. In contrast, the N4 treatment displayed its highest level of performance in these metrics in the second and third years, whereas the N8 treatment showcased its most favorable outcomes in the third and fourth years. Similar patterns were found in the number of flowers per capitulum and the number of capitula per plant. There exists a positive correlation between aboveground biomass and several factors, including the number of flowers per capitulum, the number of capitula per plant, the volume of nectar per capitulum, and the seed production per plant. A positive correlation was found between pollinator visitation and the number of flowers per capitulum or the number of capitula per plant. This implies that the addition of N affected the maintenance of plant aboveground biomass, flower trait stability, pollinator visitation, and, subsequently, the frequency of seed production and reproductive efficiency. Our results suggest that augmenting the nitrogen content in the soil may have the capacity to modify the inherent variability in seed production that is observed across various years and enhance the effectiveness of reproductive processes. These findings have the potential to enhance our comprehension of the impact of nitrogen addition on the reproductive performance of perennial herbaceous plants and the underlying mechanisms of biodiversity in the context of global environmental changes.

Keywords: reproductive success; biodiversity; plant-pollinator interactions; nitrogen applies; alpine meadow

1. Introduction

The fundamental principle underlying life-history theory revolves around the notion of trade-offs and the allocation of scarce resources to either reproduction or growth and maintenance necessary for survival [1]. Furthermore, a considerable portion of perennial species demonstrates the ability to undergo multiple cycles of growth and reproduction, enabling them to reproduce across multiple seasons. However, it is imperative to recognize that there is notable year-to-year variability in the seed production of perennial herbs in response to the prevailing growth conditions [1,2]. Plants exhibit enhanced seed production when subjected to favorable years, which encompass optimal soil moisture and nutrient levels as well as suitable weather conditions [3,4]. In light of unfavorable years, there has been a decrease in the overall yield of seeds [1,3]. Thus, there is a diverse range of interannual variations in seed production that is associated with different growth years.

The global distribution of resources is undergoing substantial changes due to increasing levels of carbon dioxide (CO_2), the deposition of nutrients, and alterations in land use. Predictive models indicate that multiple factors associated with global change have the potential to modify the interannual variation in seed output [3]. The introduction of nitrogen into the soil has a significant influence on the nutrient cycle, particularly in relation to the plant's ability to obtain readily available resources [5,6]. In theory, the introduction of nitrogen presents a challenge in allocating resources between growth and reproduction, thereby influencing the annual fluctuations in the seed production capacity of perennial plants. As a result, this phenomenon leads to an increase in their overall reproductive efficiency. Nonetheless, there is a notable deficiency in our understanding of the effects of nitrogen supply on the interannual variations in seed production patterns of perennial plants.

The influence of nitrogen on the makeup of soil nutrients controls how it affects the annual fluctuations in reproductive allocation [1,3]. During the early stages of development, plants grown in adequate-N habitats allocate a higher proportion of their resources towards the growth and development of their vegetative organs [7,8]. Subsequently, plants demonstrate an increased allocation of biomass towards their reproductive structures, resulting in an amplified production of flowers [9,10]. Additionally, the influence of nitrogen is a significant factor in the annual fluctuations of plant biomass, as it affects the storage of carbohydrates and the distribution of carbon resources [11,12]. This phenomenon possesses the capacity to induce annual variations in seed production. Additionally, the introduction of nitrogen has the potential to impact both flower production and nectar secretion [13–15]. The alterations in floral traits have the potential to influence pollinator foraging behavior, resulting in changes to the reproductive output of self-incompatibility species [9,16]. Therefore, we speculated that the addition of nitrogen would have an impact on the floral traits of perennial plants, leading to fluctuations in seed production and reproductive efficiency on an annual basis.

Based on the available empirical data, it can be deduced that there has been a significant increase in global atmospheric nitrogen inputs [16]. In Europe and North America, the nitrogen (N) addition rate commonly observed ranges from 10 to 25 kg N ha^{-1} $year^{-1}$ [17]. However, it is important to acknowledge that in China, the yearly rate of nitrogen addition is approximately 50 kg N ha^{-1} $year^{-1}$ [18]. The critical load of nitrogen (N) required to elicit a response from alpine meadow communities has been determined to be 10 kg N ha^{-1} yr^{-1} [19]. The wet deposition of nitrogen has been observed to vary across different regions, with

6.69 kg N ha^{-1} yr^{-1} on the western Tibetan Plateau and 7.55 kg N ha^{-1} yr^{-1} on the eastern Tibetan Plateau [20]. The examination of the responses exhibited by high-altitude ecosystems to elevated nitrogen deposition is highly appropriate for investigation on the Tibetan Plateau.

Saussurea nigrescens demonstrates self-incompatibility and serves as a dominant herbaceous species in alpine meadows. Our previous study showed that the abundance of honey bees evolved to reduce nectar production [21]. Recently, we observed that the production of its seeds is adversely affected by various environmental factors. Herein, we propose the hypothesis that the addition of nitrogen would have an impact on the annual fluctuations in the seed production of perennial plants, consequently leading to an improvement in their overall reproductive efficacy. In order to evaluate the veracity of this hypothesis, a multiyear field experiment was conducted utilizing *S. nigrescens* in a Tibetan meadow. The species under investigation exhibits a perennial life cycle and is dependent on honey bees for pollination [21]. The objective of this study is to investigate the impacts of nitrogen supplementation on various aspects of plant traits, including the aboveground biomass, patterns of resource allocation, production of flowers and nectar, pollinator visitation, and seed production over multiple years. The expectation was that these factors would collectively have an impact on the interannual variation in seed production. The results of our study hold promise for advancing our understanding of the effects of nitrogen addition on the reproductive efficiency of perennial herbaceous plants, as well as underlying mechanisms that regulate biodiversity in the context of global environmental changes.

2. Materials and Methods

2.1. Study Site and Species

The research was carried out within the permanent plots of the Hongyuan Alpine Meadow Ecosystem Research Station, which is affiliated with the Chinese Academy of Sciences. The station is situated on the eastern Tibetan Plateau, specifically at coordinates 32°48′–32°52′ N and 102°01′–102°33′ E, with an altitude of 3500 m. The average annual temperature is recorded at 0.9 °C, with the highest temperature occurring in July at 10.9 °C and the lowest temperature occurring in January at −10.3 °C. The yearly average precipitation amounts to 690 mm, with approximately 80% occurring from May to October. In pastures, it is common for sedges, grasses, and forb species to exhibit predominance. The overall vegetation coverage exceeds 95%, while the height of plants remains below 30 cm [21].

The grassland being studied has been subjected to yak grazing for a period of less than twenty years, specifically during the winter seasons. In addition to the practice of cattle grazing, this land does not exhibit any other agricultural uses. Since 1981, beekeepers have been transferring substantial quantities of *Apis mellifera* colonies, which consist of more than 80 million honey bees, to a specifically designated study site area for honey production. The aforementioned colonies are consistently sustained within this region annually, specifically from May to September.

The species under investigation in our study is *S. nigrescens*, which belongs to the perennial herbs of the Asteraceae family. The species frequently exhibit growth patterns within altitudinal ranges spanning from 2900 to 4300 m. The species typically attain reproductive maturity approximately three years after seed germination. The height of the plants ranges from 15 to 40 cm, and they produce 2 to 5 capitula. Each capitulum contains 20 to 55 florets, specifically mono-ovulated florets. The annular nectary is composed of a bowl-shaped tissue located at the upper region of the ovary, positioned between the ovary and the anther. The bowl's diameter ranges from 1.5 to 2.0 mm, while the style's diameter ranges from 0.3 to 0.5 mm. The observed species exhibits self-incompatibility and initiates its growth phase in the middle of May annually while undergoing senescence in the middle of September [21]. Plants frequently undergo the process of flowering during the period spanning from late July to August, followed by the subsequent development of fruits in early September. Apiculture has been consistently practiced in the alpine meadows of the Zoige Plateau, located in the eastern region of the Tibetan Plateau, since 1981. This technique involves the utilization of the honey bee species known as *Apis mellifera*. From late July to early August, honey bees engaged in foraging activities exhibited a substantial reliance on *S. nigrescens* as a primary source of nectar collection. It is worth noting that the reproductive success of *S. nigrescens* is significantly influenced by the presence of *A. mellifera* [21,22].

2.2. Nitrogen Addition Experiment

In this study, we present three levels of nitrogen (0, 4, and 8 g N m^{-2} yr^{-1}) addition experiments conducted in an alpine meadow located on the eastern Tibetan Plateau [18]. The study employed a fully randomized block design, incorporating three distinct treatments: control (0 g N m^{-2} yr^{-1}), N4 (4 g N m^{-2} yr^{-1}), and N8 (8 g N m^{-2} yr^{-1}). The three treatments were allocated randomly within a block design and replicated six times [23]. A matrix configuration comprising 18 plots (2 m $\times$ 2 m) was established. The arrangement of the plots adhered to a 3 $\times$ 6 grid pattern. The distance between adjacent plots was measured to be 3 m. The nitrogen supply was carried out using urea. In light of the frequent incidence of N wet deposition, particularly during the summer season in the specified geographical area, a solution comprising urea was administered to the plots via a sprayer, specifically in the initial week of May [18]. Following the fertilization process, the vegetation underwent thorough irrigation using water. The cumulative amount of water dispensed corresponded to approximately 2 m of precipitation. In the experimental configuration, a standardized quantity of water was evenly dispersed among the control plots [18]. Fertilization was applied to the experimental plots during the corresponding growing seasons from 2019 to 2022.

2.3. Measurements of Vegetative and Flower Trait

In late July 2018, 8–10 seedlings were randomly selected for each plot. Following that, a protective framework comprising six iron wires measuring 1 mm in diameter was constructed around each plant, which stood at a height of 5 m. During the initial flowering period, 5 plants of selection in each plot were labeled with tags. During the period of peak flowering, we measured the number of capitula per plant as well as the number of flowers per capitulum from 2019 to 2022 [21].

In order to quantify the volume and concentration of nectar, a method was employed where selected plants were subjected to a 24-h period of coverage using cylindrical metal netting [24]. This was performed with the intention of preventing pollinators' access to the plants under investigation. The volume of nectar per flower was measured using micropipettes with a capacity of 1 or 2 μL. Concurrently, a hand refractometer with a precision of 0.5% was employed to quantify the concentration of nectar (Eclipse, Stanley, and Bellingham, Basingstoke, UK). Measurements were conducted from 08:00 to 14:30 on sunny days. Based on our empirical observations, it was found that the presence of pale to white anthers in flowers was indicative of the highest level of nectar volume production [21]. We conducted measurements on a range of 3–5 flowers per capitulum. Nectar volume per capitulum was calculated as nectar volume per flower multiplied by the number of capitula per plant. The number of flowers per plant was calculated as the number of capitula per plant with multiple numbers of flowers per capitulum.

At the end of the experiment, the aerial parts of each plant were collected and divided into leaf, capitulum, and stem. The plant materials that were collected were segregated into individual paper bags, over-dried at 75 °C for 48 h, and subsequently measured using a precision balance with an accuracy of 0.001 g.

2.4. Measurements of Pollinator Visitation

Field observations were carried out during the peak blooming period of *S. nigrescens*, spanning from 2019 to 2022, with the aim of investigating the rates of pollinator visitation. The capitula number of each sampled plant in each plot was initially recorded. Honey bee visitation to the capitula was monitored by observers at a distance of 3 m. The observation periods were evenly distributed throughout the day, occurring between 9:00 and 17:00. During each hour, a single observer viewed each plant for a duration of two minutes from a stationary position. There were three observers, resulting in each plant being observed twice per hour. The observations documented the number of capitula for each plant that was visited by honey bees during the observation period [25]. The observations were exclusively conducted during sunny days and were repeated nine days per plant per year, spanning from 2019 to 2022. A cumulative duration of 144 min of field observations was recorded annually for each plant. Visitation rates (R) per capitulum per hour were calculated as the total number of visits (Nv) per hour divided by the number of capitula (Nc), i.e., R = Nv/Nc [26].

2.5. Measurements of Seed Traits

In the middle of September, ripened capitula were dissected to determine the number of ovules, unfertilized ovules, and viable seeds (which corresponded to the number of ovules per capitulum). The ratio of viable seeds to the total number of ovules was referred to as the seed set, i.e., seed set = seed number per capitulum/total ovule number per capitulum [21]. The number of viable seeds stands for the number of seeds per capitulum. The seeds per capitulum were also measured using a balance (0.001 g) as seed yield.

Seed yield per plant was calculated as seed yield per capitulum multiplied by the number of capitula per plant. Reproductive efficiency was calculated as the seed yield per plant divided by the sum of the aboveground biomass.

2.6. Data Analysis

For analysis, the data pertaining to each trait were first averaged for every plot in this study. The statistical analyses were conducted using the R software 4.2.1 (R Development Core Team 2022). The R software can be accessed at http://www.R-project.org/ (accessed on 23 June 2022).

The data were analyzed using generalized linear mixed models (GLMMs). The fixed effects in the study encompassed N addition treatments (N0, N4, and N8) and the experimental years spanning from 2019 to 2022, while the plot was considered a random effect. The above-ground biomass, nectar concentration, and pollinator visitation rate were modeled using a Gaussian model with an identity-link function. Next, we employed a Gamma model (with a Logit-link) and utilized the Laplace approximation to estimate parameters in order to model various factors, including the number of capitula per plant, number of flowers per capitulum or per plant, nectar volume per capitulum or per plant, seed set, number of seeds per capitulum or per plant, reproductive efficiency, and reproduction allocation. The parameters were estimated using the Laplace approximation method, specifically utilizing the glm.nb function in the 'lme4' package. Upon examination of the residuals, it was determined that a normal error distribution was suitable. Following the identification of a notable impact on the N addition and the experimental year, subsequent post-hoc LSD tests were employed to conduct pairwise comparisons between the N addition and the years. Calculations were performed using the glmer function from the lme4 package [27].

The study aimed to examine the association between flower traits and pollinator visitation rate at the individual plant level. This investigation utilized the automated model selection method provided by the R package glmulti. Pollinator visitation rates can be influenced by several factors, including the nectar volume per capitula, nectar concentration, number of flowers per plant, and above-ground biomass. In this study, we used a random effect model with a random intercept for each plot (random = 1 | plot) to account for potential variability between plots. The authors of the study conducted an analysis to determine the relative significance of each predictor in influencing the outcome [28]. A threshold value of 0.8 is needed in order to identify the variables deemed most significant [29].

The R package "lares" was utilized to examine the ranked cross-correlations among all plant and pollinator variables (http://laresbernardo.github.io/lares/reference/corr_cross.html, accessed on 23 June 2022).

3. Results

Interannual vegetative and floral traits of *S. nigrescens* were found to be regulated by soil nitrogen levels (Figures 1 and 2, and Tables 1 and 2). The N4 and N8 treatments exhibited the greatest aboveground biomass, stem mass, and capitulum mass in the third and fourth years. In contrast, no significant variation in aboveground biomass was observed across the years for the N0 treatment (Figure 1). The number of capitula per plant and the number of flowers per capitulum reached their highest levels during the second and third years for the N4 treatment and in the third and fourth years for the N8 treatment. The highest recorded nectar volume per capitulum was observed in the first three years for the N0 treatment, while for the N8 treatment, it was observed in the third and fourth years. On the other hand, there was no statistically significant variation in nectar volume per capitulum for the N4 treatment across different years (Figure 2). The addition of nitrogen did not result in any year-to-year variation in nectar concentration (Figure 2).

The introduction of nitrogen had an impact on both the rate of pollinator visitation and seed set (Figure 3, Table 3). Peak pollinator visitation and seed set were observed during the initial two years for the N0 treatment and in the third and fourth years for the N8 treatment. Conversely, no discernible differences in these variables were found during the years of the N4 treatment (Figure 3).

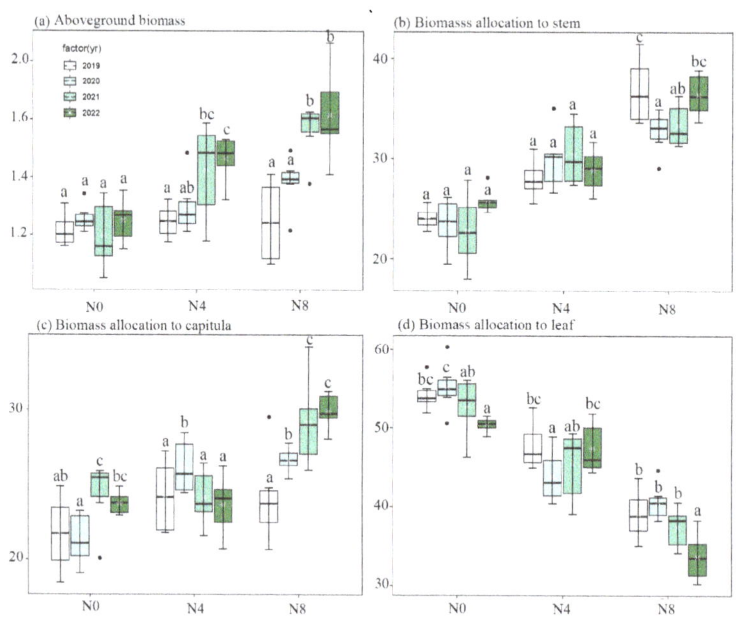

Figure 1. Comparison of aboveground biomass (**a**), biomass allocation to stem (**b**), capitulum (**c**), and leaf (**d**) across N0, N4, and N8 treatments for each year (from 2019 to 2022) of *Saussurea nigrescens*. Different letters above the boxes denote significant differences among treatments ($p < 0.05$).

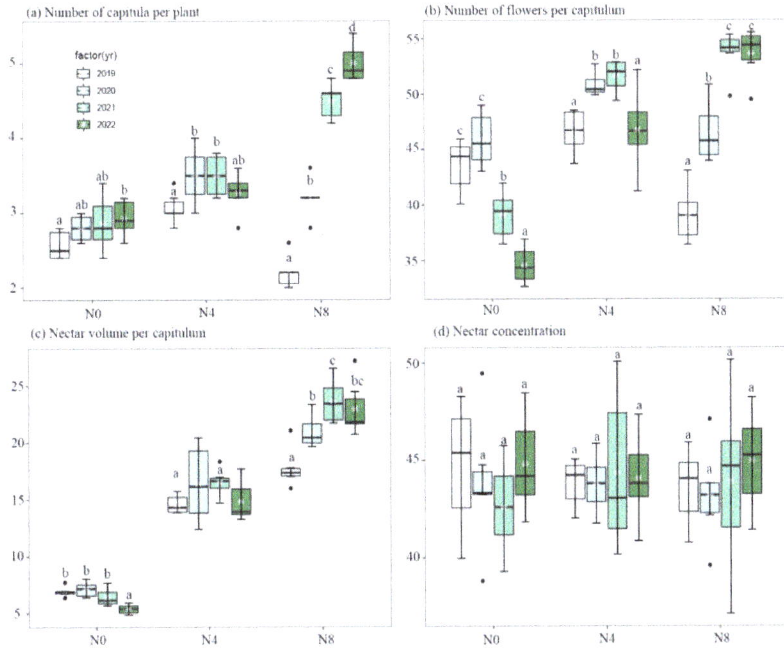

Figure 2. Comparison of the number of capitula per plant (**a**), number of flowers per capitula (**b**), nectar volume per capitula (**c**), and nectar concentration (**d**) across N0, N4, and N8 treatments for each year (from 2019 to 2022) of *Saussurea nigrescens*. Different letters above the boxes denote significant differences among treatments ($p < 0.05$).

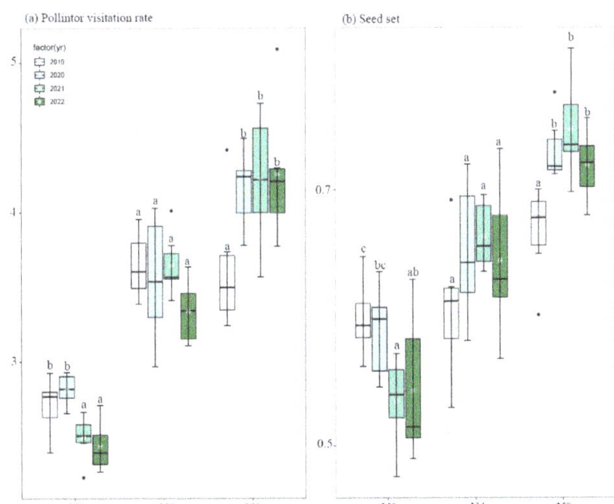

Figure 3. Comparison of the number of pollinator visitation rate (**a**) and seed set (**b**) across N0, N4, and N8 treatments for each year (from 2019 to 2022) of *Saussurea nigrescens*. Different letters above the boxes denote significant differences among treatments ($p < 0.05$). ssr—seed set; vr—pollination visitation rate (visitors capitula^{-1}·hr^{-1}).

The nitrogen supply in the soil was seen to have an impact on both reproductive efficiency and seed production per capitulum or plant (Figures 4a,b and 5; Table 3). The N0 treatment exhibited the highest reproductive efficiency and the greatest number of seeds per capitulum during the initial two years. Similarly, the N4 treatment demonstrated these characteristics in the second and third years, while the N8 treatment displayed them in the third and fourth years.

The introduction of nitrogen had varying effects on the correlations between different traits (Figure 6). The study revealed a positive correlation between seed production per plant and both flower production per capitulum and capitulum production per plant (Figure 6). There exists a positive correlation between the number of seeds per capitulum, the frequency of pollinator visits, and the quantity of nectar produced by each capitulum. The study revealed a positive correlation between above-ground biomass and various factors, including reproductive efficiency, seed set, pollinator visitation frequency, and the number of capitula per plant (Figure 6).

Both the overall quantity of capitula per plant and the number of seeds per capitulum exhibited a noteworthy influence on seed production. (Figure 4c). Both the level of nitrogen addition and the experimental year had an impact on the number of seeds that a plant produced (Figure 4c).

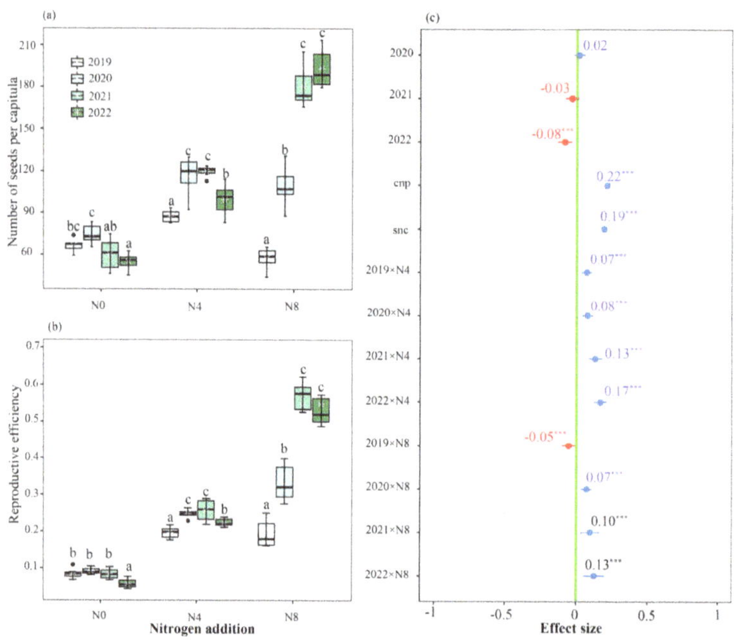

Figure 4. Interannual variation in the number of seeds produced per plant (**a**) and the reproductive efficiency (**b**) of *Saussurea nigrescens* among the N0 (0 g·m^{-2} N), N4 (4 g·m^{-2}) and N8 (8 g·m^{-2}) treatments; additionally, the effects of nitrogen addition and each experimental year on the number of seeds per plant (**c**) was analyzed using generalized linear mixed models (GLMMs). Different letters above the boxes denote significant differences among treatments ($p < 0.05$). The effect of nitrogen addition and each experimental year on the number of seeds per plant indicated by values and 95% confidence interval (CI) of estimated slopes from GLMM, with significant (95% CI does not overlap with zero, *** $p < 0.001$) positive (blue) and negative (red) effect highlighted by colored values with * and CI of the estimated slopes, and values without * and CI (overlap with zero) indicated non-significant difference across the N0, N4, and N8 treatments. The variable "cnp" represents the number of capitula per plant, while the variable "snc" represents the number of seeds per capitula.

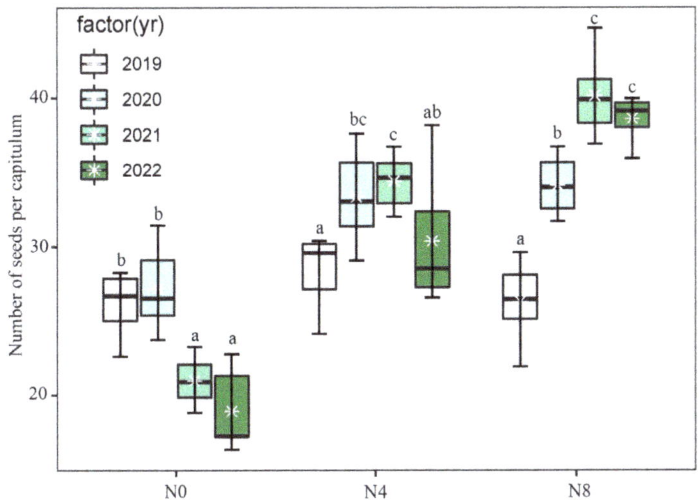

Figure 5. Comparison of the number of seeds per capitula across N0, N4, and N8 treatments for each year (from 2019 to 2022) of *Saussurea nigrescens*. Different letters above the boxes denote significant differences among treatments ($p < 0.05$).

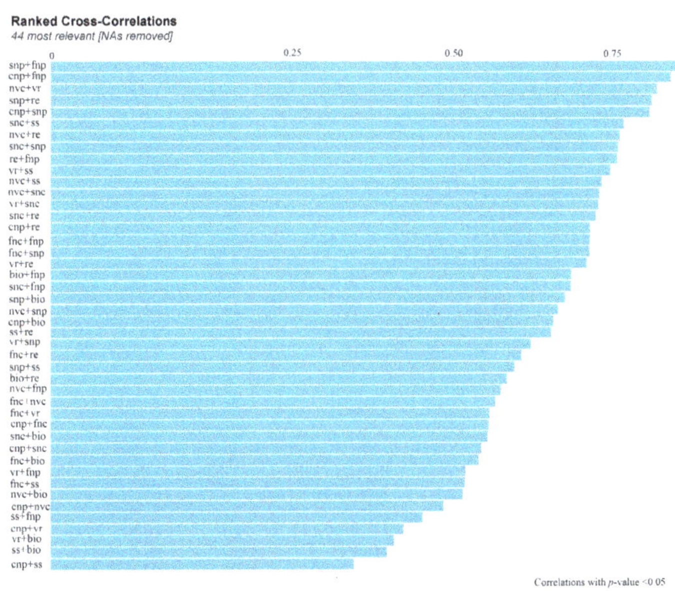

Figure 6. Ranked cross-correlations for variables of vegetative and floret traits, seed traits, and pollinator visitation rate of *Saussurea nigrescens*. Blue colors correspond to positive correlations. The size of the bar was proportional to the correlation coefficients. Forty-four most relevant reflected significant correlations of variables. NAs removed reflect the insignificant correlations that did not present. bio—above-ground biomass; cnp—number of capitula per plant; fnc—number of flowers per capitula; fnp—number of flowers per plant; nc—nectar concentration; nvc—nectar volume per capitula; nvp—nectar volume per plant; snc—number of seeds per capitula; snp—number of seeds per plant; ssr—seed set; vr—pollination visitation rate; nvc + vr represents the relationship between nectar volume per capitula and pollinator visitation rate.

The results of model selection indicated a positive relationship between pollinator visitation rates and the number of flowers and capitula on a plant (Figure 7).

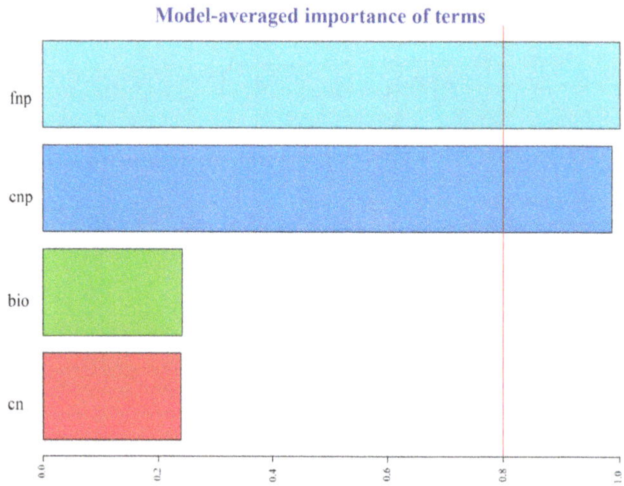

Figure 7. Relative importance of flower traits on pollinator visitation rate of *Saussurea nigrescens*. The red line on the chart indicates which terms are significant. fnp—number of flowers per plant; cnp—number of capitula per plant; bio—above-ground biomass; nc—nectar concentration. The figure shows the result of the best model selection.

Table 1. Effects of nitrogen addition and each year on vegetative and flower traits.

	Aboveground Biomass			Number of Capitulum per Plant			Number of Flowers per Capitula			Number of Flowers per Plant		
Predictors	**Estimates**	**CI**	***p***	**Estimates**	**CI**	***p***	**Estimates**	**CI**	***p***	**Estimates**	**CI**	***p***
Intercept	1.42	1.33–1.50	<0.001	2.57	2.40–2.74	<0.001	43.56	41.87–45.33	<0.001	111.68	103.47–120.54	<0.001
2020	0.04	−0.06–0.15	0.432	1.09	1.00–1.18	0.039	1.05	1.00–1.11	0.050	1.14	1.04–1.25	0.005
2021	−0.02	−0.13–0.09	0.719	1.12	1.03–1.21	0.010	0.90	0.85–0.94	<0.001	1.00	0.92–1.10	0.934
2022	0.03	−0.07–0.14	0.515	1.14	1.05–1.24	0.002	0.79	0.75–0.83	<0.001	0.91	0.83–0.99	0.034
tr [N4]	0.03	−0.08–0.14	0.568	1.20	1.10–1.30	<0.001	1.07	1.02–1.13	0.011	1.28	1.17–1.40	<0.001
tr [N8]	0.03	−0.07–0.14	0.556	0.86	0.79–0.93	<0.001	0.90	0.85–0.94	<0.001	0.77	0.70–0.84	<0.001
2020 × tr [N4]	0.01	−0.14–0.16	0.882	1.05	0.93–1.17	0.445	1.04	0.96–1.11	0.328	1.09	0.96–1.24	0.172
2021 × tr [N4]	0.20	0.05–0.35	0.011	1.02	0.91–1.15	0.704	1.23	1.15–1.33	<0.001	1.26	1.11–1.43	0.001
2022 × tr [N4]	0.18	0.03–0.33	0.018	0.93	0.83–1.05	0.227	1.26	1.18–1.36	<0.001	1.18	1.04–1.34	0.013
2020 × tr [N8]	0.09	−0.06–0.24	0.214	1.33	1.19–1.50	<0.001	1.13	1.05–1.21	0.001	1.52	1.34–1.72	<0.001
2021 × tr [N8]	0.33	0.18–0.48	<0.001	1.83	1.63–2.06	<0.001	1.53	1.42–1.64	<0.001	2.80	2.47–3.18	<0.001
2022 × tr [N8]	0.33	0.18–0.48	<0.001	1.99	1.77–2.23	<0.001	1.73	1.61–1.86	<0.001	3.44	3.03–3.90	<0.001

Table 2. Effects of nitrogen addition and each year on nectar volume and concentration.

	Nectar Concentration			Nectar Volume per Capitula			Nectar Volume per Plant		
Predictors	**Estimates**	**CI**	***p***	**Estimates**	**CI**	***p***	**Estimates**	**CI**	***p***
Intercept	44.72	42.60–46.84	<0.001	6.95	6.40–7.55	<0.001	18.10	16.18–20.25	<0.001
2020	−0.90	−3.90–2.09	0.548	1.03	0.93–1.14	0.585	1.09	0.96–1.24	0.184
2021	−2.12	−5.11–0.88	0.162	0.93	0.84–1.03	0.150	1.03	0.90–1.17	0.667
2022	0.08	−2.91–3.07	0.958	0.78	0.70–0.86	<0.001	0.87	0.76–0.99	0.030
tr [N4]	−0.89	−3.88–2.10	0.554	2.10	1.90–2.32	<0.001	2.47	2.17–2.81	<0.001
tr [N8]	−1.11	−4.10–1.88	0.461	2.56	2.31–2.83	<0.001	2.14	1.88–2.43	<0.001
2020 × tr [N4]	0.84	−3.39–5.08	0.691	1.09	0.94–1.26	0.233	1.17	0.98–1.40	0.090
2021 × tr [N4]	2.64	−1.59–6.87	0.217	1.22	1.05–1.40	0.008	1.25	1.04–1.50	0.017
2022 × tr [N4]	0.16	−4.08–4.39	0.941	1.31	1.13–1.51	<0.001	1.25	1.05–1.50	0.015
2020 × tr [N8]	0.46	−3.77–4.69	0.829	1.15	0.99–1.32	0.061	1.58	1.31–1.89	<0.001
2021 × tr [N8]	2.42	−1.82–6.65	0.258	1.43	1.24–1.65	<0.001	2.68	2.23–3.21	<0.001
2022 × tr [N8]	1.26	−2.97–5.50	0.553	1.66	1.44–1.91	<0.001	3.42	2.85–4.10	<0.001

Table 3. Effects of nitrogen addition and each year on visitation rate, seed set, number of seeds per capitula, and reproductive efficiency.

Predictors	Pollinator Visitation Rate			Seed Set			Number of Seeds per Capitula			Reproductive Efficiency		
	Estimates	CI	*p*	Estimates	CI	*p*	Estimates	CI	*p*	Estimates	CI	*p*
Intercept	2.71	2.48–2.93	<0.001	0.60	0.57–0.63	<0.001	26.14	24.33–28.09	<0.001	0.08	0.08–0.09	<0.001
2020	0.11	−0.21–0.42	0.495	0.98	0.92–1.06	0.640	1.04	0.94–1.15	0.444	1.07	0.92–1.25	0.364
2021	−0.22	−0.53–0.10	0.173	0.89	0.83–0.96	0.002	0.80	0.72–0.89	<0.001	0.98	0.84–1.14	0.789
2022	−0.28	−0.59–0.04	0.083	0.91	0.84–0.97	0.008	0.72	0.65–0.80	<0.001	0.67	0.58–0.78	<0.001
tr [N4]	0.94	0.62–1.25	<0.001	1.01	0.94–1.09	0.689	1.09	0.98–1.20	0.110	2.32	1.99–2.70	<0.001
tr [N8]	0.91	0.60–1.23	<0.001	1.11	1.04–1.20	0.004	1.01	0.91–1.11	0.921	2.28	1.96–2.65	<0.001
2020 × tr [N4]	−0.20	−0.64–0.25	0.377	1.09	0.99–1.21	0.092	1.13	0.98–1.30	0.096	1.18	0.95–1.46	0.126
2021 × tr [N4]	0.22	−0.22–0.66	0.326	1.22	1.11–1.35	<0.001	1.51	1.31–1.74	<0.001	1.34	1.08–1.65	0.009
2022 × tr [N4]	−0.03	−0.48–0.41	0.885	1.17	1.06–1.29	0.003	1.48	1.28–1.71	<0.001	1.70	1.37–2.11	<0.001
2020 × tr [N8]	0.43	−0.01–0.87	0.056	1.11	1.01–1.23	0.037	1.25	1.08–1.44	0.003	1.61	1.30–1.99	<0.001
2021 × tr [N8]	0.82	0.37–1.26	0.001	1.25	1.13–1.39	<0.001	1.90	1.65–2.20	<0.001	2.88	2.33–3.57	<0.001
2022 × tr [N8]	0.91	0.46–1.35	<0.001	1.19	1.07–1.32	0.001	2.03	1.76–2.34	<0.001	4.24	3.42–5.25	<0.001

4. Discussion

The findings indicated that increasing the nitrogen content in the soil has the potential to alter the natural fluctuations in seed production and reproductive efficiency that occur over different years of *S. nigrescens*. Nitrogen has been found to alter the interannual variations in the number of flowers per capitulum as well as the number of capitula per plant, consequently affecting the annual frequency of seed production. However, the introduction of nitrogen had an impact on the fluctuation of flower rewards from year to year. This included changes in nectar production per capitulum and the number of flowers per plant. Additionally, the patterns of pollinator visitation rate and seed set were also altered throughout the duration of the experiment. A positive correlation has been observed between the number of seeds produced per plant and several factors, such as the number of flowers per capitulum, the number of capitula per plant, the volume of nectar per capitulum, and the rate of visitation by pollinators. A significant positive correlation was identified between aboveground biomass and several factors, encompassing the allocation of biomass to the stem, the number of flowers per capitulum, the number of capitula per plant, the volume of nectar per capitulum, and the seed production per plant. This demonstrates that the incorporation of element N significantly influences the maintenance of aboveground biomass and flower traits, resulting in alterations in the frequency of pollinator visits and the interannual variability in seed production. The findings of this research have the potential to enhance our comprehension of the impacts of nitrogen supply on the reproductive efficacy of perennial herbs as well as the underlying mechanisms that influence biodiversity in alpine meadows amidst global environmental changes.

The presence of nitrogen significantly influences plant biomass through the alteration of carbohydrate reserves and the allocation of carbon resources [11,12]. Plants grown in habitats with an adequate nitrogen supply often exhibit a greater allocation of biomass towards the stem [30,31]. This allocation is necessary to provide mechanical support for the increased aboveground biomass of *S. nigrescens*. In addition, the introduction of nitrogen resulted in an increase in community height (30~35 cm). This increase in height can be attributed to the promotion of stem growth, which enables *S. nigrescens* to compete better for light resources [32,33]. Moreover, the highest aboveground biomass was observed during the third and fourth years of our study (Figure 1). We postulated that a higher percentage of underground resources experienced a transformation resulting in the production of a greater amount of aboveground biomass over the course of the latter two years [31].

Furthermore, plants frequently allocate a greater number of resources towards the development of their reproductive organs, including the production of a larger quantity of flowers and the provision of increased flower rewards [9,10,34]. This relationship is exemplified by a positive association between the aboveground biomass and several floral traits, such as the number of capitula per plant, the number of flowers per capitulum, and the amount of nectar per capitulum (Figure 6). Both the N4 and N8 treatments exhibited an increase in aboveground biomass as well as an increase in the flower mass fraction (Figure 1). This implies that an augmentation in aboveground biomass has the potential to enhance the number of flowers and nectar [13,35]. Furthermore, our study revealed a positive correlation between nectar and flower production and aboveground biomass fluctuations over consecutive years. The flower or capitulum number and nectar production of the plant and capitulum exhibit interannual variation, which is influenced by changes in aboveground biomass over the course of the experimental years. This finding suggests that the addition of nitrogen not only alters the year-to-year fluctuations in aboveground biomass [31] but also affects the allocation of resources towards flower and nectar production, consequently leading to annual changes in flower and nectar abundance.

In a research investigation examining the impacts of nutrient supply on nectar traits, the addition of nitrogen did not result in an increase in nectar secretion for *Trifolium pratense*. However, it did lead to an increase in the rate of nectar secretion for *Antirrhinum majus* [36]. Nevertheless, the augmentation in nectar secretion of *A. majus* and *Ipomopsis aggregata* was solely observed under conditions of minimal nitrogen supplementation (10 kg N ha^{-1} $year^{-1}$). At elevated levels of nitrogen addition (200 kg N ha^{-1} $year^{-1}$), there was a notable reduction in nectar secretion for both species [13,36]. This suggests that the influence of nitrogen supplementation on nectar production is dependent on both the dosage of nitrogen and the particular plant species. The nectar secretion of *S. nigrescens* was found to be enhanced by both the N4 and N8 treatments. The extent to which the dose of nitrogen reduces nectar production in *S. nigrescens* remains unknown. Indeed, numerous studies have demonstrated that the introduction of nitrogen has a positive impact on the overall concentration of amino acids [37–39]. The presence of nitrogen addition has been frequently observed to result in an increase in the levels of asparagine and glutamine among the various individual amino acids [15,37,39]. The disparity in nectar concentration (e.g., sugar content) was not identified in *S. nigrescens*. The concentration of nectar is subject to influence by microclimate factors, particularly relative humidity [40]. The available evidence indicates that the amount of water in nectar often affects how concentrated it is. The humidity gradient affects the rate of evaporation, which in turn affects the flow of water between nectar and air [41], which in turn affects the amount of water in nectar. The study site where three treatments were conducted exhibited similar levels of humidity. The plants exhibited uniform growth under equivalent relative humidity conditions, leading to nectar concentrations that were indistinguishable.

Bees derive their energy from the consumption of nectar and pollen. The provision of protein and other essential nutrients is imperative for the optimal growth and development of larvae. Plants that exhibit a high production of nectar in substantial quantities have the potential to attract a greater number of visits from pollinators as well as prolong the duration of the visits [21,42,43]. Within the designated study areas, it has been observed that honey bees play a predominant role as the principal pollinators of *S. nigrescens*. From late July to early August, domesticated honey bees demonstrate a notable reliance on *S. nigrescens* as their principal nectar-gathering resource [21]. There was a correlation between honey bee visitation and nectar production [44]. Plants that thrive in N4 and N8 habitats, characterized by the presence of nectar-rich flowers, exhibit a higher frequency of pollinator visits. Conversely, plants inhabiting the N0 habitat, which is characterized by the absence of nectar-rich flowers, tend to attract fewer pollinators (Figures 2 and 3). A positive correlation has been observed between the overall sugar concentration in nectars and specific pollinators [45]. As an illustration, honey bees exhibit a preference for nectar characterized by a diminished sugar concentration [46], while bumble bees display a preference for nectar characterized by an elevated sugar concentration [47]. The sugar concentration of *S. nigrescens* exhibited a range of 37% to 50%, aligning with the nectar concentration preferences observed in honey bees. The level of flower production has an impact on the frequency of flower visits. The findings from our model selection analysis indicated that both the quantity of nectar produced per plant and the number of capitula per plant are significant factors in determining the visitation of pollinators. It is imperative to acknowledge that floral scents have a significant impact on the attraction of pollinators [48]. Nevertheless, the impact of nitrogen supplementation on the olfactory traits of flowers and the preferences exhibited by pollinators remains uncertain. The visitation rate is contingent upon the combined factors of species richness and the abundance of pollinators [49]. The honey bee population size exhibited no discernible variation among the study sites as a result of the substantial release of honey bees by beekeepers in the designated study areas [21,22].

The introduction of nitrogen has the potential to modify the inherent variations in seed production and reproductive efficiency observed across different years in *S. nigrescens*. The presence of nitrogen has been observed to have an impact on the interannual variations in both the number of flowers per capitulum and the number of capitula per plant (Figure 6). The seed yield per plant is influenced by two factors: the number of capitula per plant and the number of seeds per capitulum (Figure 4). *S. nigrescens* exhibits self-incompatibility, with honey bees serving as the primary pollinators for fertilization [21]. This deduction is supported by the demonstrated positive correlation between seed set and pollinator visitation. The introduction of nitrogen resulted in a notable augmentation in the number of flowers and capitula, as well as an increase in nectar production and pollinator visitation [13,14,16], consequently leading to an enhanced seed output. Moreover, there is a correlation between aboveground biomass and the annual variation in seed production. A positive correlation has been identified between above-ground biomass and various reproductive traits, including seed quantity per plant, flower quantity per capitulum, and capitulum quantity per plant. This implies that the addition of nitrogen has the potential to modify the year-to-year fluctuations in aboveground biomass, resulting in changes in the number of flowers per capitulum, the number of capitula per plant, nectar production, and pollinator visitation across various years. Consequently, this can also impact interannual seed output and reproductive efficiency.

Interannual variations in seed yield exert a significant impact on the sustainability of perennial plant populations [1,31]. In light of an uncertain environment, plants undertake a thorough evaluation of the benefits and costs linked to various reproductive strategies. Under optimal circumstances, plants exhibit a tendency to produce a larger number of seeds, thereby providing benefits for the replenishment of populations. Plants demonstrate a reduction in resource allocation towards reproduction when confronted with adverse circumstances [50], opting to prioritize resource allocation towards survival. Our study suggests that over the course of the first two years of nitrogen supplementation, plants exhibited a preference for allocating resources toward vegetative tissue [51]. This allocation strategy ultimately resulted in a significant improvement in seed yields during the subsequent third and fourth years for N4 and N8 plants.

The presence of soil nutrients has the potential to influence reproductive processes [9]. For example, the presence of mycorrhizal infection in soil with high phosphorus content has the potential to enhance seed production [52]. The present study did not observe the variation in soil nitrogen (N) and the other element levels among the different nitrogen supply treatments. It was found that soil nutrients exerted a notable impact on the secretion and concentration of nectar in flowers. Consequently, these factors influenced the rates of visitation by pollinators and the production of seeds [13]. Further investigation would be required to conduct a comprehensive examination. It has been seen that adding nitrogen changes the composition and activity of nitrogen-fixing bacteria [53]. This has an effect on how seed production changes from year to year. Further investigation is warranted to empirically examine the potential impacts of this phenomenon.

In general, increasing the amount of nitrogen in the soil can change the natural differences in seed yield and improve the reproductive success of *S. nigrescens* seen in different years. The augmentation of nitrogen concentrations has a positive impact on the seed production of *S. nigrescens*, especially when combined with a dosage of 8 $g \cdot m^{-2}$ of nitrogen. However, the potential effects of nitrogen addition on the processes of seed germination and seeding establishment have yet to be investigated. The impact on plant abundance and community composition remains uncertain. The results of this study could help us learn more about how nitrogen supplementation affects the ability of perennial herbaceous plants to reproduce, as well as the basic processes that affect biodiversity in alpine meadows when the global environment changes.

Author Contributions: Conceptualization, J.M., Y.L. and C.Z.; methodology, J.M., Y.L. and C.Z.; software, J.M., Y.L. and R.W.; validation, J.M. and Y.L.; formal analysis, R.W. and Y.L.; investigation, Y.L., J.M., R.W., P.C., Y.Y., Y.C. and C.Z.; resources, Y.L.; data curation, J.M.; writing—original draft preparation, Y.L.; writing—review and editing, J.M., Y.Y. and C.Z.; visualization, Y.L.; supervision, J.M.; project administration, J.M.; funding acquisition, J.M. All authors have read and agreed to the published version of the manuscript.

Funding: This study was funded by the National Natural Science Foundation of China (31870393), by the Natural Science Foundation of Sichuan Province (22NSFSC0242), and the open project from the Ecological Security and Protection Key Laboratory of Sichuan Province (ESP1806).

Institutional Review Board Statement: Not applicable.

Informed Consent Statement: Not applicable.

Data Availability Statement: Not applicable.

Acknowledgments: We thank Jia Liu, Wenfei Dai, Hao Lin, Lijun Wu, Chuan Yuan, Hong Yang, Cuomu Baima, and Xinlin Wang for the field and lab assistance.

Conflicts of Interest: The authors declare no conflict of interest.

References

1. Friedman, J. The evolution of annual and perennial plant life histories: Ecological correlates and genetic mechanisms. *Annu. Rev. Ecol. Evol. Syst.* **2020**, *51*, 461–481. [CrossRef]
2. Herrera, C.M. Population-level estimates of interannual variability in seed production: What do they actually tell us? *Oikos* **1998**, *82*, 612–616. [CrossRef]
3. Pearse, I.S.; LaMontagne, J.M.; Koenig, W.D. Inter-annual variation in seed production has increased over time (1900–2014). *Proc. R. Soc. B* **2017**, *284*, 20171666. [CrossRef] [PubMed]
4. Dai, W.F.; Yang, Y.L.; Patch, H.M.; Grozinger, C.M.; Mu, J.P. Soil moisture affects plant-pollinator interactions in an annual flowering plant. *Philos. Trans. R. Soc. Lond. B Biol. Sci.* **2022**, *377*, 20210423. [CrossRef] [PubMed]
5. Lu, M.; Zhou, X.H.; Luo, Y.Q.; Yang, Y.H.; Fang, C.M.; Chen, J.K.; Li, B. Minor stimulation of soil carbon storage by nitrogen addition: A meta-analysis. *Agric. Ecosyst. Environ.* **2011**, *140*, 234–244. [CrossRef]
6. Deng, Q.; Hui, D.F.; Dennis, S.; Reddy, C. Responses of terrestrial ecosystem phosphorus cycling to nitrogen addition: A meta-analysis. *Glob. Ecol. Biogeogr.* **2011**, *26*, 713–728. [CrossRef]
7. Grechi, I.; Vivin, P.; Hilbert, G.; Milin, S.; Robert, T.; Gaudillère, J.P. Effect of light and nitrogen supply on internal C: N balance and control of root-to-shoot biomass allocation in grapevine. *Environ. Exp. Bot.* **2007**, *59*, 139–149. [CrossRef]
8. Metay, A.; Magnier, J.; Guilpart, N.; Christophe, A. Nitrogen supply controls vegetative growth, biomass and nitrogen allocation for grapevine (cv. Shiraz) grown in pots. *Funct. Plant Biol.* **2014**, *42*, 105–114. [CrossRef]
9. Vaudo, A.D.; Erickson, E.; Patch, H.M.; Grozinger, C.M.; Mu, J.P. Impacts of soil nutrition on floral traits, pollinator attraction, and fitness in cucumbers (*Cucumis sativus* L.). *Sci. Rep.* **2022**, *12*, 21802. [CrossRef]
10. Wang, R.; Lou, Y.; Deng, D.; Che, P.; Zhao, C.; Yang, Y.; Mu, J. Summer grazing change fitness in a Tibetan lotus. *Glob. Ecol. Conserv.* **2023**, *44*, e02505. [CrossRef]
11. Bobbink, R.; Lamers, L.P.M. Effects of increased nitrogen deposition. In *Air Pollution and Plant Life*; Bell, J.N.B., Treshow, M., Eds.; John Wiley and Sons: Chichester, UK, 2002; pp. 201–235.
12. Drewniak, B.; Gonzalez-Meler, M.A. Earth system model needs for including the interactive representation of nitrogen deposition and drought effects on forested ecosystem. *Forests* **2017**, *8*, 267. [CrossRef]
13. Burkle, L.A.; Irwin, R.E. Beyond biomass: Measuring the effects of community-level nitrogen enrichment on floral traits, pollinator visitation and plant reproduction. *J. Ecol.* **2010**, *98*, 705–717. [CrossRef]
14. Hoover, S.E.R.; Ladley, J.J.; Shchepetkina, A.A.; Tisch, M.; Gieseg, S.P.; Tylianakis, J.M. Warming, CO_2, and nitrogen deposition interactively affect a plant-pollinator mutualism. *Ecol. Lett.* **2012**, *15*, 227–234. [CrossRef] [PubMed]
15. Ceulemans, T.; Hulsmans, E.; Ende, W.V.; Honnay, O. Nutrient enrichment is associated with altered nectar and pollen chemical composition in *Succisa pratensis* Moench and increased larval mortality of its pollinator *Bombus terrestris* L. *PLoS ONE* **2017**, *12*, e0174380. [CrossRef] [PubMed]
16. David, T.I.; Storkey, J.; Stevens, C.J. Understanding how changing soil nitrogen affects plant–pollinator interactions. *Arthropod Plant Interact.* **2019**, *13*, 671–684. [CrossRef]
17. Bleeker, A.; Hicks, W.K.; Dentener, F.; Galloway, J.; Erisman, J.W. Nitrogen deposition as a threat to the world's protected areas under the convention on biological diversity (CBD). *Environ. Pollut.* **2011**, *159*, 295–303. [CrossRef]

18. Liu, X.; Zhang, Y.; Han, W.; Tang, A.; Shen, J.; Cui, Z.; Vitousek, P.; Erisman, J.W.; Goulding, K.; Christie, P.; et al. Enhanced nitrogen deposition over China. *Nature* **2013**, *494*, 459–462. [CrossRef]
19. Bowman, W.D.; Cleveland, C.C.; Halada, L.; Hreško, J.; Baron, J.S. Negative impact of nitrogen deposition on soil buffering capacity. *Nat. Geosci.* **2008**, *1*, 767–770. [CrossRef]
20. Lü, C.Q.; Tian, H.Q. Spatial and temporal patterns of nitrogen deposition in China: Synthesis of observational data. *J. Geophys. Res.* **2007**, *112*, D22S05. [CrossRef]
21. Mu, J.; Peng, Y.; Xi, X.; Wu, X.; Griffin, J.N.; Niklas, K.J.; Sun, S. Domesticated honeybees evolutionarily reduce flower nectar volume in a Tibetan lotus. *Ecology* **2014**, *95*, 3161–3172. [CrossRef]
22. Su, R.; Dai, W.; Yang, Y.; Wang, X.; Gao, R.; He, M.; Zhao, C.; Mu, J. Introduced honey bees increase host plant abundance but decrease native bumble bee species richness and abundance. *Ecosphere* **2022**, *13*, e4085. [CrossRef]
23. Xia, J.; Wan, S. Independent effects of warming and nitrogen addition on plant phenology in the Inner Mongolian steppe. *Ann. Bot.* **2013**, *111*, 1207–1217. [CrossRef] [PubMed]
24. Real, L.A.; Rathcke, B.J. Individual variation in nectar production and its effects on fitness in Kalmia latifolia. *Ecology* **1991**, *72*, 149–155.
25. Vaudo, A.D.; Patch, H.M.; Mortensen, D.A.; Grozinger, C.M.; Tooker, J.F. Bumble bees exhibit daily behavioral patterns in pollen foraging. *Arthropod Plant Interact.* **2014**, *8*, 273–283. [CrossRef]
26. Arroyo, M.T.K.; Armesto, J.J.; Primack, R.B. Community studies in pollination ecology in the high temperate Andes of central Chile. II. Effect of temperature on visitation rates and pollination possibilities. *Plant Syst. Evol.* **1985**, *149*, 187–203. [CrossRef]
27. Bates, D.M.; Maechler, M. Lme 4: Linear Mixed-Effects Models Using S4 Classes. R Package Version 0.999375-36. 2011. Available online: http://cran.r-project.org/web/packages/lme4 (accessed on 1 November 2011).
28. Chen, J.; Luo, Y.; Van Groenigen, K.J.; Hungate, B.A.; Cao, J.; Zhou, X.; Wang, R.W. A keystone microbial enzyme for nitrogen control of soil carbon storage. *Sci. Adv.* **2018**, *4*, eaaq1689. [CrossRef] [PubMed]
29. Terrer, C.; Vicca, S.; Hungate, B.A.; Phillips, R.P.; Prentice, I.C. Mycorrhizal association as a primary control of the CO_2 fertilization effect. *Science* **2016**, *353*, 72–74. [CrossRef]
30. Niu, S.; Wu, M.; Han, Y.; Xia, J.; Zhang, Z.; Yang, H.; Wan, S. Nitrogen effects on net ecosystem carbon exchange in a temperate steppe. *Glob. Chang Biol.* **2009**, *16*, 144–155. [CrossRef]
31. Yan, Z.; Eziz, A.; Tian, D.; Li, X.; Hou, X.; Peng, H.; Han, W.; Guo, Y.; Fang, J. Biomass allocation in response to nitrogen and phosphorus availability: Insight from experimental manipulations of *Arabidopsis thaliana*. *Front. Plant Sci.* **2019**, *10*, 598. [CrossRef]
32. Leith, I.D.; Hicks, W.K.; Fowler, D.; Woodin, S.J. Differential responses of UK upland plants to nitrogen deposition. *New Phytol.* **1999**, *141*, 277–289.
33. Verma, P.; Sagar, R.; Verma, H.; Verma, P.; Singh, D.K. Changes in species composition, diversity and biomass of herbaceous plant traits due to N amendment in a dry tropical environment of India. *J. Plant Ecol.* **2015**, *8*, 321–332. [CrossRef]
34. Mu, J.; Peng, Y.; Xi, X.; Wu, X.; Li, G.; Niklas, K.J.; Sun, S. Artificial asymmetric warming reduces nectar yield in a Tibetan alpine species of Asteraceae. *Ann. Bot.* **2015**, *116*, 899–906. [CrossRef] [PubMed]
35. Burkle, L.A.; Irwin, R.E. The importance of interannual variation and bottom–up nitrogen enrichment for plant–pollinator networks. *Oikos* **2009**, *118*, 1816–1829. [CrossRef]
36. Shuel, R.W. Some aspects of the relation between nectar secretion and nitrogen, phosphorus, and potassium nutrition. *Can. J. Plant Sci.* **1956**, *37*, 220–236.
37. Gardener, M.C.; Gillman, M.P. The effects of soil fertilizer on amino acids in the floral nectar of corncockle, Agrostemma githago (Caryophyllaceae). *Oikos* **2001**, *92*, 101–106. [CrossRef]
38. Gijbels, P.; Van den Ende, W.; Honnay, O. Landscape scale variation in nectar amino acid and sugar composition in a Lepidoptera pollinated orchid species and its relation with fruit set. *J. Ecol.* **2014**, *102*, 136–144. [CrossRef]
39. Gijbels, P.; Ceulemans, T.; Van den Ende, W.; Honnay, O. Experimental fertilization increases amino acid content in floral nectar, fruit set and degree of selfing in the orchid *Gymnadenia conopsea*. *Oecologia* **2015**, *179*, 785–795. [CrossRef] [PubMed]
40. Búrquez, A.; Corbet, S.A. Do flowers reabsorb nectar? *Funct. Ecol.* **1991**, *5*, 369–379. [CrossRef]
41. Castellanos, M.C.; Wilson, P.; Thomson, J.D. Dynamic nectar replenishment in flowers of *Penstemon* (Scrophulariaceae). *Am. J. Bot.* **2002**, *89*, 111–118. [CrossRef]
42. Mitchell, R.J.; Irwin, R.E.; Flanagan, R.J.; Karron, J.D. Ecology and evolution of plant-pollinator interactions. *Ann. Bot.* **2009**, *103*, 1355–1363. [CrossRef]
43. Zhao, Y.; Ren, Z.; Lázaro, A.; Wang, H.; Bernhardt, P.; Li, H.; Li, D. Floral traits influence pollen vectors' choices in higher elevation communities in the Himalaya-Hengduan Mountains. *BMC Ecol.* **2016**, *16*, 26. [CrossRef] [PubMed]
44. Sponsler, D.; Iverson, A.; Steffan-Dewenter, I. Pollinator competition and the structure of floral resources. *Ecography* **2023**, *2023*, e06651. [CrossRef]
45. Wright, G.A.; Nicolson, S.W.; Shafir, S. Nutritional physiology and ecology of honey bees. *Ann. Rev. Entomol.* **2018**, *63*, 327–344. [CrossRef] [PubMed]
46. Waller, G.D. Evaluating responses of honeybees to sugar solutions using an artificial-flower feeder. *Ann. Entomol. Soc. Am.* **1972**, *65*, 857–862. [CrossRef]
47. Crane, E. Bee products. *Bee World* **1972**, *53*, 38–39. [CrossRef]

48. Burkle, L.A.; Runyon, J.B. Floral volatiles structure plant–pollinator interactions in a diverse community across the growing season. *Funct. Ecol.* **2019**, *33*, 2116–2129. [CrossRef]
49. Mu, J.; Wu, Q.; Yang, Y.; Huang, M.; Grozinger, C.M. Plant reproductive strategies vary under low and high pollinator densities. *Oikos* **2018**, *127*, 1081–1094. [CrossRef]
50. Vico, G.; Manzoni, S.; Nkurunziza, L.; Murphy, K.; Weih, M. Trade-offs between seed output and life span—A quantitative comparison of traits between annual and perennial congeneric species. *New Phytol.* **2016**, *209*, 104–114. [CrossRef]
51. Delph, L.F.; Meagher, T.R. Sexual dimorphism masks life-history trade-offs in the dioecious plant Silene latifolia. *Ecology* **1995**, *76*, 775–785. [CrossRef]
52. Poulton, J.L.; Bryla, D.; Koide, R.T.; Stephenson, A.G. Mycorrhizal infection and high soil phosphorus improve vegetative growth and the female and male functions in tomato. *New Phytol.* **2002**, *154*, 255–264. [CrossRef]
53. Bizjak, T.; Sellstedt, A.; Gratz, R.; Nordin, A. Presence and activity of nitrogen-fixing bacteria in Scots pine needles in a boreal forest: A nitrogen-addition experiment. *Tree Physiol.* **2023**. [CrossRef]

biology

MDPI

Article

Long-Term Daytime Warming Rather Than Nighttime Warming Alters Soil Microbial Composition in a Semi-Arid Grassland

Jiayin Feng [1,2], Jingyi Ru [1,2], Jian Song [1,2], Xueli Qiu [1,2] and Shiqiang Wan [1,2,*]

1 School of Life Sciences, Hebei University, Baoding 071002, China
2 Institute of Life Science and Green Development, Hebei University, Baoding 071002, China
* Correspondence: swan@hbu.edu.cn

Simple Summary: Global mean temperature has increased by 1.07 °C since the Industrial Revolution, which arouses people's widespread concern about climate warming. Nighttime temperatures increase faster and higher than daytime temperatures around the world. Various microbial groups in the soil may respond differently to such asymmetrically diurnal warming and then change ecological functions. In this study, we used a ten-year experiment in a semi-arid grassland to examine the effects of daytime and nighttime warming on soil microbial composition. The results showed that short-term warming did not change soil microbial composition, but long-term daytime warming rather than nighttime warming decreased the fungi-to-bacteria ratio in soils. In addition, soil respiration enhanced with the decreasing fungi-to-bacteria ratio. This work implies the importance of soil microbial composition in regulating grassland C release under long-term warming, which may help us accurately assess the climate-C feedback in terrestrial ecosystems.

Abstract: Climate warming has profoundly influenced community structure and ecosystem functions in the terrestrial biosphere. However, how asymmetric rising temperatures between daytime and nighttime affect soil microbial communities that predominantly regulate soil carbon (C) release remains unclear. As part of a decade-long warming manipulation experiment in a semi-arid grassland, we aimed to examine the effects of short- and long-term asymmetrically diurnal warming on soil microbial composition. Neither daytime nor nighttime warming affected soil microbial composition in the short term, whereas long-term daytime warming instead of nighttime warming decreased fungal abundance by 6.28% ($p < 0.05$) and the ratio of fungi to bacteria by 6.76% ($p < 0.01$), which could be caused by the elevated soil temperature, reduced soil moisture, and increased grass cover. In addition, soil respiration enhanced with the decreasing fungi-to-bacteria ratio, but was not correlated with microbial biomass C during the 10 years, indicating that microbial composition may be more important than biomass in modulating soil respiration. These observations highlight the crucial role of soil microbial composition in regulating grassland C release under long-term climate warming, which facilitates an accurate assessment of climate-C feedback in the terrestrial biosphere.

Keywords: climate change; asymmetric warming; microbial community; plant cover; carbon cycling; temperate steppe

1. Introduction

Climate warming has led to an increase in global mean temperature by 1.07 °C since 1850 [1] and can profoundly influence the carbon (C) balance between plants and soils in terrestrial ecosystems, resulting in an uncertain response of soil C storage to climate warming [2,3]. Soil microorganisms, an essential component in ecological communities, are extensively involved in soil organic matter decomposition and nutrient cycles, thus playing crucial roles in maintaining ecosystem stability [4,5]. Given the high sensitivity of microbial activities to rising temperatures, climate warming may greatly change the community structure and functions of soil microorganisms [6,7]. Previous studies have suggested that warming stimulates the microbe-mediated soil organic C mineralization, thus causing decreases in soil C storage [8,9]. However, some studies show conflicting outcomes and propose our superficial cognition of warming effects on soil microbial communities and C cycling [10]. Due to the fact that soil microorganisms predominantly regulate the decomposition and formation of soil organic C, and affect soil respiration by changing their metabolic activities [10,11], understanding the variation in microbial composition and their responses to climate warming can help us make accurate predictions on the climate-C feedback in the terrestrial biosphere.

Climate warming can cause variations in soil microclimates, e.g., the increase in soil temperature and reduction in soil water content, which may stimulate the rate of organic matter decomposition or inhibit microbial growth efficiency [8–10], subsequently resulting in different levels of soil C loss. However, long-term ecosystem responses to warming may be different from transient responses [3,11]. The stimulatory effect of warming on soil respiration weakens over time, which could be possibly attributed to the depleted C availability, the thermal adaptation of the microbial community, and the enhanced microbial interactions [2,12,13]. In addition to temperature-mediated microbial activity, changes in plant communities, which provide substantial C inputs to soils and reshape microbial composition, may affect soil C release associated with rising temperatures [3,14]. For example, long-term warming stimulates C storage in forest soils by alleviating nitrogen limitation, or changes plant community composition towards low-productivity species to offset short-lived soil C loss [2,11]. However, how the temporal responses of soil microbial communities to climate warming, and their linkages with plant cover changes and soil C loss remain largely elusive.

A greater increasing rate in daily minimum temperature than that in daily maximum temperature has been found in different regions [15,16], and such asymmetric daytime and nighttime warming can differentially affect hydro-thermal factors, plant physiology, productivity, and ecosystem C cycling [15–19]. Nighttime warming-induced overcompensation of plant photosynthesis and consequent C assimilation may supply more C sources for microbial growth and activity [8,20]. In addition, the enhanced nitrogen availability and organic matter decomposition rate under nighttime warming can intensify the competition for limited nutrients between plants and soil microorganisms, leading to an increase in fungal diversity, or reductions in bacterial abundance and C use efficiency [8,21]. However, our understanding of the effects of asymmetrically diurnal warming on soil microbial communities is mainly based on some short-term (less than 5 years) field experiments, and little is known about microbial feedbacks to long-term daytime and nighttime warming.

Phospholipid fatty acid (PLFA) can be used as an efficient tool to measure microbial community composition and estimate microbial responses to environmental stress [22,23]. Although it has a low taxonomic resolution compared with current molecular methods, PLFA reflects the relative dominance of biomarkers in the soil, such as the ratios of fungi to bacteria and Gram-positive to Gram-negative bacteria [23]. As part of a 10-year (2006–2015) manipulative experiment, this study was conducted to examine the responses of soil microbial composition to short- and long-term daytime and nighttime warming in a semi-arid grassland. The following two hypotheses were specifically tested: (1) long-term warming could have greater impacts on soil microbial composition than short-term warming, likely due to the changes in soil microclimate and plant cover which can reshape the soil microbial community; (2) nighttime warming would affect more microbial groups than daytime warming because the stronger response of plants to nighttime warming can further alter the soil environment and the microbial community.

2. Materials and Methods

2.1. Site Description

The study was conducted in a semi-arid grassland in Duolun County (42°02′ N, 116°17′ E, 1324 m a.s.l.), Inner Mongolia, China. Mean annual precipitation and air temperature are 374.5 mm and 2.4 °C, respectively. Sandy soil of the experimental site is classified as Haplic Calcisols according to the Food and Agriculture Organization of the United Nations (FAO) classification standard. Surface soil bulk density and pH are 1.31 g·cm^{-3} and 7.7 [24]. Plant species in this grassland are dominated by grasses (*Agropyron cristatum*, *Cleistogenes squarrosa*, *Leymus chinensis*, and *Stipa krylovii*), forbs (*Artemisia frigida*, *Potentilla tanacetifolia*, *Heteropappus altaicus*, and *Potentilla acaulis*) and legumes (*Melilotoides ruthenica*, *Astragalus adsurgens*, and *Lespedeza bicolor*).

2.2. Experimental Design

The experiment was established in March 2006 and conducted as part of a complete random block design with 6 treatments, including control (C), daytime warming (6:00–18:00; D), nighttime warming (18:00–6:00; N), diurnal warming (24 h; W), nitrogen fertilization, and diurnal warming plus nitrogen fertilization. There were 4 replicates for each treatment, and therefore 24 3 m × 4 m plots were assigned in a 4 × 6 matrix, with a 3 m distance between adjacent plots.

The infrared radiators (165 cm × 15 cm; MSR-2420, Kalglo Electronics, Bethlehem, PA, USA) were placed 2.25 m above the ground and set at 1600W as the power output in each warming plot [19,25]. The daily mean soil temperatures were elevated by 0.91 °C and 1.18 °C under daytime and nighttime warming, respectively (Figure S1). In each control plot, a "dummy" heater with the same shape and size as the infrared heater was installed at the same height to imitate the shading effect. The asymmetric diurnal warming treatments began in April 2006, and the heaters ran continuously and automatically from 15 March to 15 November in each year. No other manipulations were performed in the experimental plots.

2.3. Soil Microclimate, PLFA Analysis, and Plant Cover Measurement

Soil temperature at the depth of 10 cm was automatically recorded every 10 min using a STM-01 Soil Temperature Measurement System Datalogger (Henan Electronic Institute, China), and soil temperatures during the growing season from May to October were used as the annual mean value [19]. Soil moisture (SM) at the same site and depth was measured by a portable soil moisture probe (Diviner-2000, Sentek, Australia). Soil moisture was analyzed three times per month from May to October each year.

A total of 2 soil cores (0–10 cm) were randomly collected using a soil auger in late August in 2006, 2007, 2011, 2012, and 2015. Soil samples in each plot were mixed together after removing visible extraneous materials and passing through a 2 mm sieve, and then stored at −20 °C until phospholipid fatty acid (PLFA) extraction. In brief, 8 g of dry soil were fractionated and quantified with a mixture containing chloroform, methanol, and phosphate buffer (1:2:0.8 $v/v/v$). The lipids were filtrated with silicic acid column and separated into phospholipids, glycolipids, and neutral lipids, and then subjected to a mild alkaline methanolysis. Then, 19:0 methyl esters were dissolved in hexane as an internal standard. Extracted fatty acid methyl esters were identified with gas chromatograph (Agilent 7890A, Agilent technologies, Santa Clara, CA, USA) and a microbial identification system (MIDI Inc., Newark, DE, USA). The relative biomass of fungi was determined by the summed concentrations of 18:1ω9c, 18:2ω6c, and 16:1ω5c. Gram-negative (G−) bacteria biomass was estimated by 16:1 2OH, 16:1ω7, 16:1ω9, 17:0 cyclo, 17:1ω8, 18:1ω5, 18:1ω7, and 19:0 cyclo ω8, while Gram-positive (G+) bacteria contained 15:0 anteiso, 15:0 iso, 16:0 iso, 16:1 iso G, 17:0 anteiso, 17:0 iso, 18:1ω7 11-methyl, 18:1ω9, and 19:1ω11. The summed concentrations of general FAME, G−, and G+ bacteria were identified as bacterial biomarkers [5]. Specific methodologies for determining microbial biomass C and soil respiration can be found in the Supplementary Materials. The percent cover of each plant species in each 1 m × 1 m plot was recorded by visual estimation in late August each year.

2.4. Data Analysis

The effects of asymmetric daytime and nighttime warming and year, and their possible interactions on soil microclimate and microbial compositions were analyzed using two-way repeated ANOVA measures with SAS V8 software (SAS Institute). Linear regressions were investigated to determine correlations among soil microbial compositions, soil microclimate, and plant functional groups. Redundancy analysis (RDA) and random forest modeling were performed to reveal environmental factors influencing microbial composition in R (v 3.4.0). Structural equation modeling (SEM) exploring the effects of abiotic and biotic factors on soil microbial composition was constructed using AMOS 24.0 (Amos Development Corporation). The goodness of model fit was tested based on the p-value (>0.05), chi-square, and root mean square error of approximation (RMSEA; <0.05). The total, direct, and indirect effects of the abiotic and biotic factors on soil microbial composition were also recorded.

3. Results

3.1. Soil Microclimate

Averaged across the 4 treatments, both soil temperature and moisture substantially fluctuated over the experimental durations (Figures 1 and S2, both $p < 0.001$, Table 1). Daily mean soil temperature was enhanced by 0.32 °C and 0.58 °C under daytime and nighttime warming averaged over the 5 years, respectively (Figure 1a, both $p < 0.001$). Soil moisture was decreased by 0.65% (absolute change, $p < 0.01$) and 0.55% ($p < 0.05$) under daytime and nighttime warming, respectively (Figure 1b). There was no interaction between daytime and nighttime warming on soil temperature or moisture.

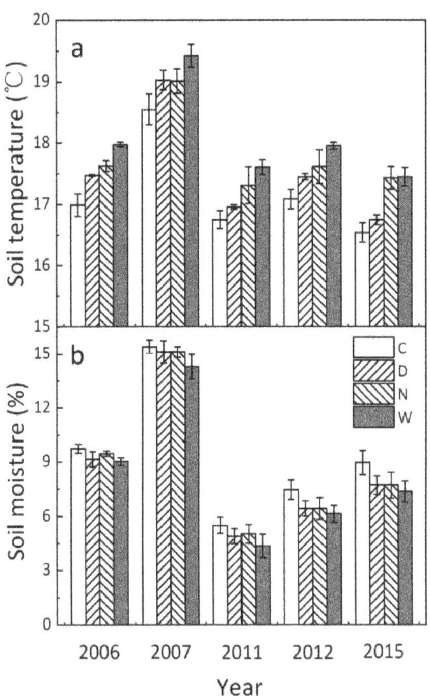

Figure 1. Annual mean soil temperature (**a**) and moisture (**b**) under the 4 treatments in 2006, 2007, 2011, 2012, and 2015 (M ± 1 SE, n = 4). C, control; D, daytime warming; N, nighttime warming; and W, diurnal warming.

When separating into the short- (2006 and 2007) and long-term (2011, 2012, and 2015) stages, daytime warming significantly increased soil temperature by 0.43° ($p < 0.001$) and marginally reduced soil moisture by 0.53% ($p < 0.1$). Nighttime warming stimulated soil temperature by 0.50 °C ($p < 0.001$) but had no effect on soil moisture ($p > 0.05$) in the short term (Figure 1 and Table 1). By contrast, daytime warming enhanced soil temperature by

0.23 °C ($p < 0.05$) and suppressed soil moisture by 0.70% ($p < 0.05$), and nighttime warming increased soil temperature by 0.64 °C ($p < 0.001$) and decreased soil moisture by 0.66% ($p < 0.05$) in the long term (Figure 1).

Table 1. Results (F-values) of repeated measures ANOVA on the effects of daytime warming (D), nighttime warming (N), and year (Y) and their interactions on soil temperature (ST), soil moisture (SM), the relative abundance of bacteria, Gram-negative (G+) bacteria, fungi, arbuscular mycorrhizal fungi (AMF), and the ratios of fungi to bacteria (F/B ratio) and G− to G+ bacteria (G−/G+ ratio) in all 5 years, short (2006 and 2007) and long (2011, 2012 and 2015) term.

Source of Variations		ST	SM	Bacteria	G+ Bacteria	Fungi	AMF	F/B Ratio	G−/G+ Ratio
All 5 years	D	18.6 ***	7.85 **	1.28	1.54	3.76 *	0.06	7.53 **	0.08
	N	64.1 ***	5.94 *	0.08	0.20	0.29	1.57	0.53	0.15
	Y	93.3 ***	233 ***	104 ***	53.1 ***	8.30 ***	17.3 ***	38.8 ***	166 ***
	D × N	0.20	0.29	1.82	0.76	0.53	0.04	0.03	0.00
	D × Y	0.70	0.05	0.52	1.01	0.60	1.01	0.47	1.38
	N × Y	0.71	0.19	1.91	1.16	0.38	0.95	0.74	1.10
	D × N × Y	0.11	0.34	0.41	0.22	0.39	1.21	0.22	0.45
Short term	D	14.3 ***	3.27 ^	1.37	0.64	0.28	0.60	2.52	0.06
	N	19.2 ***	1.67	0.25	0.48	0.93	1.84	2.02	2.55
	Y	170 ***	372 ***	4.26 *	2.17	0.05	3.37 ^	2.02	0.09
	D × N	0.20	0.10	0.80	0.43	0.19	0.17	0.01	0.12
	D × Y	0.01	0.00	0.10	1.08	0.68	0.77	0.57	1.48
	N × Y	0.39	0.33	1.08	1.48	0.10	0.04	0.17	0.39
	D × N × Y	0.02	0.32	0.02	0.24	0.11	1.96	0.06	0.80
Long term	D	6.24 *	4.75 *	0.05	1.93	4.39 *	1.12	5.11 *	0.36
	N	45.5 ***	4.27 *	2.36	0.31	0.01	0.00	0.09	0.90
	Y	9.57 ***	29.5 ***	6.35 **	3.04 ^	11.8 ***	100 ***	9.65 ***	1.14
	D × N	0.05	0.68	1.42	0.51	0.35	0.18	0.09	0.12
	D × Y	0.54	0.02	0.67	3.35*	0.47	2.13	0.66	1.94
	N × Y	0.77	0.06	5.72 **	1.15	0.37	2.48	0.52	0.20
	D × N × Y	0.19	0.22	1.80	0.09	0.75	0.79	0.41	0.36

Significance levels: ^ $p < 0.1$, * $p < 0.05$, ** $p < 0.01$, and *** $p < 0.001$.

3.2. Soil Microbial Composition

Averaged over the 5 years, daytime warming did not affect the relative abundance of bacteria, G+ bacteria, AMF, or the ratio of G− to G+ bacteria (Figure 2a,b,d,f, all $p > 0.05$, Table 1), but reduced fungi proportion by 4.47% (Figure 2c, $p < 0.05$) and the fungi-to-bacteria ratio by 5.63% (Figure 2e, $p < 0.01$). Nighttime warming had no effect on the relative abundance of any of the microbial groups or the ratios of fungi to bacteria and G− to G+ bacteria (Figure 2, all $p > 0.05$). The responses of microbial compositions to daytime and nighttime warming varied with year (all $p < 0.001$). No interaction of daytime and nighttime warming on the relative abundance of any of microbial groups or the ratios of fungi to bacteria and G− to G+ bacteria was detected.

Averaged across the four treatments, the relative abundance of bacteria and the G− to G+ bacteria ratio were higher in the long- than in the short-term stage (Figure 3a, all $p < 0.001$). However, the relative abundance of G+ bacteria ($p < 0.001$) and fungi ($p < 0.05$) as well as the fungi-to-bacteria ratio ($p < 0.001$) in the long term were lower than those in the short term (Figure 3a). Nighttime warming did not change the composition of any of the microbial groups in either short or long terms (all $p > 0.05$, Table 1), whereas daytime warming reduced the relative abundance of fungi by 6.28% ($p < 0.05$) and the ratio of fungi to bacteria by 6.76% ($p < 0.01$) in the long term only (Figure 3b).

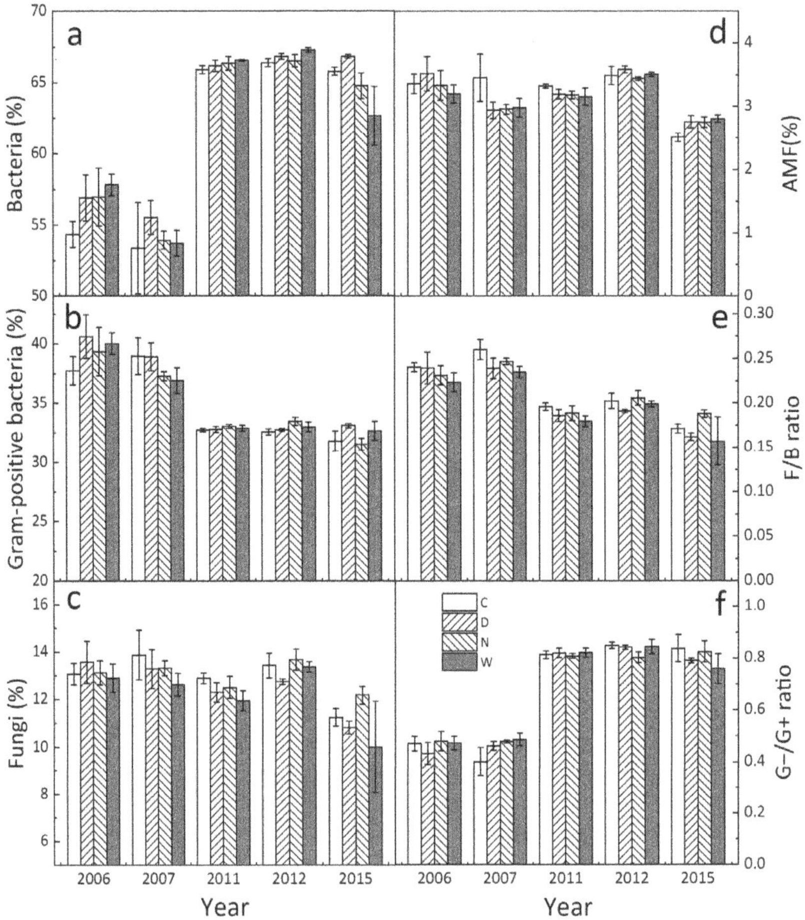

Figure 2. Annual mean relative abundance of bacteria (**a**), Gram-positive bacteria (**b**), fungi (**c**), arbuscular mycorrhizal fungi (AMF, (**d**)), fungi-to-bacteria ratio (F/B ratio) (**e**), and Gram-negative bacteria to Gram-positive bacteria ratio (G−/G+ ratio) (**f**) under the 4 treatments in 2006, 2007, 2011, 2012, and 2015 (M ± 1 SE, n = 4). See Figure 1 for abbreviations.

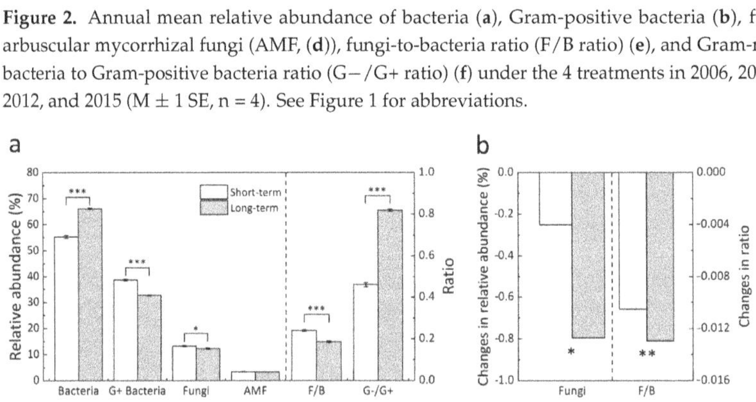

Figure 3. Comparison of microbial compositions between short- (2006 and 2007) and long-term (2011, 2012, and 2015) warming (**a**) and daytime warming-induced changes in the relative abundance of fungi and the fungi-to-bacteria ratio (F/B) (**b**). Each blank column in Panel a represents the mean value of 2006 and 2007, and each gray column in *Panel a* represents the mean value of 2011, 2012, and 2015. Significance levels: * $p < 0.05$, ** $p < 0.01$, and *** $p < 0.001$.

3.3. Relationships of Microbial Composition with Soil Microclimate and Plant Cover

Grass cover was not affected by daytime warming but was significantly reduced by 19.2% (Figure S3a, $p < 0.05$, Table S1) under nighttime warming in the long term. Forb cover was suppressed by 8.89% and 22.5% (both $p < 0.05$) under daytime warming in the short and long term, respectively, but remained unchanged under nighttime warming (Figure S3b). Legume cover was increased by 82.8% ($p < 0.01$) under daytime warming in the short term but was not changed under nighttime warming (Figure S3c).

The relative abundance of G+ bacteria positively depended on soil moisture (Figure S4a, $R^2 = 0.32$, $p < 0.001$), but negatively related to grass cover (Figure S4b, $R^2 = 0.34$, $p < 0.001$), whereas G− bacteria linearly reduced with increasing soil moisture (Figure S4e, $R^2 = 0.60$, $p < 0.001$) and legume cover (Figure S4h, $R^2 = 0.21$, $p < 0.001$), but enhanced with increasing grass cover (Figure S4f, $R^2 = 0.39$, $p < 0.001$). The relative abundance of fungi showed negative linear dependence on grass cover (Figure 4a, $R^2 = 0.16$, $p < 0.001$) over the 5 years, but was not related to forb or legume cover (Figure 4b,c, both $p > 0.05$). The ratio of fungi to bacteria linearly decreased with grass cover (Figure 4d, $R^2 = 0.45$, $p < 0.001$) and increased with legume cover (Figure 4f, $R^2 = 0.09$, $p < 0.01$), but was not correlated with forb cover (Figure 4e, $p > 0.05$). Both the relative abundance of fungi (Figure S6a, $R^2 = 0.12$, $p < 0.001$) and fungi-to-bacteria ratio (Figure S6c, $R^2 = 0.37$, $p < 0.001$) positively depended upon soil temperature, but only the fungi-to-bacteria ratio was positively related to soil moisture (Figure S6d, $R^2 = 0.39$, $p < 0.001$).

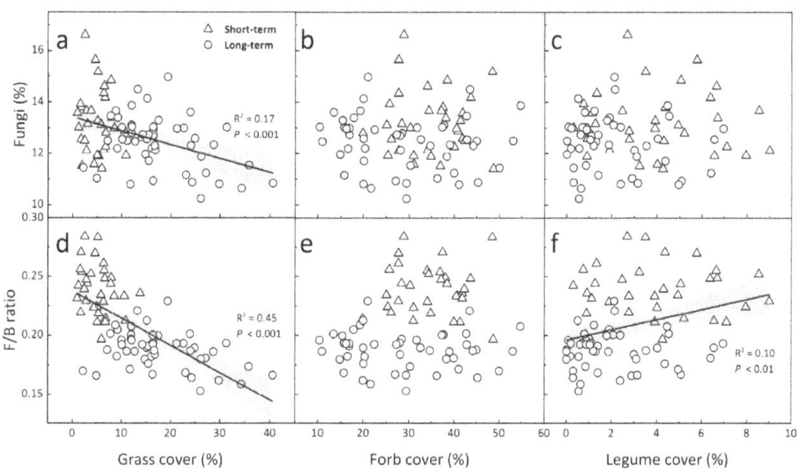

Figure 4. Dependences of the relative abundance of fungi (**a–c**) and the ratios of fungi to bacteria (**d–f**) on plant cover of grass (**a**,**d**), forb (**b**,**e**), and legume (**c**,**f**). Each data point represents the mean annual value in each plot in the short (2006 and 2007) and long term (2011, 2012, and 2015). The solid lines and shaded areas describe linear regressions and 95% CIs.

3.4. Controls of Abiotic and Biotic Factors on Soil Microbial Composition

Redundancy analysis (RDA) revealed the correlations among different treatments, soil microclimate, and plant functional groups, with the first and second canonical axis explaining 87.6% and 2.3% of the variance, respectively (Figure 5a). Grass cover, soil temperature, and moisture were the dominant factors influencing microbial variations, accounting for 8.91% ($p < 0.01$), 7.40% ($p < 0.05$), and 4.81% ($p < 0.05$) of the increase in mean square error (InMSE) based on random forest model analysis (Figure 5b).

The results of SEM analysis showed that under daytime warming, the increased soil temperature directly changed microbial composition as revealed by the fungi-to-bacteria ratio (standardized direct effect was 0.472, Figure 6a), and indirectly affected it through the changes in soil moisture and grass cover (standardized indirect effect was 0.024, Figure 6a).

Under nighttime warming, the ratio of fungi to bacteria was significantly correlated with nighttime soil temperature only (standardized total effect was 0.412, Figure 6b).

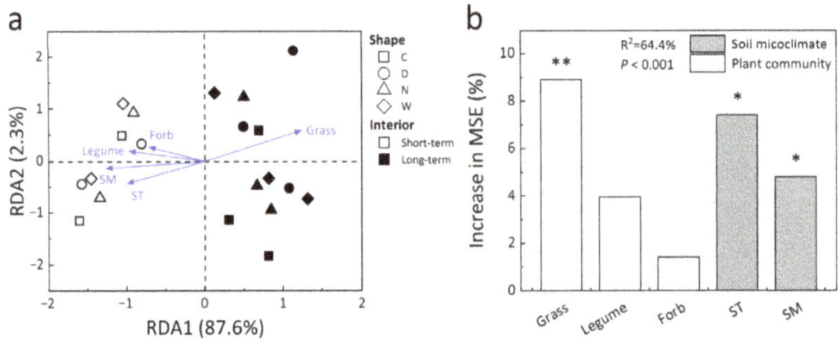

Figure 5. Redundancy analysis (RDA) (**a**) among soil temperature (ST), soil moisture (SM), plant (grass, forb, and legume) cover and different treatments, and random forest analysis (**b**) for important predictors of the fungi-to-bacteria ratio. ST, soil temperature; SM, soil moisture. Significance levels: * $p < 0.05$, ** $p < 0.01$. See Figure 1 for abbreviations.

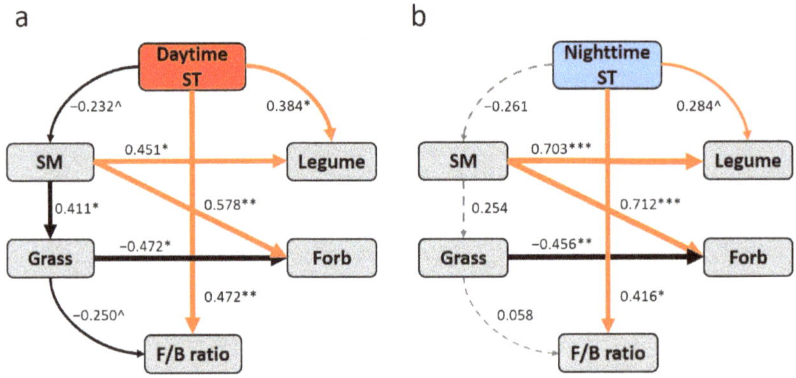

Figure 6. Structural equation models of the direct and indirect effects of daytime soil temperature on the fungi-to-bacteria ratio under daytime warming (**a**) and nighttime soil temperature on the fungi-to-bacteria ratio under nighttime warming (**b**). The solid and dashed arrows represent significant ($p < 0.1$) and non-significant ($p > 0.1$) paths. Orange and black arrows represent positive and negative paths. The width of the arrow indicates the strength of the relationship. Numbers adjacent to arrows are standardized path coefficients and are indicative of the effect size of the relationships. The final model fits the data well, as suggested by the chi-square and RMSEA values ((**a**), $\chi^2 = 4.622$, $p = 0.706$, RMSEA = 0, $df = 7$; (**b**), $\chi^2 = 4.035$, $p = 0.776$, RMSEA = 0, $df = 7$). Significance levels: ^ $p < 0.1$, * $p < 0.05$, ** $p < 0.01$, *** $p < 0.001$. See Figures 1 and 5 for abbreviations.

4. Discussion

4.1. Short-and Long-Term Warming on Soil Microbial Composition

The obvious interannual variation in soil microclimate and some microbial groups (e.g., G+ and G− bacteria) found in our study can be mainly attributed to the changes in precipitation amounts, the natural succession of plant communities, and the potential interactions among different microbial components (Figures S2–S5). Our results that long-term rather than short-term warming decreases fungal abundance and alters microbial composition revealed by the fungi-to-bacteria ratio agree with previous findings that warming effects on soil microbial communities mainly depend on treatment duration [9,26], with unchanged microbial biomass, diversity, and composition observing in several short-term (e.g., 1–6 years) warming experiments [9,26,27]. This can be explained by several possible reasons: First, the moderate elevation of soil temperatures (<1 °C) during the first two years found in our study might not be sufficient to induce a significant response in the microbial community due to the wide range of microbial adaptation [9,26]. Second, although soil moisture, forb, and legume cover vary with short-term warming, they may have legacy effects on the soil microbial community because of the strong resistance and functional redundancy [8,9]. Third, the assembly and stabilization of the soil microbial community experienced considerable long-term environmental fluctuations, which partially explains the weak response of soil microorganisms to short-term warming [28].

In addition, the reductions in fungal abundance and consequent fungi-to-bacteria ratio under long-term warming can be attributed to the changes in soil microclimate and increased cover of grasses (Figures 4, 5, S6 and S8), which agrees with our Hypothesis 1. Previous studies have found the suppressed competitive ability of some grasses decreased microbial biomass and C use efficiency under warming in this temperate steppe [8,19], but no linkages between above- and below-ground components were examined. The dominant grass species in this grassland, such as *Achnatherum Sibiricum*, *Agropyron Cristatum*, and *Stipa Krylovii*, typically have taller stature but shallower root distributions than forbs and legumes; therefore, the presence of these species may lead to lower soil temperature because of the shading effect and have subsequent impacts on microbial activity in surface soils. Moreover, warming can reduce organic C inputs through grass litter [29] which decreases substrate availability, thereby decreasing fungal activity and saprotrophic fungal proportions [7]. Overall, our study not only suggests the close association between soil microbial composition and grass cover but also indicates that fungi respond to long-term warming more sensitively than bacteria in this semi-arid grassland.

4.2. *Asymmetrically Diurnal Warming on Soil Microbial Composition*

The observations of decreased fungal abundance and the fungi-to-bacteria ratio under daytime warming in the long term do not agree with our Hypothesis 2 and previous findings that nighttime warming increases fungal diversity and changes microbial composition and C use efficiency [8,19,21]. Two potential mechanisms may account for the daytime warming-induced reduction in fungal abundance. First, the enhancement in microbial metabolism and water consumption under daytime warming might lead to rapid exhaustion of C substrates, thus inducing an inhibition of microbial activity. Given that soil fungi, especially the saprotrophic lineages, are important decomposers of soil organic matter [30], the decreased water can significantly suppress the abundance of the fungal community. Second, because of the key role of plant growth in triggering microbial variations, the competition between root absorption and microbial activity for nutrients in soils may limit the growth of fungi which usually colonize plant rhizospheres [8,31]. Nighttime warming-induced overcompensation of plant photosynthesis also improves the substrates' utilization of plant roots in the daytime and further reduces fungal abundance [8,15,19,20].

The stronger impacts of warming on fungi than bacteria have been widely reported [32,33]. One possibility is that most fungi have higher C assimilation efficiency and metabolic ability than bacteria, hence the warming-induced decrease in soil water availability could significantly inhibit the growth and activity of fungi, showing greater suppression on the biomass and diversity of fungi, especially in our experimental grassland which is largely water limited [23]. A reduction in plant litter residues with higher cellulose contents under warming can also lead to a decline in saprotrophic fungal abundance [34,35]. In addition, although AMF is important in facilitating plant nutrient absorption and usually shows sensitive responses to warming [36], the unchanged AMF abundance observed in our study might contribute to their different temperature optima [37] and the direct nutrient acquisition process because they are obligate symbionts that can receive growth substrates from host plants.

4.3. *Implications*

Climate warming can result in soil C loss (release more CO_2 to the atmosphere) by stimulating microbial respiration and improving microbe-mediated mineralization of soil

C stocks [35,38]. As the important component of soil respiration, microbial respiration is always thought to be closely associated with their biomass in most studies. However, the insignificant correlation between soil respiration and microbial biomass C, and the negative dependence of soil respiration on the fungi-to-bacteria ratio observed in our study suggest that microbial composition may be more important than biomass in regulating soil respiration under long-term warming (Figure 7). These findings indicate that changes in microbial composition in response to climate warming are critical for ecosystem C cycling in the semi-arid grassland and can provide new insights to predict the climate-C feedback in terrestrial ecosystems.

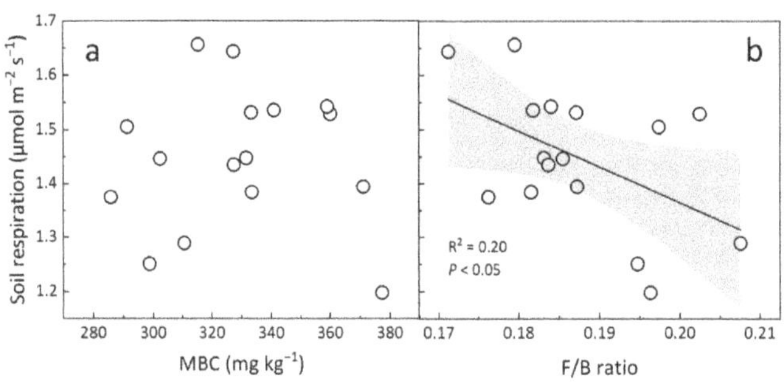

Figure 7. Dependence of soil respiration on microbial biomass C (MBC, (**a**)) and the fungi-to-bacteria ratio (F/B ratio, (**b**)) across spatial scales. Each data point represents the 5-year mean value of each plot. The solid lines and shaded areas describe linear regression and 95% CIs.

5. Conclusions

Climate warming has both direct and indirect effects on soil microbial composition with implications for soil respiration in the temperate steppe. Our study shows that long-term rather than short-term warming has significant impacts on soil microbial composition, which can be attributed to the changes in soil microclimate and grass cover. The decreases in fungal abundance and the fungi-to-bacteria ratio under long-term daytime warming instead of nighttime warming observed in this study are also mainly driven by the increased soil temperature, reduced soil moisture, and improved grass cover. Our findings indicate that microbial composition may be more important than microbial biomass in regulating soil respiration under long-term climate warming, which will facilitate an accurate assessment of climate warming-C feedback in terrestrial ecosystems.

Supplementary Materials: The following supporting information can be downloaded at: https://www.mdpi.com/article/10.3390/biology12050699/s1, Figure S1: Warming-induced changes in soil temperature from 0:00–23:00 on each day (a) and from 1 May to 31 October during the growing season in each year (b) at a depth of 10 cm. Mean soil temperature (M ± 1 SE, n = 4) during short term, long term, and all 5 years (c). C, control; D, daytime warming; N, nighttime warming; and W, diurnal warming; Figure S2: Dependences of soil moisture on precipitation amount in August in 2006, 2007, 2011, 2012, and 2015. Each data point represents the mean annual value in each plot. The solid lines and shaded areas describe linear regressions and 95% CIs; Figure S3: Annual mean grass (a), forb (b), and legume cover (c) under the 4 treatments in 2006 2007, 2011, 2012, and 2015 (M ± 1 SE, n = 4). See Figure S1 for abbreviations; Figure S4: Dependences of the relative abundance of Gram-positive (G+; a–d) and Gram-negative (G−; e–h) bacteria on soil moisture (a,e) and plant cover of grasses (b,f), forbs (c,g), and legumes (d,h). Each data point represents the mean annual value in each plot. The solid lines and shaded areas describe linear regressions and 95% CIs; Figure S5: Dependences of the relative abundance of fungi on bacteria (a), G+ bacteria (b), G− bacteria (c), and G−/G+ ratio (d). Each data point represents the mean annual value in each plot. The solid lines and shaded areas describe linear regressions and 95% CIs; Figure S6: Dependences of the relative abundance of fungi (a,b) and the ratio of fungi to bacteria (c,d) on soil temperature (a,c) and moisture (b,d). Each data point represents the mean annual value in each plot in the short (2006 and 2007) and long term

(2011, 2012, and 2015). The solid lines and shaded areas describe linear regressions and 95% CIs; Figure S7: Dependences of grass (a,b), forb (c,d), and legume cover (e,f) on soil temperature (a,c,e) and moisture (b,d,f). Each data point represents the mean annual value in each plot in the short (2006 and 2007) and long term (2011, 2012, and 2015). The solid lines and shaded areas describe linear regressions and 95% CIs; Figure S8: Dependences of warming-induced changes in the relative abundance of fungi (a–e) and the ratio of fungi to bacteria (f–j) on warming-induced changes in grass cover (a,f), forb cover (b,g), legume cover (c,h), soil temperature (d,i), and moisture (e,j). Each data point represents the mean annual value in each plot in the short (2006 and 2007) and long term (2011, 2012, and 2015). The solid lines and shaded areas describe linear regressions and 95% CIs; Table S1: Results (*F*-values) of repeated measures ANOVA on the effects of daytime warming (D), nighttime warming (N), and year (Y) on plant functional community (grass, forb, and legume) cover in the short (2006 and 2007) and long (2011, 2012, and 2015) term.

Author Contributions: Conceptualization, S.W.; methodology, S.W.; data analysis, J.F. and X.Q.; writing—original draft preparation, J.F.; writing—review and editing, J.R., J.S. and S.W.; visualization, J.F.; supervision, S.W.; funding acquisition, S.W. and J.F. All authors have read and agreed to the published version of the manuscript.

Funding: This research was funded by the National Natural Science Foundation of China (31830012), the Hebei Natural Science Foundation (C2022201042), and the High-level Talent Research Funding Project of Hebei University (521000981186 and 521000981405).

Institutional Review Board Statement: Not applicable.

Informed Consent Statement: Not applicable.

Data Availability Statement: The data presented in this study are available upon request from the corresponding author.

Acknowledgments: Special thanks to Chengde Yu, Cancan Zhao, Qian Zhang, Naili Zhang, and Lili Zhu for their excellent work on the experimental measurements for this article.

Conflicts of Interest: The authors declare no conflict of interest.

References

1. IPCC. Summary for Policymakers. In *Climate Change 2021: The Physical Science Basis*; Cambridge University Press: Cambridge, UK, 2021.
2. Melillo, J.; Steudler, P.; Aber, J.; Newkirk, K.; Lux, H.; Bowles, F.; Catricala, C.; Magill, A.; Ahrens, T.; Morrisseau, S. Soil warming and carbon-cycle feedbacks to the climate system. *Science* **2002**, *298*, 2173. [CrossRef] [PubMed]
3. Yan, Y.; Wang, J.; Tian, D.; Luo, Y.; Xue, X.; Peng, F.; He, J.; Liu, L.; Jiang, L.; Wang, X.; et al. Sustained increases in soil respiration accompany increased carbon input under long-term warming across global grasslands. *Geoderma* **2022**, *428*, 116157. [CrossRef]
4. Zhou, Z.; Wang, C.; Luo, Y. Meta-analysis of the impacts of global change factors on soil microbial diversity and functionality. *Nat. Commun.* **2020**, *11*, 3072. [CrossRef] [PubMed]
5. Feng, J.; Li, Z.; Hao, Y.; Wang, J.; Ru, J.; Song, J.; Wan, S. Litter removal exerts greater effects on soil microbial community than understory removal in a subtropical-warm temperate climate transitional forest. *For. Ecol. Manag.* **2022**, *505*, 119867. [CrossRef]
6. Romero-Olivares, A.; Allison, S.; Treseder, K. Soil microbes and their response to experimental warming over time: A meta-analysis of field studies. *Soil Biol. Biochem.* **2017**, *107*, 32–40. [CrossRef]
7. Che, R.; Wang, S.; Wang, Y.; Xu, Z.; Cui, X. Total and active soil fungal community profiles were significantly altered by six years of warming but not by grazing. *Soil Biol. Biochem.* **2019**, *139*, 107611. [CrossRef]
8. Zhang, N.; Xia, J.; Yu, X.; Ma, K.; Wan, S. Soil microbial community changes and their linkages with ecosystem carbon exchange under asymmetrically diurnal warming. *Soil Biol. Biochem.* **2011**, *43*, 2053–2059. [CrossRef]
9. Gao, Y.; Ding, J.; Yuan, M.; Chiariello, N.; Yang, Y. Long-term warming in a mediterranean-type grassland affects soil bacterial functional potential but not bacterial taxonomic composition. *NPJ Biofilms Microbiomes* **2021**, *7*, 17. [CrossRef]
10. Yu, Y.; Liu, L.; Wang, J.; Zhang, Y.; Xiao, C. Effects of warming on the bacterial community and its function in a temperate steppe. *Sci. Total Environ.* **2021**, *792*, 148409. [CrossRef]
11. Saleska, S.; Shaw, M.; Fischer, M.; Dunne, J.; Still, C.; Holman, M.; Harte, J. Plant community composition mediates both large transient decline and predicted long-term recovery of soil carbon under climate warming. *Global Biogeochem. Cycles* **2002**, *16*, 1055. [CrossRef]
12. Guo, X.; Feng, J.; Shi, Z.; Zhou, X.; Yuan, M.; Tao, X.; Hale, L.; Yuan, T.; Wang, J.; Qin, Y.; et al. Climate warming leads to divergent succession of grassland microbial communities. *Nat. Clim. Chang.* **2018**, *8*, 813–818. [CrossRef]
13. Yuan, M.; Guo, X.; Wu, L.; Zhang, Y.; Zhou, J. Climate warming enhances microbial network complexity and stability. *Nat. Clim. Chang.* **2021**, *11*, 343–348. [CrossRef]

14. Wang, F.; Li, Z.; Su, F.; Guo, H.; Wang, P.; Guo, J.; Zhu, W.; Wang, Y.; Hu, S. Sensitive groups of bacteria dictate microbial functional responses to short-term warming and N input in a semiarid grassland. *Ecosystems* **2021**, *25*, 1346–1357. [CrossRef]
15. Xia, J.; Han, Y.; Zhang, Z.; Wan, S. Effects of diurnal warming on soil respiration are not equal to the summed effects of daytime and nighttime warming in a temperate steppe. *Biogeosciences* **2009**, *6*, 1361–1370. [CrossRef]
16. Peng, S.; Piao, S.; Ciais, P.; Myneni, R.; Chen, A.; Chevallier, F.; Dolman, A.; Janssens, I.; Peñuelas, J.; Zhang, G.; et al. Asymmetric effects of daytime and night-time warming on Northern Hemisphere vegetation. *Nature* **2013**, *501*, 88–92. [CrossRef]
17. Xia, J.; Chen, J.; Piao, S.; Ciais, P.; Luo, Y.; Wan, S. Terrestrial carbon cycle affected by non-uniform climate warming. *Nat. Geosci.* **2014**, *7*, 173–180. [CrossRef]
18. Zhang, Y.; Dong, S.; Gao, Q.; Ganjurjav, H.; Wang, X.; Geng, W. "Rare biosphere" plays important roles in regulating soil available nitrogen and plant biomass in alpine grassland ecosystems under climate changes. *Agric. Ecosyst. Environ.* **2019**, *279*, 187–193. [CrossRef]
19. Wang, J.; Zhang, Q.; Song, J.; Ru, J.; Zhou, Z.; Xia, J.; Dukes, J.; Wan, S. Nighttime warming enhances ecosystem carbon-use efficiency in a temperate steppe. *Funct. Ecol.* **2020**, *34*, 1721–1730. [CrossRef]
20. Wan, S.; Xia, J.; Liu, W.; Niu, S. Photosynthetic overcompensation under nocturnal warming enhances grassland carbon sequestration. *Ecology* **2009**, *90*, 2700–2710. [CrossRef]
21. Liu, J.; Long, Z.; Zhang, J.; Chen, C. Effects of nighttimely warming and nitrogen application on the diversity of soil fungi in winter wheat in the lower reach of the Yangtze River. *Arch. Agron. Soil Sci.* **2022**, *68*, 838–851. [CrossRef]
22. Halász, J.; Kotroczó, Z.; Szabó, P.; Kocsis, T. Biomonitoring and assessment of dumpsites soil using phospholipid fatty acid analysis (PLFA) method—Evaluation of possibilities and limitations. *Chemosensors* **2022**, *10*, 409. [CrossRef]
23. Yergeau, E.; Bokhorst, S.; Kang, S.; Zhou, J.; Greer, C.; Aerts, R.; Kowalchuk, G. Shifts in soil microorganisms in response to warming are consistent across a range of Antarctic environments. *ISME J.* **2012**, *6*, 692–702. [CrossRef] [PubMed]
24. Song, J.; Xia, J.; Hui, D.; Zheng, M.; Wang, J.; Ru, J.; Wang, H.; Zhang, Q.; Yang, C.; Wan, S. Plant functional types regulate non-additive responses of soil respiration to 5-year warming and nitrogen addition in a semi-arid grassland. *Funct. Ecol.* **2021**, *35*, 2593–2603. [CrossRef]
25. Xia, J.; Chen, S.; Wan, S. Impacts of day versus night warming on soil microclimate: Results from a semiarid temperate steppe. *Sci. Total Environ.* **2010**, *408*, 2807–2816. [CrossRef] [PubMed]
26. Melillo, J.; Frey, S.; DeAngelis, K.; Werner, W.; Bernard, M.; Bowles, F.; Pold, G.; Knorr, M.; Grandy, A. Long-term pattern and magnitude of soil carbon feedback to the climate system in a warming world. *Science* **2017**, *358*, 101. [CrossRef]
27. DeAngelis, K.; Pold, G.; Topçuoglu, B.; van Diepen, L.; Varney, R.; Blanchard, J.; Melillo, J.; Frey, S. Long-term forest soil warming alters microbial communities in temperate forest soils. *Front. Microbiol.* **2015**, *6*, 104. [CrossRef]
28. Cabrol, L.; Poly, F.; Malhautier, L.; Pommier, T.; Lerondelle, C.; Verstraete, W.; Lepeuple, A.; Fanlo, J.; Le Roux, X. Management of microbial communities through transient disturbances enhances the functional resilience of nitrifying gas-biofilters to future disturbances. *Environ. Sci. Technol.* **2016**, *50*, 338–348. [CrossRef]
29. Luo, C.; Xu, G.; Wang, Y.; Wang, S.; Lin, X.; Hu, Y.; Zhang, Z.; Chang, X.; Duan, J.; Su, A.; et al. Effects of grazing and experimental warming on DOC concentrations in the soil solution on the Qinghai-Tibet plateau. *Soil Biol. Biochem.* **2009**, *41*, 2493–2500. [CrossRef]
30. Yuste, J.; Penuelas, J.; Estiarte, M.; Garcia-Mas, J.; Mattana, S.; Ogaya, R.; Pujol, M.; Sardans, J. Drought-resistant fungi control soil organic matter decomposition and its response to temperature. *Glob. Chang. Biol.* **2011**, *17*, 1475–1486. [CrossRef]
31. Ball, B.; Convey, P.; Feeser, K.; Nielsen, U.; Van Horn, D. Environmental harshness mediates the relationship between aboveground and belowground communities in Antarctica. *Soil Biol. Biochem.* **2022**, *164*, 108493. [CrossRef]
32. García-Palacios, P.; Vandegehuchte, M.; Shaw, E.; Dam, M.; Post, K.; Ramirez, K.; Sylvain, Z.; Milanode Tomasel, C.; Wall, D. Are there links between responses of soil microbes and ecosystem functioning to elevated CO_2, N deposition and warming? A global perspective. *Glob. Chang. Biol.* **2015**, *21*, 1590–1600. [CrossRef] [PubMed]
33. Seaton, F.; Reinsch, S.; Goodall, T.; White, N.; Jones, D.; Griffiths, R.; Creer, S.; Smith, A.; Emmett, B.; Robinson, D. Long-term drought and warming alter soil bacterial and fungal communities in an upland heathland. *Ecosystems* **2021**, *25*, 1279–1294. [CrossRef]
34. Morrison, E.; Pringle, A.; Linda, T.; Grandy, A.; Melillo, J.; Frey, S. Warming alters fungal communities and litter chemistry with implications for soil carbon stocks. *Soil Biol. Biochem.* **2019**, *132*, 120–130. [CrossRef]
35. Verbrigghe, N.; Meeran, K.; Bahn, M.; Canarini, A.; Fransen, E.; Fuchslueger, L.; Ingrisch, J.; Janssens, I.; Richter, A.; Sigurdsson, B.; et al. Long-term warming reduced microbial biomass but increased recent plant-derived C in microbes of a subarctic grassland. *Soil Biol. Biochem.* **2022**, *167*, 108590. [CrossRef]
36. Tatsuhiro, E.; Katsuharu, S. How do arbuscular mycorrhizal fungi handle phosphate? New insight into fine-tuning of phosphate metabolism. *New Phytol.* **2018**, *220*, 1116–1121.
37. Barrett, G.; Campbell, C.; Hodge, A. The direct response of the external mycelium of arbuscular mycorrhizal fungi to temperature and the implications for nutrient transfer. *Soil Biol. Biochem.* **2014**, *78*, 109–117. [CrossRef]
38. Crowther, T.; van den Hoogen, J.; Wan, J.; Mayes, M.; Keiser, A.; Mo, L.; Averill, C.; Maynard, D. The global soil community and its influence on biogeochemistry. *Science* **2019**, *365*, 6455. [CrossRef]

biology

MDPI

Article

Introduced Western Honeybees Dramatically Reduce the Abundance of Wild Bees in Alpine Meadows, Eastern Tibet Plateau

Ruimin An and Shucun Sun *

Department of Biology, School of Life Sciences, Nanjing University, Nanjing 210093, China; dz1930006@smail.nju.edu.cn
* Correspondence: shcs@nju.edu.cn

Simple Summary: Western honeybees have been introduced across China, yet their effects on native pollinators in alpine meadows remain unclear. We compared native bee abundance and diversity close to and far from apiaries on the eastern Tibetan Plateau, analyzing plant–bee networks and niche overlap (apparent competition) between honeybees and native species. The results show that native bee abundance in distant plots and their niche overlap with honeybees explain interspecific variations in species abundance changes.

Abstract: Over the past few decades, the western honeybee (*Apis mellifera*) has been widely introduced throughout China. Introduced honeybees have often been observed to pose a significant threat to native bee species diversity through competition for floral resources. However, the specific impact on native pollinator communities is not well understood, especially in alpine meadows, where bee diversity is particularly high. In this study, we determined the difference in species abundance and diversity of native bees between nearby and distant plots relative to apiaries in alpine meadows on the eastern Tibetan Plateau. We constructed a plant–bee network and calculated perceived apparent competition (i.e., the feeding niche overlap) between the introduced honeybees and each of the native bee species. Furthermore, we determined the relationship between the relative change in species abundance and the perceived apparent competition and species abundance across bee species. Among the 15 native bee species, 9 bee species were significantly lower in abundance in the nearby plots compared to the distant plots, and, in particular,, 5 rare species were not found in the nearby plots. For the other six species, the abundance difference was statistically non-significant. Data analysis reveals that the species abundance of native bees in distant plots, along with the feeding niche overlap between introduced bees and native bees in distant plots, explains the variation in the relative change in species abundance across bee species. However, the feeding niche overlap between introduced bees and native bees in nearby plots does not account for this variation. Our findings demonstrate that rare native bee species with greater feeding niche overlap have been significantly impacted by the introduced western honeybees. These results highlight important implications for pollinator management in natural ecosystems and the conservation of wild bees.

Keywords: alpine meadow; *Apis mellifera*; native bee; niche overlap; plant–pollinator interactions; pollination network; rare species; Tibetan Plateau

1. Introduction

The western honeybee (*Apis mellifera*) is one of the earliest domesticated insects. It is the primary species maintained by beekeepers worldwide for its honey production and pollination services [1]. Through human assistance, the western honeybee has been introduced to every continent except Antarctica [2]. It was introduced to China in the late 19th century and has since become a common species in beekeeping. Several subspecies, such as *Apis mellifera ligustica* and *A. mellifera carnica*, are prevalent. Currently, China has approximately 6.8 million colonies of western honeybees and over three hundred thousand beekeepers. Beekeepers migrate hives from southern to northern regions and from lowland to high-altitude areas across the country [3,4]. While providing pollination services to crops and producing honey and other hive products for commercial markets, the western honeybee also has negative impacts on wild pollinators [5,6].

As a eusocial insect, the western honeybee is recognized as a highly impactful invasive species [7]. This is due to its ability to rapidly exploit resources, such as nectar and pollen, with a coordinated group of foragers in its introduced range. This exploitation leads to significant ecological impacts on wild pollinators through competition for food resources in natural ecosystems [5]. For instance, in North America, the introduction of *Apis mellifera* has caused intense competition with native bumblebees (*Bombus* spp.), resulting in population declines as resources are depleted for native bee species [8]. Additionally, the introduced honeybees can have indirect effects on native pollinators. In South America, for example, the presence of *Apis mellifera* has altered the pollination dynamics of native plants, favoring those that are more attractive to honeybees. This shift can reduce the diversity of plant species that rely on native pollinators, potentially altering community composition and function.

However, not all native pollinator species experience the same degree of abundance decline in honeybee-invaded ecosystems. Several factors contribute to this variation. First, differences in ecological niches play a role, as some native pollinators specialize in specific plants or resources that are less preferred by honeybees, thereby reducing competition [9–11]. For example, native pollinators that specialize in deep, tubular flowers, are less affected by introduced honeybees, which prefer more accessible flowers [10]. Additionally, native pollinators with different activity patterns, such as those that forage earlier in the morning or later in the evening when honeybees are less active, may experience reduced competition [12]. Second, differences in species abundance influence the extent of impact. Rare pollinators with lower densities are typically more sensitive to the dominant invaders compared to common pollinators with higher densities [13,14]. This is likely because rare pollinators often interact with specific resources, which are often a subset of those utilized by generalist honeybees, leading to intensified competition [15]. Furthermore, the introduced honeybees may alter plant community composition and structure, potentially decreasing the abundance of plants preferred by some rare pollinators [16]. In contrast, common pollinators, with broader ecological niches, are often more adaptable to changes in resource availability caused by introduced honeybees [11]. With this in mind, we can hypothesize that (1) native bee species with greater niche overlap with introduced honeybees would experience greater declines in abundance in the presence of introduced bees, and that (2) rare species with lower densities would experience greater declines in abundance compared to common species with higher densities.

Native pollinators can also adapt to the presence of introduced western honeybees by shifting their ecological niches [17]. For example, native pollinators may reduce direct competition with honeybees by altering their foraging times, diet preferences, or habitat use. Behavioral shifts, such as foraging at different times of the day or focusing on alternative underutilized floral resources, can allow native pollinators to coexist with introduced honeybees [11,18,19]. Over generations, native pollinators that successfully shift their ecological niches may gain a survival advantage, making their populations more resilient to the presence of introduced honeybees. However, previous studies have largely overlooked the importance of niche shifts in shaping native pollinator population dynamics. These studies typically predict changes in native pollinator abundance based on the niche overlap (perceived apparent competition) between introduced honeybees and native pollinators in the introduced areas, rather than considering the niche overlap between introduced honeybees in introduced areas and native pollinators in their native areas [11,16]. This approach may underestimate the niche overlap between introduced and native bees, thereby reducing the ability to predict the invasive honeybee's impact on native pollinators. Consequently, we can hypothesize that (3) niche overlap between invasive honeybees in introduced areas and native pollinators in their native habitats is a stronger predictor of invasive honeybees' impact on native pollinators than niche overlap within invaded areas.

In this study, we investigated the abundance response of native bees to the introduced western honeybee in alpine meadows on the eastern Tibetan Plateau, where native bee species diversity is particularly high [20]. We compared the species abundance of 15 native bee species between sites close to beekeeper hives (with high densities of introduced bees) and sites far from beekeeper hives (with low densities of introduced bees). According to the aforementioned hypotheses, we predicted that the abundance decline would be greater in the native bee species with greater niche overlap with introduced honeybees and in the rare species with lower densities compared to their counterparts. Additionally, considering the possibility of niche shift in native bee species in response to the introduced honeybees, we predicted that niche overlap between invasive honeybees in introduced areas and native pollinators in their native areas would better predict the invasive honeybee's impact on native pollinators compared to niche overlap within the invaded areas.

2. Materials and Methods

2.1. Study Site and Natural History

This study was conducted in an alpine meadow located on the Qinghai–Tibet Plateau in Sichuan Province, China (32°48′ N, 102°33′ E, 3500 m a.s.l.). The region experiences a typical continental plateau climate, characterized by short, cool summers and autumns, and a long winter. The annual mean air temperature is 1.7 °C, and annual precipitation ranges from 450 mm to 900 mm, with most precipitation occurring during the growing season from late May to September [21].

The study site is dominated by sedges (*Kobresia setchwanensis, Carex* spp.), grasses (*Deschampsia caespitosa, Festuca ovina* and *Elymus nutans*), and forb species (*Saussurea nigrescens, Polygonum viviparum, Carum carvi, Pedicularis kansuensis* and *Anemone trullifolia* var. *linearis*). The primary vegetation type is an alpine meadow, with total vegetation coverage exceeding 95% and plant height averaging ~30 cm [22].

The grassland serves as a summer or winter pasture. The winter pasture, used as our study site, is grazed by livestock during the winter only. Additionally, the study site experiences intense apiculture during the summer due to the abundance of nectar-rich species, including Asteraceae, Lamiaceae, Boraginaceae, Umbelliferae, and Fabaceae [23].

Since the 1980s, the number of domesticated honeybees (*Apis mellifera*) has increased significantly in Hongyuan County. Honeybees cannot overwinter on the plateau, so beekeepers move hives from other regions to the meadows during the growing season (May to September). It is estimated that there are currently ~300 beekeepers and approximately 80 million honeybees every summer on the winter-grazing pasture, covering about

1500 km^2 of the county. The estimated annual honey production ranges between 40 and 60 tons [24].

2.2. Field Sampling

In 2023, we selected three apiaries as reference sites. Each apiary has been operational for over 30 years, with at least 40 hives per year [24]. We chose three plots near the apiaries and three plots distant from them, each plot covering >1 ha. Apiaries were spaced >2 km apart, with paired experimental plots (nearby vs. distant relative to an apiary) separated by >10 km. *A. mellifera* becomes extremely rare when sites are >6 km from the apiary, no *A. mellifera* was observed or captured in the distant plots during any survey. Furthermore, the plant community was dominated by Asteraceae in each plot.

Field sampling involved the use of light muslin sweep nets (38 cm diameter, 180 cm handles) to survey the abundance of introduced and native bees. We conducted five surveys at approximately two-week intervals for each plot. For each survey, sweeping was performed three times; each sweeping was conducted at a constant speed for 50 nets. All sweepings were carried out in the central area of each plot. Sampling occurred between 11:00 am and 3:00 pm on sunny days. All bees collected in the nets were identified and recorded based on morphological traits. Unidentified species were taken to the laboratory for taxonomic identification.

We calculated the sum and average abundance values across the five surveys for each plot and determined the relative species abundance of native bees for both nearby and distant plots using the average data from the three sites. During the flowering season (late June to early September), we conducted ten surveys at approximately ten-day intervals for each community. For each survey, we collected all flower-visiting bees (Hymenoptera: Apoidea) by walking along three prearranged transects (100 m in length, 2 m in width) within each plot. Surveys were conducted between 9:30 am and 4:30 pm on sunny days. We captured >15 individuals for each bee species in each plot, and these samples were taken to the laboratory for palynological analysis. A total of 656 bee individuals were collected, and 2203 bee individuals were recorded during sweeping. Among native bees in the distant plots, *Andrena tarsata* was the most abundant species (22.54% relative abundance), while *Sphecodes* sp. was the rarest (0.35% relative abundance). Nine bee species had relative abundances below 5%, so they were classified as rare.

In addition, we surveyed the flowering plant community five times with two-week intervals during the flowering season (late June to early September) in each plot. For each survey, we randomly placed twenty-five 1 m × 1 m quadrats in each plot. We recorded the number of flowers or inflorescences for each plant species in each quadrat, and calculated flower abundance for each plant species.

2.3. Identification of Pollen Carried by Bee Individuals

For palynological analysis, we sampled pollen grains from all collected bee individuals. Each insect specimen was bathed in 95% ethanol in a centrifuge tube using an ultrasonic cleaner for ten minutes. Each bathed bee was removed and preserved in a second co-referenced tube in 95% alcohol. The pollen samples remaining in the first tube were centrifuged at 5000 rpm for 10 min, and the supernatants were decanted. Each pollen sample was then placed on slides with a micropipette. We identified the pollen contained on the slides at 100 and 400× magnifications using a light microscope (Nikon, E600) by comparing them to a pollen reference library constructed from field-collected and identified flowers.

To construct the pollen library, we collected 95 flowering plant species from the alpine meadow throughout the flowering season. The dehisced anthers from each flower were suspended in 95% ethanol and photographed under a microscope using the same method as described above.

An insect specimen was considered to be a potential pollinator of the plant species if a minimum of 3 grains of the same pollen morphotype were observed. We used a slightly lower number of pollen grains than other authors to define interactions [25], because some plant species may have less pollen on individuals of the same bee species in the interactions that have been identified (e.g., some *Asteraceae*, personal observation by Ruimin An).

2.4. Perceived Apparent Competition

The collected individuals were identified to 15 different species, including the introduced A. mellifera, 4 Apidae species, 1 Megachilidae species, 3 Andrenidae species, and 7 Halictidae species (Table A1). Ten native bee species were found in both the nearby and distant plots, and five species were found in the distant plots only. Moreover, among the bee species observed in both nearby and distant plots, 6 species are common, having a higher abundance, whereas 9 species are rare with a lower abundance.

We constructed quantitative pollen transport networks (bee species in rows and plant species in columns) for each plot. The strength of each interaction was calculated as the number of individuals collected carrying pollen of the visited plants [25]. The plant and bee species that had no interaction partners were removed from the pollen transport networks. We combined ten surveys in each plot and constructed 6 networks from these six plots.

We calculated perceived apparent competition (PAC) for each network using the bipartite package [26,27] in R version 4.2.1 (R Core Team, 2022). Perceived apparent competition (PAC) estimates the degree of niche overlap between two species using Müller's index [28,29]. In this study, we only considered the PAC between *A. mellifera* and each native bee species. For all pairwise comparisons of *A. mellifera* against each of native bee species, we calculated Müller's index as

$$d_{ij} = \sum_{k}\left[\frac{\alpha_{ik}}{\sum_{l}\alpha_{il}} \times \frac{\alpha_{jk}}{\sum_{m}\alpha_{mk}}\right], \tag{1}$$

where α_{ik} presents the number of interactions between pollinator i and plant k, α_{il} represents the number of interactions by pollinator i across all plants I, α_{jk} represents the number of interactions between pollinator j and plant k, and α_{mk} represents visits to plant k from all pollinators m.

Müller's index is sensitive to the relative abundance of competing species. Therefore, we used the number of bee specimens involved in constructing the pollen transport networks to adjust the strength of each interaction before we calculated PAC. We calculated the new number of each interaction as

$$N_{xy} = P_{xy}/A_x, \tag{2}$$

where A_x represents the individuals of pollinator x involved in constructing each network, and P_{xy} represents the number of interactions between pollinator x and plant y in each past network. As for the *A. mellifera*, A_x represents the average number of individuals involved in constructing three networks. P_{xy} represents the number of interactions between *A. mellifera* and plant y in each network. Both N_{xy} and P_{xy} are the same in six networks.

For PAC calculations between a native bee species and *Apis mellifera*, we first developed a pollination network matrix exclusively comprising these two species. We then applied the second formula to this matrix to calculate interaction strengths under conditions of equal abundances. Finally, these interaction strengths were substituted into the first formula to determine the PAC metric. We calculated two PAC values for each native bee species using the reconstructed interaction frequencies (N_{xy}). One is PAC_C, representing the niche overlap between the introduced and native bee species in the nearby plots, and the other is PAC_D, representing the niche overlap between the introduced bee and native bee species in the distant plots.

2.5. Data Analysis

All analyses were conducted using R 4.4.1 (R Core Team, 2024). The networks were visualized using the plotweb function from the bipartite R package v.2.16. Prior to modeling, we examined the distributions of the variables.

We determined whether the presence of A. mellifera was associated with changes in native bee abundance using separate generalized linear models (GLMs) fitted with a negative binomial distribution for each native bee species. We included plot type (nearby vs. distant) as a fixed predictor with the abundance of native bee species as the response variable.

We calculated the relative change in abundance for each native bee species as the abundance difference between nearby and distant sites divided by the bee abundance at distant sites. To compare the predictive performance of PAC_C and PAC_D for the relative change in bee abundance, we assessed whether PAC_C or PAC_D was associated with the relative change in bee abundance using separate generalized linear models (GLMs) fitted with a Gaussian distribution for two types (nearby vs. distant) of plots. Moreover, to determine whether rare species are more sensitive to the introduction of western honeybees, we assessed whether the abundance of native bee species in the distant plots was associated with the relative change in bee abundance using generalized linear models (GLM) fitted with a gaussian distribution.

To determine to what extent species abundance and PAC_D affected the abundance change induced by the introduced honeybees, we assessed whether species abundance in the distant plot, PAC_D, and their interactions were associated with the relative change in native bee abundance using generalized linear models (GLM) fitted with a Gaussian distribution. We included bee abundance in distant plots, PAC_D, and their interactions as predictors and the relative change in abundance as a response variable.

To examine whether native bees changed their feeding niches, we conducted a permutational analysis of variances (PERMANOVA) on the Bray–Curtis dissimilarity matrix including the diet preference of the six common species. Bee diet preference was calculated as the proportion of links for each plant to the total number of links. We combined adjacent samples to ensure consistency of variables. The PERMANOVA was run with 999 permutations using the "adonis2" function in the R package vegan v. 2.6–4 [30]. The dissimilarity in native bees' niches between nearby and distant plots was also computed using non-metric multidimensional scaling (NMDS) with Bray–Curtis distance and the "metaMDS" function in the R package vegan v. 2.6–4 [30].

To further investigate the similarity in flowering plant species composition across six communities, we performed a permutational analysis of variances (PERMANOVA) on the Bray–Curtis dissimilarity matrix including the flower abundance for each species. The PERMANOVA was run with 999 permutations using the "adonis2" function in the R package vegan v. 2.6–4 [30]. Additionally, we assessed the dissimilarity in flowering plant species composition across six communities using non-metric multidimensional scaling (NMDS) with Bray–Curtis distance. The NMDS analysis was carried out using the "metaMDS" function in the R package vegan v. 2.6–4 [30].

3. Results

3.1. The Plant–Bee Interaction Network

Using pollen analysis on the pollen loads carried by the bees, we determined 858 interactions between individual bees and plants. There was a total of 133 species pair interactions in total (Figure 1). Among all native bee species, the most generalist pollinator was *Andrena tarsata*, visiting 22 plant species (41.5% of floral taxa recorded), whereas the extreme specialist *Sphecodes* sp. exclusively foraged on a single plant species (*Angelica dahurica*). Additionally, the PERMANOVA and NMDS analyses indicated no significant differences in flowering plant species composition among the six communities (Figure A1).

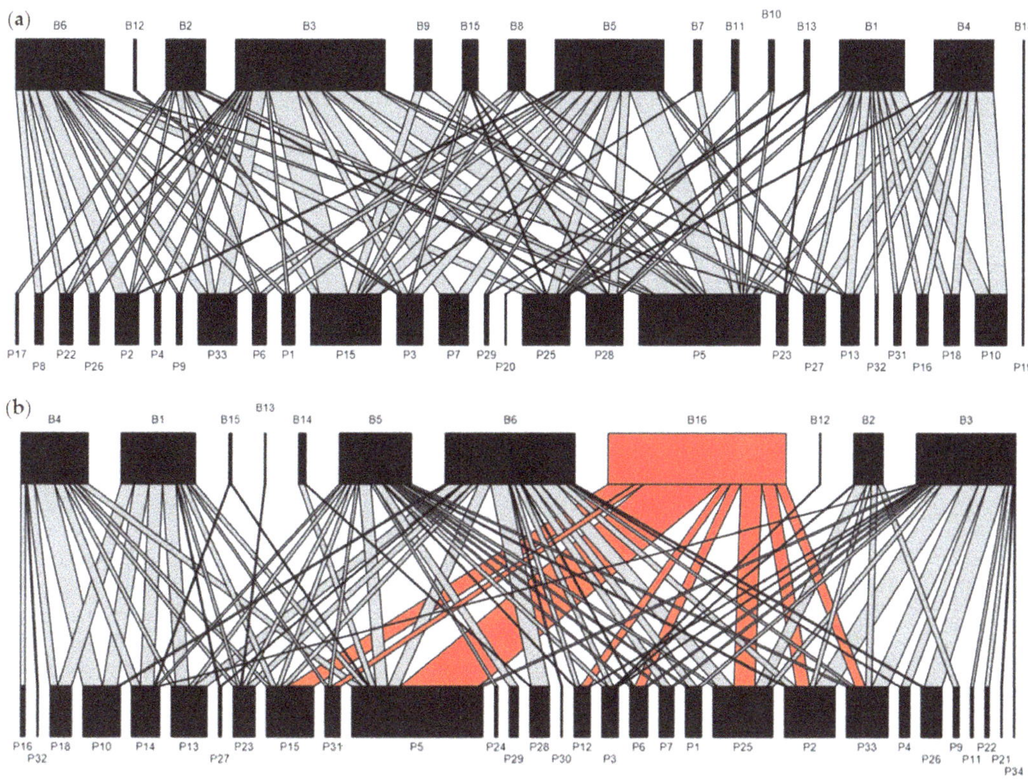

Figure 1. Illustration of a plant–bee network based on the examination of pollen carried by bees in (**a**) distant plots and (**b**) nearby plots. Top nodes represent pollinators, and bottom nodes represent plants. The width is proportional to the frequency of particular pollinators visiting plants. The introduced *A. mellifera* (B16) is shown in red. The bee and plant species identities are listed in the Appendix A.

3.2. The Difference in Native Bee Species Abundance Between Nearby and Distant Plots

For 6 common species, the presence of *A. mellifera* significantly decreased the abundance of the four species. There was no statistically significant difference in the abundance of the two common species between the nearby and distant plots (Figure 2a). Five rare bee species were found only in distant plots, and the other four bee species remained unchanged by the introduced honeybees (Figure 2b).

3.3. PAC and Species Abundance on Relative Change in Bee Abundance

Across all species that were found in both nearby and distant plots, PAC_D showed a significant positive relationship with the relative change in bee abundance decline (Figure 3a), whereas PAC_C exhibited no significant relationship with the relative change (Figure 3b).

Across all species, species abundance in the distant plots showed a significant negative relationship with the relative change in bee abundance (Figure 3c).

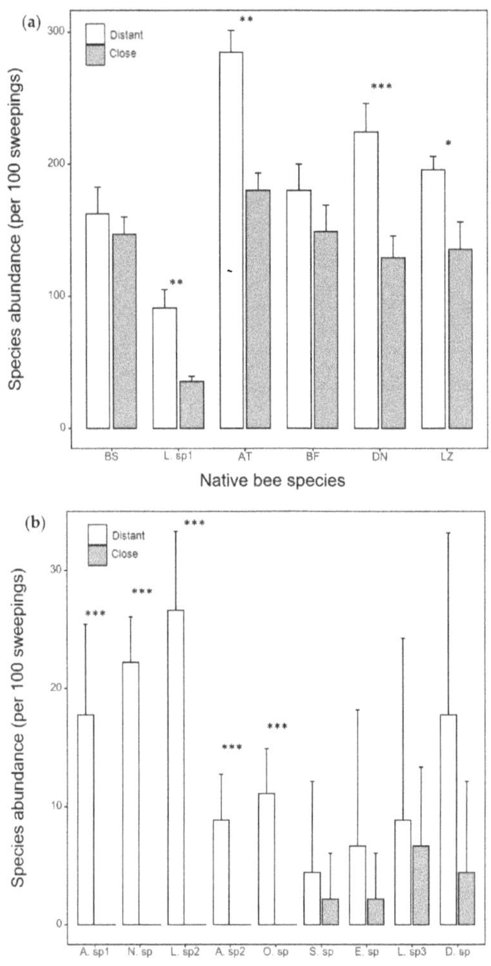

Figure 2. The difference in species abundance of (**a**) 6 common bee species and (**b**) 9 rare species (5 of which were found in the distant plots only) between nearby and distant plots. Error bars represent standard deviation. * $p < 0.05$; ** $p < 0.01$; *** $p < 0.001$. Species names are provided in the Appendix A.

In addition, both species, abundance in the distant plot and PAC_D, significantly affected the relative change in bee abundance, but their interaction effect was non-significant (Table 1).

Table 1. Effects of species abundance, PAC_D, and their interaction on the relative change in species abundance for 15 native bees (residual standard error: 0.1575 on 11 degrees of freedom; Adjusted R^2: 0.778). * $p < 0.05$. N = 15.

Response	Predictor	Estimate	df	p
Relative change of species abundance	Species abundance	−0.024	1	0.036 *
	PAC_D	1.476	1	0.002 *
	Interaction	0.036	1	0.432

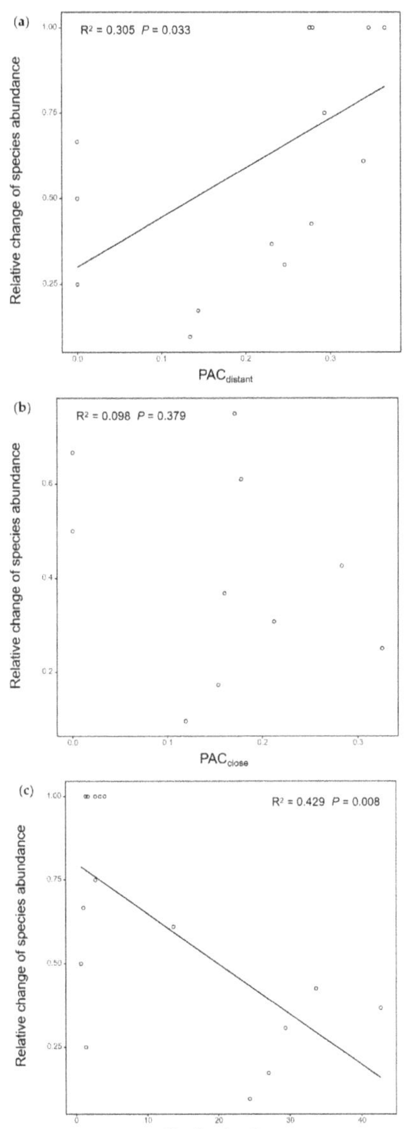

Figure 3. The relationship between relative change in bee abundance vs. (**a**) PAC_D, (**b**) PAC_C and (**c**) bee species abundance in the distant plot across native bee species. N = 15 for both (**a**,**c**); N = 10 for (**b**) because of species loss in the nearby plots.

3.4. The Shift in the Feeding Niche of Native Bees

The PERMANOVA and NMDS analysis showed that significant changes in feeding niche occurred in *Lasioglossum* sp1 and *Andrena tarsata* (Figure A2).

4. Discussion

We have shown that among the 15 native bee species we investigated, 9 exhibited significantly lower abundance in distant plots compared to nearby plots where western honeybees (*Apis mellifera*) had been introduced to the alpine meadows of the Tibetan Plateau. Notably, five rare native bee species were not detected in the introduced areas. Given the similarity in flowering plant species composition between nearby and distant plots, our findings suggest that the introduced honeybees had a significantly negative impact on native pollinators in natural ecosystems. The differences in this negative effect among species are likely due to variations in species abundance and the degree of niche overlap between native bees and introduced honeybees. These results highlight the critical importance of understanding ecological interactions involving invasive species, as they can have unintended consequences on local biodiversity.

Floral resources are often a key limiting factor for bee reproduction [31–33] and population growth [34–36], although parasitism and nest site availability are sometimes more important limiting factors [37,38]. The introduced honeybees have been demonstrated to be highly efficient pollen and nectar foragers and are able to outcompete many native bees by constraining pollen collection and offspring provisioning [39,40]. Consistently, our data show that most of the studied bee species were significantly and negatively impacted by the introduced honeybees. Moreover, some native bees, even if they shift their diet, they still suffer abundance decline, because they are forced to forage on less nutritious plants, spend more time searching for resources that are unoccupied or have not yet been depleted, or foraging further from their nests [11,41–45].

However, the effect of the introduced honeybees was different among native bees. As predicted, perceived apparent competition, a measure of niche overlap, can be used to predict the population dynamics of native bees. This is similar to the finding that the performance of the native generalist megachilid *Osmia pumilia*, which had a higher resource overlap with the introduced honeybees (*A. mellifera*), was more detrimentally affected than other native *Osmia* species having a lower resource overlap [46]. Consistently, in the study, the degree of niche overlap between native bees and the introduced honeybees was positively correlated with the degree of decline of native bee abundance. Specifically, the presence of the *A. mellifera* had no significant effect on the abundance of the two bumble bees, which have the least niche overlap with the introduced honeybees. This is because only bumble bees can utilize the plant species with complex floral morphologies (e.g., tubular, funnel-shaped, and campanulate corollas), as found in temperate, arctic, and alpine zones of the Northern Hemisphere [9,10]. The depth of the floral tube in these plants has been selected to correspond to the proboscis length of bumble bee species, leading to specialized mutualisms with specific bumble bee species [10]. In the current study, they did have a lower PAC with *A. mellifera* compared to other native bees, although bumble bees might have competed with *A. mellifera* by visiting abundant Asteraceae plants (e.g., *Saussurea nigrescens*).

It is worth noting that PAC_D, but not PAC_C, can successfully predict the interspecific difference in abundance decline among different native bee species. This is presumably because some native bees might have shifted their feeding niche in invaded sites, which might broaden their diet to improve their survival and reproduction, or narrow their diet breath to reduce population size. Indeed, the rare honeybee species are also observed to experience diet change in response to the introduced honeybees. For example, the species *Andrena tarsata* became more specialized at sites with *A. mellifera*, and it visited a lower proportion of available plant species in the invaded sites. Similarly, the other species *Lasioglossum* sp1 also showed a reduction in the proportion of links to plants visited primarily by *A. mellifera*, and an increase in the proportion of links to plants not primarily visited by *A. mellifera*, in response to competition for resources by *A. mellifera*. Such a diet shift is similar to those observed in other studies [15,47].

The difference in the effect of introduced honeybees can also be explained by species abundance. As predicted, the abundance of rare species disproportionally more rapid than that of common species. Notably, five endemic species were found to have disappeared in the invaded sites. One potential cause leading to the disappearance is that the rare species occupy similar and little differential feeding niches to superior competitors of common species including the introduced species. Indeed, except for few exclusive floral resources, these rare and endemic species are observed to visit common plant species including *Saussurea nigrescens*, *Allium chrysanthum*, *Taraxacum sikkimense*, *Angelica apaensis*, which are also visited by the introduced honeybees. Consistent with previous studies, rare pollinators with lower densities are typically more sensitive to the dominant invaders compared to common pollinators with higher densities [13,14]. This is likely because rare pollinators often interact with specific resources, which are often a subset of those utilized by generalist honeybees, leading to intensified competition [15]. Once the abundance of these rare species declined due to competition with the generalist introduced honeybees, they likely experienced local species loss rapidly because of environmental and demographic stochasticity [48].

It must be noted that, theoretically, our methodology—comparing nearby and distant plots—cannot exclude the possibility that the observed differences in native bee abundance were due to initial differences in plant and pollinator communities. This is because we lack pre- and post-apiculture data for the same sites [16,49]. However, in the current study, the flowering plant species composition was similar across all plots, and the soil within each plot is classified as meadow soil according to the Chinese soil classification system [50]. Based on these factors, we attribute the observed decline in native bee abundance to the introduced honeybees.

5. Conclusions

Our study demonstrates that, among 15 native bee species investigated, 9 species had a significantly lower abundance in distant plots than close plots, where western honeybees (*Apis mellifera*) had been introduced in alpine meadows of the Tibetan Plateau. In particular, five of these nine species were not detected in the introduced areas. Our findings suggest that the introduced honeybees likely outcompete many native bee species, potentially leading to local species loss, particularly among rare native bee species. Importantly, our results indicate that the interspecific difference in the abundance decline among different native pollinator species are well explained by their species abundance and niche overlap with the introduced honeybees. This adds to current knowledge that apiculture may inadvertently introduce superior competitive invasive pollinators, causing ecological damage to local pollinators and even protective species, though it is traditionally recognized to be important to agricultural production and economic development. To this end, we recommend limiting apicultural practice in nature ecosystems, especially where native honeybees are rich and rare.

Author Contributions: Conceptualization, S.S.; investigation, R.A.; writing—original draft preparation, R.A.; writing—review and editing, S.S.; funding acquisition, S.S. All authors have read and agreed to the published version of the manuscript.

Funding: This research was funded by the National Natural Science Foundation of China, grant numbers 32071605 and 32301391.

Institutional Review Board Statement: The study was conducted in accordance with the policy of the Ethics Committee of Nanjing University, China.

Data Availability Statement: Not applicable.

Acknowledgments: We thank Z. Huang and S. He for field and laboratory assistance, and the Sichuan Zoige Alpine Wetland Ecosystem National Observation and Research Station for platform support.

Conflicts of Interest: The authors declare no conflicts of interest.

Appendix A

Table A1. List of native honeybee species and their relative abundance in both close and distant sites. The text in the first column corresponds the species code in Figure 1; the abbreviation corresponds the species code in Figure 2.

ID	Family	Species	Abbreviation
B1	Apidae	*Bombus supremus*	BS
B2	Halictidae	*Lasioglossum* sp1	L. sp1
B3	Andrenidae	*Andrena tarsata*	AT
B4	Apidae	*Bombus filchnerae*	BF
B5	Halictidae	*Dufourea novaeangliae*	DN
B6	Halictidae	*Lasioglossum zonulum*	LZ
B7	Andrenidae	*Andrena* sp1	A. sp1
B8	Apidae	*Nomada* sp.	N. sp.
B9	Halictidae	*Lasioglossum* sp2	L. sp2
B10	Andrenidae	*Andrena* sp2	A. sp2
B11	Megachilidae	*Osmia* sp.	O. sp.
B12	Halictidae	*Sphecodes* sp.	S. sp.
B13	Apidae	*Epeolus* sp.	E. sp.
B14	Halictidae	*Lasioglossum* sp3	L. sp3
B15	Halictidae	*Dufourea* sp.	D. sp.

Table A2. List of plant species. The text in the first column corresponds the species code in Figure 1.

ID	Family	Species
P1	Gentianaceae	*Gentiana abaensis*
P2	Ranunculaceae	*Trollius farreri*
P3	Apiaceae	*Angelica dahurica*
P4	Ranunculaceae	*Anemone rivularis*
P5	Asteraceae	*Saussurea nigrescens*
P6	Apiaceae	*Angelica apaensis*
P7	Apiaceae	*Semenovia malcolmii*
P8	Rosaceae	*Potentilla anserina*
P9	Rosaceae	*Potentilla discolor*
P10	Orobanchaceae	*Pedicularis kansuensis*
P11	Apiaceae	*Carum carvi*
P12	Geraniaceae	*Geranium pylzowianum*
P13	Amaryllidaceae	*Allium sikkimense*
P14	Fabaceae	*Oxytropis kansuensis*

Table A2. *Cont.*

ID	Family	Species
P15	Amaryllidaceae	*Allium chrysanthum*
P16	Gentianaceae	*Halenia elliptica*
P17	Gentianaceae	*Comastoma pulmonarium*
P18	Orobanchaceae	*Pedicularis longiflora*
P19	Asteraceae	*Leontopodium souliei*
P20	Rosaceae	*Dasiphora fruticosa*
P21	Rosaceae	*Geum japonicum*
P22	Ranunculaceae	*Ranunculus tanguticus*
P23	Gentianaceae	*Gentiana formosa*
P24	Geraniaceae	*Geranium pratense*
P25	Asteraceae	*Taraxacum sikkimense*
P26	Ranunculaceae	*Anemone trullifolia*
P27	Fabaceae	*Oxytropis glabra*
P28	Asteraceae	*Ajania przewalskii*
P29	Asteraceae	*Anaphalis flavescens*
P30	Asteraceae	*Saussurea stella*
P31	Fabaceae	*Hedysarum tanguticum*
P32	Orobanchaceae	*Pedicularis* sp.
P33	Asteraceae	*Aster alpinus*
P34	Polygonaceae	*Polygonum viviparum*

Appendix B

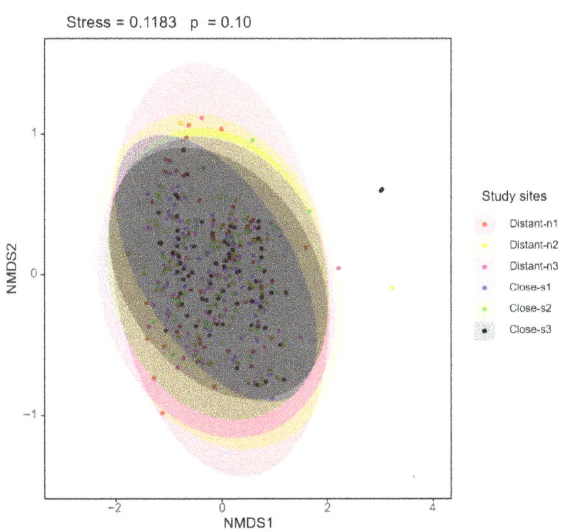

Figure A1. Results of non-metric multidimensional scaling (NMDS) showing plant species community similarity among six field sampling plots. Dissimilarity (Bray–Curtis) was estimated using the abundance of flowers or inflorescences.

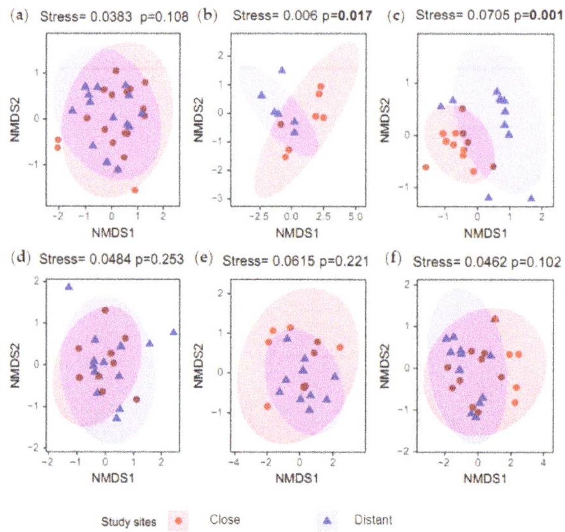

Figure A2. Results of non-metric multidimensional scaling (NMDS) showing whether the feeding niche of native bees differs between nearby (pink) and. distant (gray) plots for six common species. Dissimilarity (Bray–Curtis) was estimated using the diet preference of each sampling. (**a**) *Bombus supremus*, (**b**) *Lasioglossum* sp1, (**c**) *Andrena tarsata*, (**d**) *Bombus filchnerae*, (**e**) *Dufourea novaeangliae*, and (**f**) *Lasioglossum zonulum*. *p*-values are shown in bold as < 0.05.

References

1. Crane, E. Recent research on the world history of beekeeping. *Bee World* **1999**, *80*, 174–186. [CrossRef]
2. Requier, F.; Garnery, L.; Kohl, P.L.; Njovu, H.K.; Pirk, C.W.W.; Crewe, R.M.; Steffan-Dewenter, I. The Conservation of Native Honey Bees Is Crucial. *Trends Ecol. Evol.* **2019**, *34*, 789–798. [CrossRef] [PubMed]
3. Yang, G. Harm of introducing the western honeybee *Apis mellifera L.* to the Chinese honeybee *Apis cerana F.* and its ecolocjcal impact. *Acta Entomol. Sin.* **2005**, *42*, 401–406. [CrossRef]
4. Chen, C.; Liu, Z.; Pan, Q.; Chen, X.; Wang, H.; Guo, H.; Liu, S.; Lu, H.; Tian, S.; Li, R.; et al. Genomic Analyses Reveal Demographic History and Temperate Adaptation of the Newly Discovered Honey Bee Subspecies *Apis mellifera sinisxinyuan* n. ssp. *Mol. Biol. Evol.* **2016**, *33*, 1337–1348. [CrossRef] [PubMed]
5. Goulson, D. Effects of introduced bees on native ecosystems. *Annu. Rev. Ecol. Evol. Syst.* **2003**, *34*, 1–26. [CrossRef]
6. Paini, D.R. Impact of the introduced honey bee (*Apis mellifera*) (Hymenoptera: Apidae) on native bees:: A review. *Austral Ecol.* **2004**, *29*, 399–407. [CrossRef]
7. Roubik, D.W.; Villanueva-Gutierrez, R. Invasive Africanized honey bee impact on native solitary bees: A pollen resource and trap nest analysis. *Biol. J. Linn. Soc.* **2009**, *98*, 152–160. [CrossRef]
8. Thomson, D.M. Detecting the effects of introduced species: A case study of competition between *Apis* and *Bombus*. *Oikos* **2006**, *114*, 407–418. [CrossRef]
9. Goulson, D. *Bumblebees: Bahavior, Ecology, and Conservation*; Online edn; Oxford Academic: Oxford, UK, 2023. [CrossRef]
10. Suzuki, K.; Dohzono, I.; Hiei, K. Evolution of pollinator generalization in bumblebee-pollinated plants. *Plant Species Biol.* **2007**, *22*, 141–159. [CrossRef]
11. Page, M.L.; Williams, N.M. Evidence of exploitative competition between honey bees and native bees in two California landscapes. *J. Anim. Ecol.* **2023**, *92*, 1802–1814. [CrossRef]
12. Stenc, J.; Janosik, L.; Freudenfeld, M.; Matouskova, E.; Hadrava, J.; Mikat, M.; Dankova, K.; Hadravova, T.; Rysan, T.; Simonova, J.; et al. Pollen presentation mitigates competition for pollinators due to diurnal stratification of pollen transfer. *Perspect. Plant Ecol. Evol. Syst.* **2025**, *67*, 125868. [CrossRef]
13. Thomas, C.D.; Mallorie, H.C. Rarity, species richness and conservation-butterflies of the Atlas Mountains in Morocco. *Biol. Conserv.* **1985**, *33*, 95–117. [CrossRef]
14. Kunin, W.E.; Gaston, K.J. The biology of rarity-patterns, causes and consequences. *Trends Ecol. Evol.* **1993**, *8*, 298–301. [CrossRef]
15. Valido, A.; Rodriguez-Rodriguez, M.C.; Jordano, P. Honeybees disrupt the structure and functionality of plant-pollinator networks. *Sci. Rep.* **2019**, *9*, 4711. [CrossRef]

16. Magrach, A.; Gonzalez-Varo, J.P.; Boiffier, M.; Vila, M.; Bartomeus, I. Honeybee spillover reshuffles pollinator diets and affects plant reproductive success. *Nat. Ecol. Evol.* **2017**, *1*, 1299–1307. [CrossRef]
17. Wang, L.-L.; Huang, Z.Y.; Dai, W.-F.; Yang, Y.-P.; Duan, Y.-W. Mixed effects of honey bees on pollination function in the Tibetan alpine grasslands. *Nat. Commun.* **2024**, *15*, 8164. [CrossRef] [PubMed]
18. Montero-Castano, A.; Vila, M. Influence of the honeybee and trait similarity on the effect of a non-native plant on pollination and network rewiring. *Funct. Ecol.* **2017**, *31*, 142–152. [CrossRef]
19. Thomson, D.M. Local bumble bee decline linked to recovery of honey bees, drought effects on floral resources. *Ecol. Lett.* **2016**, *19*, 1247–1255. [CrossRef] [PubMed]
20. Macior, L.W.; Ya, T.; Zhang, J. Reproductive biology of Pedicularis (Scrophulariaceae) in the Sichuan Himalaya. *Plant Species Biol.* **2001**, *16*, 83–89. [CrossRef]
21. Hu, X.; Zhou, W.; Li, X.; Niklas, K.J.; Sun, S. Changes in Community Composition Induced by Experimental Warming in an Alpine Meadow: Beyond Plant Functional Type. *Front. Ecol. Evol.* **2021**, *9*, 569422. [CrossRef]
22. Wu, X.; Duffy, J.E.; Reich, P.B.; Sun, S. A brown-world cascade in the dung decomposer food web of an alpine meadow: Effects of predator interactions and warming. *Ecol. Monogr.* **2011**, *81*, 313–328. [CrossRef]
23. Peng, Y.; Dong, Y.; Xu, H.; Xi, X.; Niklas, K.J.; Sun, S. Domesticated honeybees facilitate interspecific hybridization between two (Taraxacum) congeners. *J. Ecol.* **2018**, *106*, 1204–1216. [CrossRef]
24. Mu, J.; Peng, Y.; Xi, X.; Wu, X.; Griffin, J.N.; Niklas, K.J.; Sun, S. Domesticated honey bees evolutionarily reduce flower nectar volume in a Tibetan lotus. *Ecology* **2014**, *95*, 3161–3172. [CrossRef]
25. Zhao, Y.-H.; Lazaro, A.; Ren, Z.-X.; Zhou, W.; Li, H.-D.; Tao, Z.-B.; Xu, K.; Wu, Z.-K.; Wolfe, L.M.; Li, D.-Z.; et al. The topological differences between visitation and pollen transport networks: A comparison in species rich communities of the Himalaya-Hengduan Mountains. *Oikos* **2019**, *128*, 551–562. [CrossRef]
26. Dormann, C.F.; Fründ, J.; Blüthgen, N.; Gruber, B. Indices, Graphs and Null Models: Analyzing Bipartite Ecological Networks. *Open Ecol. J.* **2009**, *2*, 7–24. [CrossRef]
27. Dormann, C.F. How to be a specialist? Quantifying specialisation in pollination networks. *Netw. Biol.* **2011**, *1*, 1–20.
28. Morris, R.J.; Lewis, O.T.; Godfray, H.C.J. Apparent competition and insect community structure: Towards a spatial perspective. *Ann. Zool. Fenn.* **2005**, *42*, 449–462.
29. Müller, C.B.; Adriaanse, I.C.T.; Belshaw, R.; Godfray, H.C.J. The structure of an aphid–parasitoid community. *J. Anim. Ecol.* **1999**, *68*, 346–370. [CrossRef]
30. Oksanen, J.; Blanchet, F.G.; Friendly, M.; Kindt, R.; Legendre, P.; McGlinn, D. R Package, Version 2.5–4; Vegan: Community Ecology Package. Available online: https://CRAN.R-project.org/package=vegan (accessed on 14 August 2025).
31. Carvell, C.; Bourke, A.F.G.; Dreier, S.; Freeman, S.N.; Hulmes, S.; Jordan, W.C.; Redhead, J.W.; Sumner, S.; Wang, J.; Heard, M.S. Bumblebee family lineage survival is enhanced in high-quality landscapes. *Nature* **2017**, *543*, 547–549. [CrossRef]
32. Goodell, K. Food availability affects *Osmia pumila* (Hymenoptera: Megachilidae) foraging, reproduction, and brood parasitism. *Oecologia* **2003**, *134*, 518–527. [CrossRef]
33. Williams, N.M.; Kremen, C. Resource distributions among habitats determine solitary bee offspring production in a mosaic landscape. *Ecol. Appl.* **2007**, *17*, 910–921. [CrossRef]
34. Crone, E.E.; Williams, N.M. Bumble bee colony dynamics: Quantifying the importance of land use and floral resources for colony growth and queen production. *Ecol. Lett.* **2016**, *19*, 460–468. [CrossRef] [PubMed]
35. Malfi, R.L.; Crone, E.; Williams, N. Demographic benefits of early season resources for bumble bee (*B. vosnesenskii*) colonies. *Oecologia* **2019**, *191*, 377–388. [CrossRef] [PubMed]
36. Thomson, D.M.; Page, M.L. The importance of competition between insect pollinators in the Anthropocene. *Curr. Opin. Insect Sci.* **2020**, *38*, 55–62. [CrossRef]
37. Forrest, J.R.K.; Chisholm, S.P.M. Direct benefits and indirect costs of warm temperatures for high-elevation populations of a solitary bee. *Ecology* **2017**, *98*, 359–369. [CrossRef] [PubMed]
38. Steffan-Dewenter, I.; Schiele, S. Do resources or natural enemies drive bee population dynamics in fragmented habitats? *Ecology* **2008**, *89*, 1375–1387. [CrossRef]
39. Thomson, D. Competitive interactions between the invasive European honey bee and native bumble bees. *Ecology* **2004**, *85*, 458–470. [CrossRef]
40. Hudewenz, A.; Klein, A.-M. Red mason bees cannot compete with honey bees for floral resources in a cage experiment. *Ecol. Evol.* **2015**, *5*, 5049–5056. [CrossRef]
41. Di Pasquale, G.; Salignon, M.; Le Conte, Y.; Belzunces, L.P.; Decourtye, A.; Kretzschmar, A.; Suchail, S.; Brunet, J.-L.; Alaux, C. Influence of Pollen Nutrition on Honey Bee Health: Do Pollen Quality and Diversity Matter? *PLoS ONE* **2013**, *8*, e72016. [CrossRef]
42. Fruend, J.; Dormann, C.F.; Holzschuh, A.; Tscharntke, T. Bee diversity effects on pollination depend on functional complementarity and niche shifts. *Ecology* **2013**, *94*, 2042–2054. [CrossRef]

43. Roulston, T.H.; Cane, J.H. Pollen nutritional content and digestibility for animals. *Plant Syst. Evol.* **2000**, *222*, 187–209. [CrossRef]
44. Spiesman, B.J.; Gratton, C. Flexible foraging shapes the topology of plant-pollinator interaction networks. *Ecology* **2016**, *97*, 1431–1441. [CrossRef]
45. Zurbuchen, A.; Cheesman, S.; Klaiber, J.; Mueller, A.; Hein, S.; Dorn, S. Long foraging distances impose high costs on offspring production in solitary bees. *J. Anim. Ecol.* **2010**, *79*, 674–681. [CrossRef]
46. Goodell, K. The Impact of Introduced Honey Bees on Native Solitary Bees: Competition and Indirect Effects. Ph.D. Thesis, State University of New York at Stony Brook, Stony Brook, NY, USA, 2000.
47. Walther-Hellwig, K.; Fokul, G.; Frankl, R.; Buechler, R.; Ekschmitt, K.; Wolters, V. Increased density of honeybee colonies affects foraging bumblebees. *Apidologie* **2006**, *37*, 517–532. [CrossRef]
48. Lande, R. Risks of population extinction from demographic and environmental stochasticity and random catastrophes. *Am. Nat.* **1993**, *142*, 911–927. [CrossRef]
49. Magrach, A.; Holzschuh, A.; Bartomeus, I.; Riedinger, V.; Roberts, S.P.M.; Rundlof, M.; Vujic, A.; Wickens, J.B.; Wickens, V.J.; Bommarco, R.; et al. Plant-pollinator networks in semi-natural grasslands are resistant to the loss of pollinators during blooming of mass-flowering crops. *Ecography* **2018**, *41*, 62–74. [CrossRef]
50. Gong, Z.T. *Chinese Soil Taxonomy: Theory, Method and Application*; Science Press: Beijing, China; New York, NY, USA, 1999; pp. 1–903.

biology

MDPI

Article

The Influences of Rainfall Intensity and Timing on the Assemblage of Dung Beetles and the Rate of Dung Removal in an Alpine Meadow

Wenxiao Sun, Wenting Tang, Yashi Wu, Shuaibing He and Xinwei Wu *

Department of Ecology, College of Life Science, Nanjing University, Nanjing 210023, China; swenxiao430@163.com (W.S.); wttang@whu.edu.cn (W.T.); wuyashi200009@163.com (Y.W.); 15670523502@163.com (S.H.)

* Correspondence: wuxw@nju.edu.cn; Tel.: +86-25-89681213

Simple Summary: Changes in precipitation patterns can significantly alter biological processes and associated ecosystem functions. This study aimed to investigate the impacts of rainfall intensity and timing on the structure of dung beetle assemblage and dung mass loss rate in a Tibetan alpine meadow. The results revealed that exposure to rainfall within a 2 h timeframe, regardless of rainfall intensity, significantly reduced the rate of dung mass loss. However, this decrease was not attributed to the activity of dung beetles. Earlier rainfall tended to decrease species richness of dung beetles, while heavier rainfall significantly decreased beetle abundance. These findings suggest that changes in precipitation patterns can influence both the assemblage of dung beetles and the rate of dung decomposition, but may also decouple their relationship under a certain circumstance. The variability in local biological processes and ecosystem functions deserves greater attention within a global climate change scenario.

Abstract: Changes in precipitation patterns, including rainfall intensity and rainfall timing, have been extensively demonstrated to impact biological processes and associated ecosystem functions. However, less attention has been paid to the effects of rainfall intensity and rainfall timing on the assembly of detritivore communities and the decomposition rate of detritus such as animal dung. In a grazed alpine meadow on the eastern Qinghai–Tibet Plateau, we conducted a manipulative experiment involving two levels of rainfall intensity (heavy rainfall, 1000 mL/5 min; light rainfall, 100 mL/5 min) and five levels of rainfall timing (0, 2, 4, 24, and 48 h after yak dung deposition). The aim was to determine the effects of rainfall intensity, timing, and their interaction on the assemblage of dung beetles and dung removal rate during the early stage (i.e., 96 h after yak dung deposition) of dung decomposition. Light rainfall significantly increased species richness in the treatment of 48 h after dung pats were deposited. Heavy rainfall significantly decreased beetle abundance in both the 0 h and 48 h treatments while light rainfall had no effect on beetle abundance. Dung mass loss was significant lower in the 2 h treatment compared to other treatments regardless of rainfall intensity. The structural equation model further revealed that the species richness of dung beetles and dung mass loss were significantly affected by rainfall timing but not by rainfall intensity. However, no significant relationships were observed between any variables examined. These findings suggest that changes in precipitation patterns can influence both the structure of dung beetles and the rate of dung decomposition but may also decouple their relationship under a certain circumstance. Therefore, it is crucial to pay greater attention to fully understand local variability between the biological processes and ecosystem functions within a global climate change scenario.

Keywords: climate change; detritus; detritivore; dung decomposition; Qinghai–Tibet Plateau; rainfall pattern

1. Introduction

Extensive evidence has demonstrated the profound impact of global climate change on biological processes and associated ecosystem functions in both terrestrial and aquatic ecosystems [1,2]. Precipitation patterns, a key driver of global climate change, have already undergone significant changes and are predicted to undergo further alterations in the future [3]. Specifically, there has been a widespread increase in rainfall intensity across various regions worldwide [4]. The pronounced seasonal variation in rainfall timing draws attention to extreme events caused by the uneven distribution of rainfall time and changes in rainfall intervals [5,6]. These changes in rainfall patterns have exerted influences on metabolic processes [7,8], population dynamics [9], species interactions [10], assemblages [11], the distribution ranges of natural plants and animals [12,13], as well as ecosystem functions and services [14]. Nevertheless, compared with other drivers such as climate warming, our comprehensive understanding of the impacts resulting from changes in precipitation patterns remains incomplete, particularly regarding their effects on the assemblage of detritivore community and the decomposition rates of non-living organic matter.

The decomposition of dead organic matter is a biogeochemical process that plays a fundamental role in nutrient, carbon, and energy cycling within and among ecosystems, as well as the exchange between the biosphere and atmosphere [15]. The decomposition rates of dead organic matter are largely influenced by the diversity and population dynamics of detritivores responsible for breaking down non-living organic matter into simpler forms to gain energy and matter for their own biomass maintenance [16]. Importantly, the effect sign and strength of these soil micro-, meso- and macro-organisms on decomposition can be modified by rainfall patterns [17]. For instance, a 50% reduction in rainfall significantly decreased the decomposition rate of leaf litter through influencing the activities of soil microorganisms and soil animals [18]. A microcosm experiment investigating leaf litter decomposition revealed that microbial-driven decomposition was positively correlated with cumulative rainfall quantity, while isopod-driven decomposition was strongly controlled by rainfall frequency [19]. However, it should be noted that these studies have primarily focused on the decomposition of plant detritus rather than animal-based materials such as feces or carcasses.

Ignoring the decomposition of animal detritus would be inappropriate [20]. For example, terrestrial plants annually produce approximately 120 billion tons of organic carbon [21], with up to 90% entering the detritus pool as plant litter [22]. Even of the remaining 10% consumed by herbivores, only very little is assimilated for growth and development, while the majority eventually returns to the detritus pool as excreta and carrion [23]. Although these animal detritus components constitute only a small percentage compared to plant detritus within the total natural detritus pool [24,25], they display a disproportionate role in assembling and ecosystem functioning [26]. Firstly, animal detritus provides energy and habitat for a suite of highly dependent species [27]. For example, the whole life history of a variety of dung beetle species depends completely on excreta from large animals. Secondly, compared to plant detritus, animal detritus is an ephemeral resource but very rich in nutrients. The nitrogen content of dung is two to three times higher than that of plant litter [18,28]. Moreover, the rates of nutrient turnover and organic matter decomposition of animal detritus are undoubtedly quicker than plant detritus [24]. This energy pulse of animal detritus is crucial to the maintenance of biodiversity and the normal functioning of many ecosystems.

Like the decomposition of plant detritus, the decomposition of animal detritus also depends largely on the population dynamics, species diversity, and community structure of detritivores [15,29], all of which are sensitive to environmental change [30]. Changes in rainfall patterns are suggested to influence decomposition by modifying physiological traits, feeding behaviors, and species interactions among decomposing arthropods [31–34]. For example, decreased rainfall significantly reduces the abundance of dung beetles, leading to a decline in dung decomposition rates [32]. Although the onset of rainfall has been linked to the emergence of dung beetle populations for many years, available field data about the effect of the changes in rainfall patterns on the richness, abundance, and ecological functions of dung beetle species are scarce [31–34]. Given the great substantial variability in global rainfall patterns, further research is necessary for a comprehensive understanding of the relationship between changes in rainfall patterns and animal detritus decomposition.

As the "Third Pole of the Earth", the Qinghai–Tibet Plateau is a biodiversity hotspot and is particularly sensitive to global climate change [35,36]. Specifically, there has been an increase in rainfall amounts and intensity, frequency of heavy rainy events, and the daily maximum of rainfall during the period from 1961 to 2017 [37]. Additionally, precipitation duration has decreased since 2000 [37], with this trend predicted to persist until the end of the century [38]. Moreover, the Qinghai–Tibet Plateau houses China's largest area of grassland and rangeland, which supports billions of wild animals and livestock, including yaks (*Bos grunniens* Linnaeus, 1766), which are predominantly reared by local people. The estimated density of yak dung pats can reach up to 5900 per hectare, covering more than 20% area of the total rangeland [39]. The timely decomposition of yak dung is thus important to nutrient turnover and primary productivity. Dung beetles (Coleoptera: Scarabaeoidea) are the major group that contributes substantially to dung decomposition on the Qinghai–Tibet Plateau. Notably, because the species richness and abundance of dung beetles always peak within the first two to three days after dung deposition [39], the early stage of dung decomposition is the best period to examine the population dynamics and assemblage structure of dung beetles. Rainfall events might affect the assemblage and dung decomposition rates by influencing volatile substances released from dung as well as dung beetle behavior. Given the importance of rainfall patterns on detritivores and detritivores on decomposition, we hypothesize here that changes in rainfall patterns will alter decomposition rates through affecting detritivore community assemblage. Specifically, we aim to test whether high-intensity rainfall events that occurred earlier would lead to reduced richness and abundance of dung beetles, which in turn would decrease dung decomposition rates.

2. Materials and Methods

2.1. Study Site

The experiment was conducted in an alpine meadow in Hongyuan County (32°48′ N, 102°33′ E) in the northwest of Sichuan Province, located in the eastern edge of the Qinghai–Tibet Plateau, China. The average altitude is about 3500 m. The climate is characterized by long, cold winters and short, cool summers. According to the Hongyuan County Climate Station (a 4 km distance to our study site) from 1961 to 2019, the mean annual temperature was 1.7 °C, with the maximum and minimum monthly means being 13.4 °C and −15.1 °C in July and January, respectively. Precipitation mostly (>80%) occurred from May to August. The mean annual precipitation was about 760 mm with significant fluctuations across years, mostly ranging from 450 mm to 900 mm [40]. Details on the species composition of plant community at the study site can be found in previous studies [39]. The soil is characterized by high organic content and low nitrogen and phosphorus content. The meadow has been under intensive grazing (by yaks) for decades. The study site was a typical summer-grazing pasture, which was under grazing from May to September by local people. According to an independent pre-experimental survey, the estimated population density is about 10 yaks per hectare.

The decomposer community responsible for yak dung decomposition in the Tibetan alpine meadows is as complicated as in other areas [41]. In addition to the microfauna (bacteria and fungi), macro-invertebrates such as beetles and flies are dominant decomposer groups especially in the early stages of dung decomposition. In our study area, *Calliphora vicina* (Robineau-Desvoidy, 1830) is usually the most abundant fly species. The coprophagous beetle species include *Eupleurus subterraneus* (Linnaeus, 1758), *Aphodius erraticus* (Linnaeus, 1758), *A. rectus* (Motschulsky, 1866),

A. prodromus (Brahm,1790), *A. elegans* (Allibert, 1847), *Geotrupes ibericus* (Baraud, 1958), *Onthophagus uniformis* (Heyden, 1886), etc. Predaceous beetles include *Philonthus rubripennis* (Stephens, 1832), *Quedius (Microsaurus) liangshanensis* (Zheng and Zheng, 2006), and *Sphaerites* sp. In the very late successional stage, there may be spiders (e.g., *Araneus marmoreus* (Clerk, 1757)), centipedes, and beetle larvae emerging within and underneath the dung [42].

2.2. Experimental Design and Sampling

The experiment was conducted using a two-factor full design, where rainfall intensity with two levels (heavy rainfall, 1000 mL rainwater in five mins; light rainfall, 100 mL rainwater in five mins), and rainfall timing with five levels (0, 2, 4, 24 and 48 h after fresh dung pats were deposited). The heavy rainfall was a little beyond the usual maximum rainfall. Notably, however, the classification of the heavy and light rainfall here was not a meteorological definition. There were thus ten treatments, each with five replicates, and each replicate was an artificial fresh dung pat. There were 50 dung pats in total.

The experiment was conducted in a flat pasture with an area of about two hectares where the vegetation cover is over 80%. Before the start of the experiment, enough rainwater was collected and stored in buckets. In the early morning of a sunny day (20 August 2021), enough fresh yak dung free of any macro-invertebrates was collected in a stall of a Tibet family. After thoroughly mixing in large buckets, the dung was built into small artificial pats using plastic modules with a diameter of 18 cm and a height of 4 cm, respectively. The fresh weight of each pat was 1000 ± 20 g (250 ± 5 g in dry weight). The pats were then randomly placed in the pasture, with at least an interval of 5 m between any two pats to reduce possible mutual interference with each other. Finally, rainwater was sprayed evenly on the surface of each pat using watering pots according to the experimental design. In addition, two additional pats were arranged as candidate compensates for each treatment. Pats damaged by grazing yaks or rodents during the experimental period would have been randomly replaced by these candidates. In addition, the temperature was measured using thermometers (model DS1921G, Maxim Integrated Products, Sunnyvale, CA, USA). Over the course of the experiment, the mean was 12.21 °C based on daily average (over 24 h of measurements made every 30 min), with the highest and lowest values being 28.5 °C and 0 °C, respectively.

On the fourth day (i.e., 96 h after deposition of yak dung) since the beginning of the experiment, dung beetles and any other visible decomposers (if any) within both the dung pats and the soil beneath the pats were collected by hand-sorting and counted. Dung beetles were identified to the species level. Furthermore, the residual dung was also collected and dried at 75 °C for 72 h to assess the rate of dung mass loss. It is important to note that there was no rainfall throughout the entire experimental period.

2.3. Data Analysis

Two-way ANOVA was used to test the effects of rainfall intensity, rainfall timing, and their interaction on dung mass loss, species richness and abundance (log-transformed) of dung beetles. Once a significant factor effect was detected, post hoc Tukey's HSD tests were used to further determine the differences among treatments. The relative density of each species was calculated by dividing the abundance of the specie in the treatment by the total abundance of all beetles in the treatment. Then, the similarity percentage (SIMPER) analysis based on the Bray–Curtis similarity matrix was employed to determine the species contribution among different treatments [40]. In addition, a piecewise structural equation model (SEM) was constructed to examine the causal relations among rainfall, dung beetles, and dung mass loss. An initial SEM model was constructed based on our hypothesis that dung decomposition would be affected either directly by rainfall patterns or indirectly by rainfall patterns through affecting dung beetle assemblages (Figure S1). In doing this, collinearity between variables in the model was firstly assessed through a variance inflation factor (VIF) test using the car package [43]. Because our terms had more than 1 degree of freedom (df), generalized variance-inflation factors [44] were calculated. Secondly, the model was performed with the piecewiseSEM package using the maximum likelihood estimation method [45]. The conceptual model was evaluated based on goodness-of-fit statistics and the selection of optimal model relied on the lowest AIC value. In the model, both main and interactive effects of rainfall intensity and timing were considered. All these analyses were performed in R 4.1.1 [46].

3. Results

3.1. Responses of Beetle Assemblage and Dung Decomposition

Rainfall timing significantly affected the species richness and abundance of dung beetles (Table 1). Rainfall intensity and its interaction with rainfall timing significantly affected beetle abundance (Table 1). Specifically, light rainfall led to a significant increase in species richness in the 48 h treatment (Figure 1A). On the other hand, heavy rainfall resulted in a significant decrease in beetle abundance for both the 0 h treatment and the 48 h treatment, while light rainfall had no effect on beetle abundance (Figure 1B).

Table 1. Results of a two-way ANOVA showing the effects of rainfall timing (timing), rainfall intensity (intensity), and their interaction on the species richness, abundance of dung beetles, and dung mass loss.

	df	Richness		Abundance		Dung Mass Loss	
		SS	F	SS	F	SS	F
Timing	4	0.06	3.43 *	0.35	3.83 **	14,350	7.70 ***
Intensity	1	0.00	0.55	0.12	5.37 *	61	0.13
Timing × Intensity	4	0.02	1.31	0.38	4.27 **	2401	1.29
Residuals	40	0.18		0.90		18,627	

*, $p < 0.05$; **, $p < 0.01$, ***, $p < 0.001$.

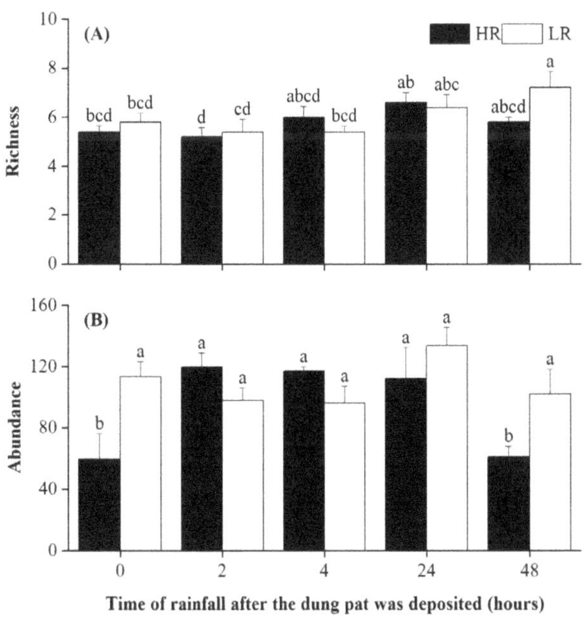

Figure 1. Species richness (**A**) and abundance (**B**) of dung beetles in different treatments at the end of the experiment. The letters above the error bars denote statistically significant differences among treatments at a significant level of $p = 0.05$, as revealed by two-way ANOVAs followed by post hoc Tukey's test for multiple comparisons. Error bars are ±SE.

A total of 5067 dung beetle individuals were collected, belonging to three families and 14 species (Table S1). The average dissimilarity among different rainfall timings were 13.62% and 16.07% in the heavy rainfall and light rainfall conditions, respectively. In the heavy rainfall conditions, the dissimilarity was primarily driven by the contribution of seven species (>2%), which accounted for a significant proportion (95%) of the overall dissimilarity. Notably, *E. subterraneus* contributed approximately 36% to the dissimilarity (Figure 2A). Similarly, in the light rainfall conditions, the dissimilarity was mainly influenced by the presence of eight species that collectively accounted for 96.85% of the overall dissimilarity. Among these species, *E. subterraneus* displayed a prominent role with an approximate contribution of about 41% (Figure 2B). Furthermore, it is worth mentioning that there were no evident differences in relative density of *E. subterraneus* across different treatments (Figure 2).

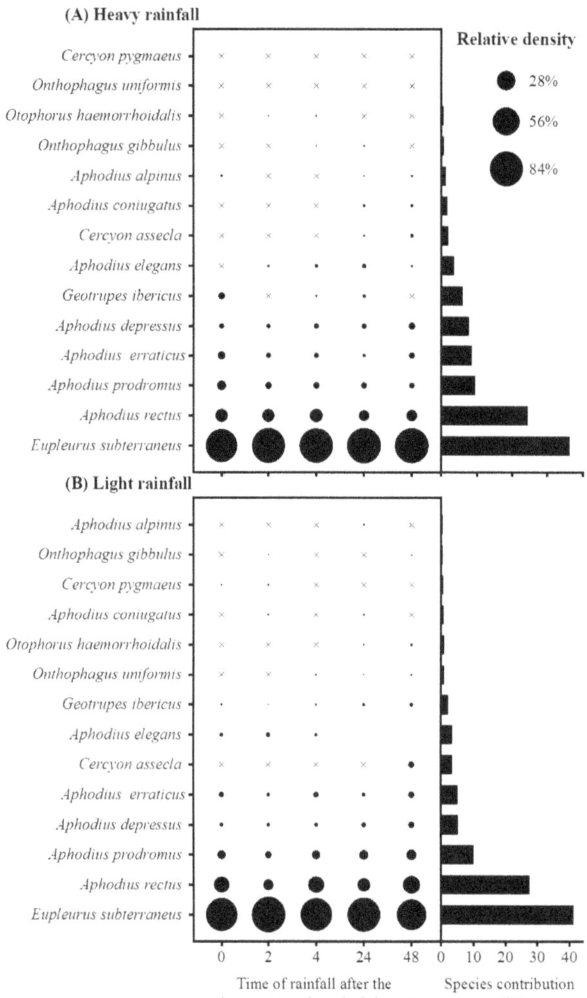

Figure 2. The relative density of different species in different treatments, and species with their corresponding contributions in SIMPER analysis of heavy rainfall (**A**) and light rainfall (**B**). "×" means the relative density of the specie is zero in the treatment.

In addition, rainfall timing significantly affected dung mass loss (Table 1). The dung mass loss was lower in the 2 h treatment compared to other treatments regardless of rainfall intensity (Figure 3).

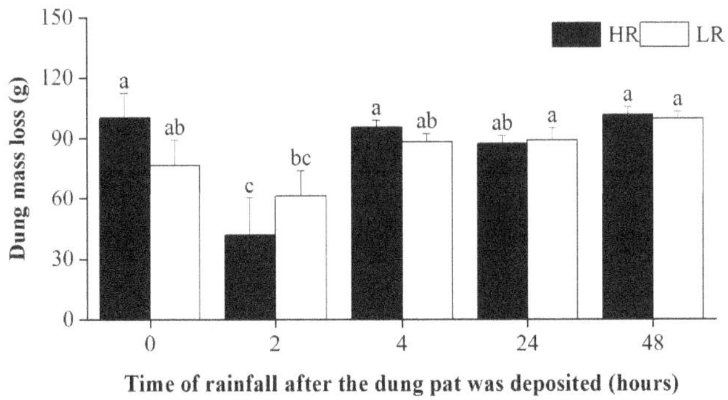

Figure 3. Dung mass loss in different treatments at the end of the experiment. The letters above the error bars denote statistically significant differences among treatments at a significant level of $p = 0.05$, as revealed by two-way ANOVAs followed by post hoc Tukey's test for multiple comparisons. Error bars are ±SE.

3.2. Relationships among Rainfall, Beetle Assemblage, and Dung Decomposition

The results of the SEM showed that species richness of dung beetles ($\beta = 0.69$) as well as the dung mass loss ($\beta = 0.50$) were significantly and positively influenced by rainfall timing rather than rainfall intensity or their interaction. However, no significant relationships were observed between any other variables (Figure 4).

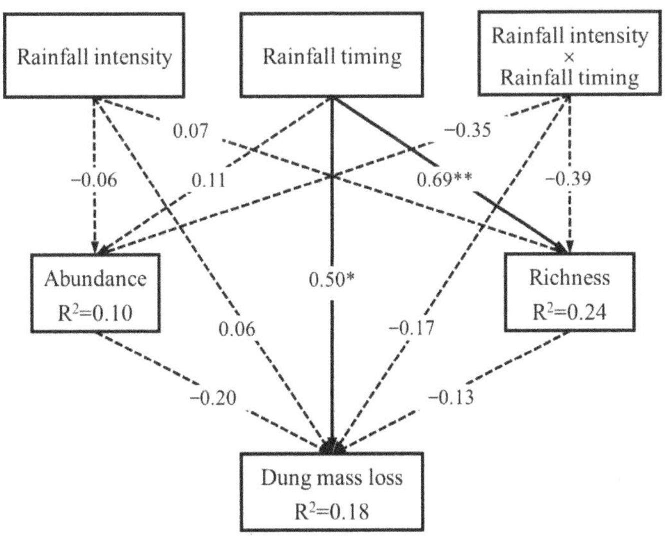

Figure 4. Structural equation model exploring the casual relationship among rainfall, dung beetles, and dung mass loss (Fisher' C = 2.91, $p = 0.23$, AIC = 36.91; a low AIC value associated with Fisher's C test indicates a strong fit of the model to data). Arrows represent unidirectional relationships between variables, with standardized path coefficients (β) indicated on each arrow. Solid arrows represent significant relationships, while dashed arrows represent non-significant relationships. **, $p < 0.01$; *, $p < 0.05$.

4. Discussion

Compared to plant litter, the organic matter in animal dung generally decomposes at a faster rate due to the presence of a more diverse and abundant decomposer community [28]. Nevertheless, it is evident that the composition of dung decomposers varies significantly among different regions or ecosystems. For example, earthworms play a primary role in degrading cattle dung in temperate climatic areas [47,48], while termites contribute significantly in arid [49] or seasonally dry areas [50]. In tropical, temperate and highland grassland ecosystems, dung beetles are typically the most important dung-feeding guild responsible for dung decomposition and nutrient cycling [39,51]. Although maggots have also been reported to be one important dung decomposer group in the Tibetan alpine grasslands [39], it is curious that we did not observe their presence within dung pats during our experimental period. We speculate that this might be due to the low temperature, which could have limited the flying and egg-laying as well as egg-hatching activities of flies during the experiment. Therefore, we here focused primarily on the responses of dung beetles. Previous studies have revealed that changes in environmental conditions can influence the community structure of dung beetles and subsequently impact dung decomposition rates [32,51]. However, we did not find any clear evidence supporting our hypothesis that rainfall influenced dung decomposition rates by altering the structure of dung beetles. Indeed, we found that both rainfall intensity and rainfall timing significantly affected the structure of dung beetles, and also a significant effect of rainfall timing on the rate of dung mass loss. However, changes in rainfall patterns simultaneously decoupled the links between dung beetles and dung decomposition. These results validate the fact that local-scale environmental changes can alter the inherent relationships between biological processes and associated ecosystem functions.

Our results showed that rainfall timing significantly influenced the species richness of dung beetles, that is to say, the rain played an important role in determining whether beetles visited a fresh dung pat successfully. The species composition of the beetle assemblage within dung pats usually exhibits a distinct successional pattern during the early stage of dung decomposition [41], which is believed to be closely linked to the emission and component of volatile organic compounds (VOCs) from fresh dung pats [52]. These VOCs, driven by microbes, provide dung beetles with multifaceted information including the location, type, and condition of the potential resources [53]. Therefore, any changes in environments that disrupt the emission of VOCs may disturb the selection by and visitation of dung beetles of their target resource. In our study, it can be inferred that rainfall might have affected the normal release of volatiles from dung pats [54], thereby impeding dung beetle's ability to locate target dung pats and reducing the species diversity of dung beetles. Moreover, it appears that a later (e.g., 48 h) rainfall had a weaker effect on the species richness. This could largely be attributed to the crust formed on the surface of dung pat which may have offset part of the rainfall effect. Interestingly, rainfall intensity did not have a significant effect on the species richness of dung beetles. Nevertheless, heavy rainfall, particularly when it occurred earlier (i.e., 0 h after pats were deposited), resulted in a significant decline in beetle abundance. This could potentially be because that a heavy rainfall might have made the surface of dung pats viscous enough for beetles to crawl into. Once the crust was successfully formed, the impact of heavy rainfall on beetle abundance became negligible as observed in the 2-h, 4 h and 24 h treatments. In addition, the flight of beetles may have been hindered by heavy rainfall, particularly during an actual rainfall event. It is curiously that beetle abundance in the 48 h treatment was observed to decrease under the heavy rainfall condition. We speculate that this could be attributed to the higher water content within dung pats induced by heavy rainfall, which exerted stress on dung beetles. Notably, the species *E. subterraneus* always predominated the beetle assemblage regardless of changes in rainfall patterns. This probably suggests that this particular species may have a low sensitivity to rainfall events. In this end, we boldly speculate that with regard to rainfall patterns, rainfall timing can partly determine the number of beetle species attracted to a dung pat, while rainfall intensity determines the number of beetle individuals can dwell within a pat.

Given the important role of dung beetles in the decomposition process of dung, changes in species diversity and the abundance of dung beetles should lead to corresponding responses in dung decomposition rates. However, our findings indicate that there was no significant correlation between the pattern of dung mass loss and the richness and abundance of dung beetles. This finding was inconsistent with results reported in many other studies [32,55]. This may be because we cannot separate the rainfall effect from the beetle effect due to our experimental design. Differences in the geographic location may also have an important influence. Indeed, rainfall timing was observed to significantly impact on dung mass loss, i.e., dung mass loss was significantly lower when fresh dung pats that were deposited two hours later, regardless of rainfall intensity. In case of the fresh dung pats (e.g., pats in the 0 h treatment), their compact surface may prevent rainfall from penetrating into the pats. However, as the dung pat aged, the crust on its surface, liquid within it, along with holes created by beetles, may have collectively facilitated a leaching effect [56]. Notably, however, once the dung pat became drier (e.g., after depositing for 4 h), the leaching effect might become negligible.

As a special type of natural resource, dung from large vertebrates has been recognized as an important ecological unit for its high biodiversity, intriguing successional process, and important role in ecosystem functioning [41]. In particular, the status of dung pats during the early stage (i.e., the first two to three days after dung is dropped) of dung decomposition can largely determine the community structure of dung decomposers, which in turn influences the rates of organic matter decomposition and nutrient release. Several deficiencies and specialties of our experiment need to be clarified. Firstly, it should be noted that the present study employed a manipulated experiment to simulate rain conditions on a specific resource, which deviated from natural rainfall patterns occurring over a broad geographical region. This disparity may elicit distinct effects on dung beetles. For instance, while the humidity level of the resource might not necessarily impact the flight activity of beetles, intense precipitation can have an effect. Secondly, the observed responses of beetle adults and dung decomposition rate may not fully capture the overall effects resulted from changes in rainfall patterns on dung decomposition because this study solely focused on the early stage of dung decomposition processes, and it was lacking a real experiment control. Actually, the complete breakdown of organic matter and nutrient release from dung requires an extended period spanning several months or even 1 to 2 years on the Qinghai–Tibet Plateau [39]. In particular, dung beetle larvae contribute largely to the loss of organic matter during the middle and late stages of decomposition. Thirdly, the species composition and diversity of decomposers was simplistic, reflecting the decomposer community on the Qinghai–Tibet Plateau. For example, the species *E. subterraneus* occupied for more than 80% of the total beetle abundance. The experiment was conducted in the late August, a period characterized by a relatively low temperature, which could result in the absence of some decomposer groups such as flies and certain species of dung beetles. Furthermore, this study was carried out in a geographically specific climatic area particularly characterized by high amounts and a strong intensity of precipitation. Therefore, the results should be cautiously generalized to other areas. In this end, considering the increasing variability in rainfall patterns on the Qinghai–Tibet Plateau [37,38] as well as other grazing ecosystems, more local research is urgently needed to fully comprehend the relationship between the biological processes and detritus decomposition.

5. Conclusions

In this study, we revealed that rainfall patterns, particularly rainfall timing, exerted a significantly impact on the structure of dung beetles and dung mass loss rate; meanwhile, the relationship between dung beetles and dung mass loss appeared to be decoupled in a Tibetan alpine meadow. These findings provide clear evidence that changes in precipitation patterns can influence the correlation between the species diversity of dung beetles and the decomposition rate of yak dung. This may contribute to our understanding of the relationship between biodiversity–ecosystem function regarding detrital systems under the environmental change conditions.

Supplementary Materials: The following supporting information can be downloaded at: https://www.mdpi.com/article/10.3390/biology12121496/s1, Table S1: The number of different beetle species in different treatments during the experiment period; Figure S1: Diagram of possible pathways that rainfall patterns and dung beetles affect the dung mass loss.

Author Contributions: Conceptualization, X.W.; data curation, W.S.; formal analysis, W.S.; funding acquisition, X.W.; investigation, W.S.; methodology, W.S., W.T. and Y.W.; resources, W.S. and X.W.; software, W.S. and S.H.; supervision, X.W.; validation, W.S., W.T., Y.W. and S.H.; visualization, W.S.; writing—original draft, W.S.; writing—review and editing, X.W. All authors have read and agreed to the published version of the manuscript.

Funding: This research was funded by National Natural Science Foundation of China, grant number 31971442.

Institutional Review Board Statement: Not applicable.

Informed Consent Statement: Not applicable.

Data Availability Statement: The data presented in this study are available upon request from the corresponding author.

Acknowledgments: We thank Qinghai–Tibetan Research Base of Southwest University for Nationalities for providing research facilities.

Conflicts of Interest: The authors declare no conflict of interest.

References

1. Wang, Y.; Gu, J. Ecological Responses, Adaptation and Mechanisms of Mangrove Wetland Ecosystem to Global Climate Change and Anthropogenic Activities. *Int. Biodeterior. Biodegrad.* **2021**, *162*, 105248. [CrossRef]
2. Hoffmann, A.A.; Sgro, C.M. Climate Change and Evolutionary Adaptation. *Nature* **2011**, *470*, 479–485. [CrossRef] [PubMed]
3. Zhang, B.; Cao, J.; Bai, Y.; Zhou, X.; Ning, Z.; Yang, S.; Hu, L. Effects of Rainfall Amount and Frequency on Vegetation Growth in a Tibetan Alpine Meadow. *Clim. Chang.* **2013**, *118*, 197–212. [CrossRef]
4. Gordon, H.B.; Whetton, P.H.; Pittock, A.B.; Fowler, A.M.; Haylock, M.R. Simulated Changes in Daily Rainfall Intensity Due to the Enhanced Greenhouse Effect: Implications for Extreme Rainfall Events. *Clim. Dyn.* **1992**, *8*, 83–102. [CrossRef]
5. Westra, S.; Fowler, H.J.; Evans, J.P.; Alexander, L.V.; Berg, P.; Johnson, F.; Kendon, E.J.; Lenderink, G.; Roberts, N.M. Future Changes to the Intensity and Frequency of Short-Duration Extreme Rainfall. *Rev. Geophys.* **2014**, *52*, 522–555. [CrossRef]
6. Qiu, T.; Song, C.; Clark, J.S.; Seyednasrollah, B.; Rathnayaka, N.; Li, J. Understanding the Continuous Phenological Development at Daily Time Step with a Bayesian Hierarchical Space-Time Model: Impacts of Climate Change and Extreme Weather Events. *Remote Sens. Environ.* **2020**, *247*, 111956. [CrossRef]
7. Henen, B.T.; Peterson, C.C.; Wallis, I.R.; Berry, K.H.; Nagy, K.A. Effects of Climatic Variation on Field Metabolism and Water Relations of Desert Tortoises. *Oecologia* **1998**, *117*, 365–373. [CrossRef]
8. Szarek, S.R.; Ting, I.P. Seasonal Patterns of Acid Metabolism and Gas Exchange in Opuntia Basilaris. *Plant Physiol.* **1974**, *54*, 76–81. [CrossRef]
9. Previtali, M.A.; Lima, M.; Meserve, P.L.; Kelt, D.A.; Gutiérrez, J.R. Population Dynamics of Two Sympatric Rodents in a Variable Environment: Rainfall, Resource Availability, and Predation. *Ecology* **2009**, *90*, 1996–2006. [CrossRef]
10. Pires, A.P.F.; Marino, N.A.C.; Srivastava, D.S.; Farjalla, V.F. Predicted Rainfall Changes Disrupt Trophic Interactions in a Tropical Aquatic Ecosystem. *Ecology* **2016**, *97*, 2750–2759. [CrossRef]
11. Sánchez, A.M.; Peralta, A.M.L.; Luzuriaga, A.L.; Prieto, M.; Escudero, A. Climate Change and Biocrust Disturbance Synergistically Decreased Taxonomic, Functional and Phylogenetic Diversity in Annual Communities on Gypsiferous Soils. *Oikos* **2022**, *2022*, 08809. [CrossRef]
12. Amissah, L.; Mohren, G.M.J.; Bongers, F.; Hawthorne, W.D.; Poorter, L. Rainfall and Temperature Affect Tree Species Distribution in Ghana. *J. Trop. Ecol.* **2014**, *30*, 435–446. [CrossRef]
13. Bongers, F.; Poorter, L.; Rompaey, R.S.A.R.; Parren, M.P.E. Distribution of Twelve Moist Forest Canopy Tree Species in Liberia and Côte D'ivoire: Response Curves to a Climatic Gradient. *J. Veg. Sci.* **1999**, *10*, 371–382. [CrossRef]
14. Abalos, D.; Sanz-Cobena, A.; Andreu, G.; Vallejo, A. Rainfall Amount and Distribution Regulate Dmpp Effects on Nitrous Oxide Emissions under Semiarid Mediterranean Conditions. *Agric. Ecosyst. Environ.* **2017**, *238*, 36–45. [CrossRef]

15. Wardle, D.A.; Bardgett, R.D.; Klironomos, J.N.; Setala, H.; van der Putten, W.H.; Wall, D.H. Ecological Linkages between Aboveground and Belowground Biota. *Science* **2004**, *304*, 1629–1633. [CrossRef] [PubMed]
16. Bradford, M.A.; Berg, B.; Maynard, D.S.; Wieder, W.R.; Wood, S.A. Understanding the Dominant Controls on Litter Decomposition. *J. Ecol.* **2016**, *104*, 229–238. [CrossRef]
17. Cole, L.; Staddon, P.L.; Sleep, D.; Bardgett, R.D. Soil Animals Influence Microbial Abundance, but Not Plant-Microbial Competition for Soil Organic Nitrogen. *Funct. Ecol.* **2004**, *18*, 631–640. [CrossRef]
18. Salamanca, E.F.; Kaneko, N.; Katagiri, S. Rainfall Manipulation Effects on Litter Decomposition and the Microbial Biomass of the Forest Floor. *Appl. Soil Ecol.* **2003**, *22*, 271–281. [CrossRef]
19. Joly, F.X.; Weibel, A.K.; Coulis, M.; Throop, H.L. Rainfall Frequency, Not Quantity, Controls Isopod Effect on Litter Decomposition. *Soil Biol. Biochem.* **2019**, *135*, 154–162. [CrossRef]
20. DeVault, T.L.; Rhodes, J.; Olin, E.; Shivik, J.A. Scavenging by Vertebrates: Behavioral, Ecological, and Evolutionary Perspectives on an Important Energy Transfer Pathway in Terrestrial Ecosystems. *Oikos* **2003**, *102*, 225–234. [CrossRef]
21. Beer, C.; Reichstein, M.; Tomelleri, E.; Ciais, P.; Jung, M.; Carvalhais, N.; Rodenbeck, C.; Arain, M.A.; Baldocchi, D.; Bonan, G.B.; et al. Terrestrial Gross Carbon Dioxide Uptake: Global Distribution and Covariation with Climate. *Science* **2010**, *329*, 834–838. [CrossRef] [PubMed]
22. Gessner, M.O.; Swan, C.M.; Dang, C.K.; McKie, B.G.; Bardgett, R.D.; Wall, D.H.; Hättenschwiler, S. Diversity Meets Decomposition. *Trends Ecol. Evol.* **2010**, *25*, 372–380. [CrossRef] [PubMed]
23. Cebrian, J.; Lartigue, J. Patterns of Herbivory and Decomposition in Aquatic and Terrestrial Ecosystems. *Ecol. Monogr.* **2004**, *74*, 237–259. [CrossRef]
24. Parmenter, R.R.; Macmahon, J.A. Carrion Decomposition and Nutrient Cycling in a Semiarid Shrub–Steppe Ecosystem. *Ecol. Monogr.* **2009**, *79*, 637–661. [CrossRef]
25. Barton, P.S.; Evans, M.J.; Foster, C.N.; Pechal, J.L.; Bump, J.K.; Quaggiotto, M.M.; Benbow, M.E. Towards Quantifying Carrion Biomass in Ecosystems. *Trends Ecol. Evol.* **2019**, *34*, 950–961. [CrossRef]
26. Barton, P.S.; Cunningham, S.A.; Lindenmayer, D.B.; Manning, A.D. The Role of Carrion in Maintaining Biodiversity and Ecological Processes in Terrestrial Ecosystems. *Oecologia* **2013**, *171*, 761–772. [CrossRef]
27. Moore, J.C.; Berlow, E.L.; Coleman, D.C.; Ruiter, P.C.; Dong, Q.; Hastings, A.; Johnson, N.C.; McCann, K.S.; Melville, K.; Morin, P.J.; et al. Detritus, Trophic Dynamics and Biodiversity. *Ecol. Lett.* **2004**, *7*, 584–600. [CrossRef]
28. Wang, Y.; Li, F.; Liu, Y.; Cheng, J.; Wang, Y.; Liu, J.; Wang, X.; Li, Y. Herbivore Dung Promotes Plant Litter Decomposition Rate in a Semi-Arid Grassland Ecosystem. *Ecosystems* **2023**, *26*, 661–674. [CrossRef]
29. Cole, L.; Bardgett, R.D.; Ineson, P.; Adamson, J.K. Relationships between Enchytraeid Worms (Oligochaeta), Climate Change, and the Release of Dissolved Organic Carbon from Blanket Peat in Northern England. *Soil Biol. Biochem.* **2002**, *34*, 599–607. [CrossRef]
30. Noriega, J.A.; Santos, A.M.C.; Calatayud, J.; Chozas, S.; Hortal, J. Short- and Long-Term Temporal Changes in the Assemblage Structure of Amazonian Dung Beetles. *Oecologia* **2021**, *195*, 719–736. [CrossRef]
31. Clein, J.S.; Schimel, J.P. Reduction in Microbial Activity in Birch Litter Due to Drying and Rewetting Events. *Soil Biol. Biochem.* **1994**, *26*, 403–406. [CrossRef]
32. Dickinson, C.H.; Underhay, V.S.H.; Ross, V. Effect of Season, Soil Fauna and Water Content on the Decomposition of Cattle Dung Pats. *New Phytol.* **1981**, *88*, 129–141. [CrossRef]
33. Medina, A.M.; Lopes, P.P. Seasonality in the Dung Beetle Community in a Brazilian Tropical Dry Forest: Do Small Changes Make a Difference? *J. Insect Sci.* **2014**, *14*, 123. [CrossRef] [PubMed]
34. Novais, S.M.; Evangelista, L.A.; Reis-Júnior, R.; Neves, F.S. How Does Dung Beetle (Coleoptera: Scarabaeidae) Diversity Vary Along a Rainy Season in a Tropical Dry Forest? *J. Insect Sci.* **2016**, *16*, 81. [CrossRef] [PubMed]
35. Salunke, P.; Jain, S.; Mishra, S.K. Performance of the Cmip5 Models in the Simulation of the Himalaya-Tibetan Plateau Monsoon. *Theor. Appl. Climatol.* **2018**, *137*, 909–928. [CrossRef]
36. Duan, A.; Wu, G.; Zhang, Q.; Liu, Y. New Proofs of the Recent Climate Warming over the Tibetan Plateau as a Result of the Increasing Greenhouse Gases Emissions. *Chin. Sci. Bull.* **2006**, *51*, 1396–1400. [CrossRef]
37. Feng, X.; Shen, H.; Li, W.; Wang, Q.; Duan, L.; Li, H. Spatiotemporal Changes for Extreme Precipitation in Wet Season over the Qinghai-Tibetan Plateau and the Surroundings During 1961–2017. *Plateau Meteorol.* **2020**, *39*, 694–705. [CrossRef]
38. Zhang, R.; Su, F.; Jiang, Z.; Gao, X.; Guo, D.; Ni, J.; You, Q.; Cuo, L.; Zhou, B. An Overview of Projected Climate and Environmental Changes across the Tibetan Plateau in the 21st Century. *Chin. Sci. Bull.* **2015**, *60*, 3036–3047. [CrossRef]
39. Wu, X.; Sun, S. The Roles of Beetles and Flies in Yak Dung Removal in an Alpine Meadow of Eastern Qinghai-Tibetan Plateau. *Ecoscience* **2010**, *17*, 146–155. [CrossRef]
40. Kou, L.; Sun, W.; Wei, X.; Wu, X.; Sun, S. Drainage Increases Species Richness and Density of Soil Macro-Invertebrates in the Zoige Peatland of Eastern Tibetan Plateau. *Pedobiologia* **2021**, *89*, 150773. [CrossRef]
41. Mohr, C.O. Cattle Droppings as Ecological Units. *Ecol. Monogr.* **1943**, *13*, 275–298. [CrossRef]
42. Wu, X.; Li, G.; Sun, S. Effect of Rainfall Regimes on the Decomposition Rate of Yak Dung in an Alpine Meadow of Northwest Sichuan Province, China. *Acta Ecol. Sin.* **2011**, *31*, 28–36.
43. Fox, J.; Weisberg, S. *An R Companion to Applied Regression*, 3rd ed.; Sage: Thousand Oaks CA, USA, 2019; ISBN 978-154-433-647-3.
44. Fox, J.; Monette, G. Generalized Collinearity Diagnostics. *J. Am. Stat. Assoc.* **1992**, *87*, 178–183. [CrossRef]

45. Lefcheck, J.S. Piecewisesem: Piecewise Structural Equation Modelling in R for Ecology, Evolution, and Systematics. *Methods Ecol. Evol.* **2016**, *7*, 573–579. [CrossRef]
46. RR Core Team. R: A Language and Environment for Statistical Computing. Available online: https://www.R-project.org/ (accessed on 3 September 2023).
47. Hoter, P. Effect of Dung-Beetles (*Aphodius* spp.) and Earthworms on the Disappearance of Cattle Dung. *Oikos* **1979**, *32*, 393–402. [CrossRef]
48. Freymann, P.B.; Buitenwerf, R.; Desouza, O.; Olff, H. The Importance of Termites (Isoptera) for the Recycling of Herbivore Dung in Tropical Ecosystems: A Review. *Eur. J. Entomol.* **2008**, *105*, 165–173. [CrossRef]
49. Herrick, J.E.; Lal, R. Dung Decomposition and Pedoturbation in a Seasonally Dry Tropical Pasture. *Biol. Fertil. Soils* **1996**, *23*, 177–181. [CrossRef]
50. Janzen, D.H. Seasonal Change in Abundance of Large Nocturnal Dung Beetles (Scarabaeidae) in a Costa Rican Deciduous Forest and Adjacent Horse Pasture. *Oikos* **1983**, *41*, 274–283. [CrossRef]
51. Noriega, J.A.; March-Salas, M.; Castillo, S.; García, Q.H.; Hortal, J.; Santos, A.M.C. Human Perturbations Reduce Dung Beetle Diversity and Dung Removal Ecosystem Function. *Biotropica* **2021**, *53*, 753–766. [CrossRef]
52. Weithmann, S.; von Hoermann, C.; Schmitt, T.; Steiger, S.; Ayasse, M. The Attraction of the Dung Beetle Anoplotrupes Stercorosus (Coleoptera: Geotrupidae) to Volatiles from Vertebrate Cadavers. *Insects* **2020**, *11*, 476. [CrossRef]
53. Dormont, L.; Epinat, G.; Lumaret, J.P. Trophic Preferences Mediated by Olfactory Cues in Dung Beetles Colonizing Cattle and Horse Dung. *Environ. Entomol.* **2004**, *33*, 370–377. [CrossRef]
54. Vallat, A.; Gu, H.; Dorn, S. How Rainfall, Relative Humidity and Temperature Influence Volatile Emissions from Apple Trees in Situ. *Phytochemistry* **2005**, *66*, 1540–1550. [CrossRef] [PubMed]
55. Gotcha, N.; Cuthbert, R.N.; Machekano, H.; Nyamukondiwa, C. Density-Dependent Ecosystem Service Delivery under Shifting Temperatures by Dung Beetles. *Sci. Total Environ.* **2022**, *807*, 150575. [CrossRef]
56. Underhay, V.H.S.; Dickinson, C.H. Water, Mineral and Energy Fluctuations in Decomposing Cattle Dung Pats. *Grass Forage Sci.* **1978**, *33*, 189–196. [CrossRef]

biology

MDPI

Article

Hidden Threats: The Unnoticed Epidemic System of Pine Wilt Disease Driven by Sexually Mature *Monochamus* Beetles and Asymptomatic Trees

Kazuyoshi Futai [1,*] and Hideaki Ishiguro [2]

[1] Graduate School of Agriculture, Kyoto University, Kyoto 606-8502, Japan
[2] Ishiguro Tree Doctor Office, Suzuka 513-0012, Japan; info@ishiguro.cc
* Correspondence: kazun4696060@gmail.com; Tel.: +81-090-3032-6855

Simple Summary: Based on robust results from in-cage vector beetle release experiments and pine seedling inoculation trials, we demonstrate that sexually mature *Monochamus* beetles, attracted to withering trees for reproduction, can transmit pathogenic nematodes to surrounding healthy trees. This leads to the formation of new asymptomatic carrier trees, which in turn facilitates the spread of pine wilt disease and may nullify the effectiveness of current control efforts.

Abstract: Pine wilt disease, caused by the nematode *Bursaphelenchus xylophilus*, poses a significant threat to pine forests worldwide. Understanding the dynamics of its spread is crucial for effective disease management. In this study, we investigated the involvement of asymptomatic carrier trees in the expansion of pine wilt disease through a series of experiments. Cage-releasing experiments revealed that sexually immature Japanese pine sawyer beetles, *Monochamus alternatus*, feeding on healthy pine branches drops only a minimal number of nematodes (primary infection). However, sexually mature beetles, still harboring numerous nematodes, fly to weakened trees for breeding and extend their feeding activities to healthy pines around weakened trees, infecting them with nematodes and thus spreading the disease further. Inoculation experiments on field-planted black pine seedlings demonstrated that even a small number of nematodes can lead to a high occurrence of asymptomatic carrier trees. Our findings suggest that nematode infections transmitted by sexually mature *Monochamus* beetles significantly contribute to the expansion of pine wilt damage and play a crucial role in the persistence of asymptomatic carrier trees. This conclusion is based on cage-release experiments demonstrating nematode transmission by mature beetles and inoculation experiments highlighting the conditions leading to asymptomatic carrier trees.

Keywords: pine wilt disease; asymptomatic infected tree; *Monochamus* beetle; *Bursaphelenchus xylophilus*; infection cycle; sexually mature beetle

1. Introduction

Pine wilt disease (hereafter abbreviated as PWD), responsible for significant damage to pine forests throughout Japan except for Hokkaido, peaked around 1980, resulting in an annual loss of over 2 million cubic meters of pine trees nationwide. Following this peak, the disease appeared to gradually subside. However, this apparent decline can be attributed to the significant reduction in pine forests susceptible to the disease. Presently, pine wilt disease is spreading to the last remaining forests of Japanese red pine, *Pinus densiflora* Sieb.

et Zucc. in cold high-altitude areas and high-latitude regions in Japan [1]. Additionally, due to the increased global movement of materials, this disease, which has devastated Japanese forests, spread to other East Asian countries [2,3] and reached Portugal in Europe by the end of the 20th century [4], later spreading to Spain [5] and Armenia [6].

The established infection cycle of pine wilt disease (PWD) has been understood as follows: Pine sawyer beetles with immature reproductive organs emerge from pine trees that died due to PWD the previous year [7]. To mature their reproductive organs, beetles fly to healthy pines and vigorously feed on the bark of young branches, creating feeding scars. This period of sexual immaturity lasts about 5–15 days [8] for males and 16–30 days [9] for females, during which the feeding activity creates scars on the bark. Nematodes retained within the beetle's body enter the pine tissues through these feeding scars, initiating infection (primary infection). As the host pine becomes diseased, it emits volatile compounds such as ethanol and terpenes, attracting sexually mature beetles that fly to weakened pines for mating and oviposition. Beetles overwinter as matured larvae within the dead pine wood, pupate the following spring, and emerge as adults carrying numerous nematodes from the dead wood. These adults then fly to healthy pines, perpetuating the infection cycle (Figure 1).

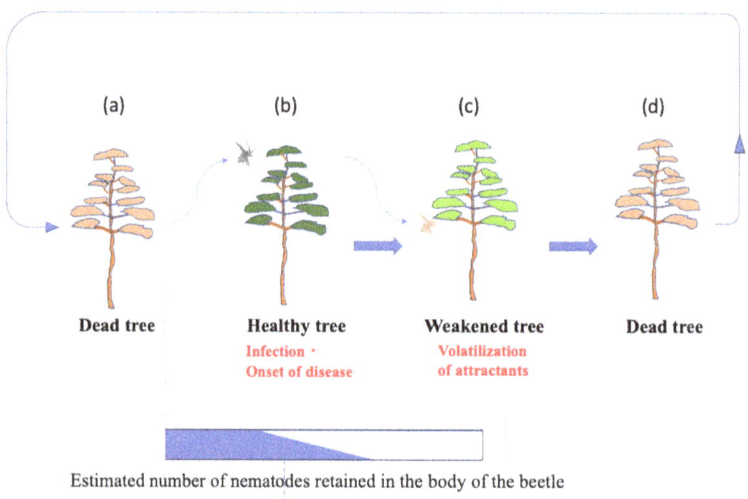

Figure 1. Established theory on the infection chain of PWD. The rectangular bar at the bottom of the figure shows the gradually decreasing percentage of nematodes (blue) retained in the body of the *Monochamus* beetle. Descriptions in red refer to the disease status of the host tree.

Based on this understanding of the infection cycle, two primary methods have been utilized to control this forest epidemic in Japan. One strategy entails the preventive application of insecticides on host pine trees to deter the vector, the Japanese pine sawyer beetle, *Monochamus alternatus* Hope (Japanese pine sawyer beetle, hereafter referred to as *Monochamus* beetle, or beetle), from feeding on young branches. This prevents the pathogen, the pine wood nematode *Bursaphelenchus xylophilus* (Steiner & Buhrer) Nickle (hereafter abbreviated as PWN or Bx nematode), harbored within the beetle's body from invading the tree tissues via the beetle's feeding scars. The other method involves the eradication of dead trees, where both the nematodes and beetle larvae breed, through felling and removing by burning or fumigation. Despite the understanding that thorough implementation of these methods should prevent damage, in reality, even after felling and removing all dead trees in the forest in early spring, a few wilting trees due to PWD occur surrounding the stumps of removed dead trees in early summer [10]. To understand the reasons behind these puzzling patterns of damage occurrence, one of the authors conducted a thorough investigation over four years focusing on a total of 72 Korean pine trees (*Pinus koraiensis* Sieb. et Zucc.), each approximately 40 years old and planted in an arboretum [11]. This investigation involved tracking the locations of dead trees while continuously monitoring resin exudation [12], a physiological indicator of pine health, in all trees starting from the year when two dead trees first appeared due to pine wilt. It is important to note that the damage in this forest continued to increase, even though all dead wood had been completely removed by early summer, prior to the appearance of the beetles. This removal was intended to prevent damage from occurring and spreading further throughout the forest. As a key to solving this mystery, we incorporated the concept of "**asymptomatic infected trees**". These are trees that become infected with nematodes in a given year and experience a decline in resin exudation but survive without visible symptoms until the following year or later. As a result, they are overlooked during the eradication of dead trees by early summer and later develop visible symptoms, contributing to disease progression [11].

To reassess existing theories on the PWD infection cycle and clarify the role of asymptomatic carrier trees in disease spread, it is essential to quantify the number of nematodes carried by beetles arriving at these trees for reproduction. Additionally, a detailed investigation into the relationship between their feeding behavior during reproduction and nematode transmission to healthy trees is required. Therefore, in this study, we initially observed the behavior of beetles towards both asymptomatic carrier trees and healthy trees nearby and assessed nematode infection in the healthy trees through cage release experiments.

Furthermore, to elucidate the conditions under which asymptomatic carrier trees arise as breeding targets for beetles, we inoculated 3- or 5-year-old Japanese black pines, *Pinus thunbergii* Parl, in nurseries, and potted 3-year-old Japanese black pines with a small number of nematodes in early and late summer. We then observed the progression of disease until the following summer. Additionally, we regularly harvested inoculated seedlings and isolated nematodes from various parts of the seedlings to determine their distribution within the host plant and its association with their physiological and external symptoms.

2. Materials and Methods

The following in-cage experiments and the field inoculation experiments were implemented in Suzuka City, Mie Prefecture (34°52′ N, 136°35′ E)

1. In-Cage Insect release experiments: Investigation of Beetle Reproductive Behavior and Transition of Retained Nematodes to Surrounding Healthy Trees

As some of the results of the in-cage experiments have already been published in another Japanese journal [13], a summary of the methods and results is given here to aid understanding of the overall purpose of this experiment. For statistical details of the results of the in-cage experiments, see the original paper [13].

The experiments were conducted in a mesh cage measuring 2 m in height, width, and depth. About a month before the cage release experiments, one externally healthy three-year-old black pine seedling was designated as the oviposition target tree —essentially an artificially created asymptomatic carrier tree. To induce disease onset, indicated by the cessation of resin exudation, the seedling was pre-inoculated with 5000 pathogenic nematodes. This seedling was placed at the center of the cage, surrounded by one healthy black pine seedling in each of the four directions. Positioned in two corners of the cage were polystyrene foam boxes (dimensions: height × width × depth = 15 × 18 × 20 cm). Within each box, four wooden sticks, each 20 cm in length, were erected to facilitate natural takeoff for the beetles (Figure 2).

Figure 2. Mesh cage used for the *Monochamus* beetle release experiments (left). Five potted pine seedlings were used in each experiment: the seedling placed at the center was an artificially latently infected tree inoculated with nematodes (top right). The arrangement of the seedlings within the cage is shown in the schematic diagram (bottom right). An artificially made latent carrier seedling was pre-inoculated with 5000 *Bx* nematodes, and its cessation of resin exudation was confirmed 2 days prior to each experiment, then placed in the center. Insect release boxes were placed in two corners of the cage. In each, three male or three female sawyer beetles were placed. (Modified and reproduced from Ishiguro and Futai [13]).

Pine logs that had been killed due to pine wilt in Ueno Forest Park, 43 km away, were transported and stocked in a 1.5 × 2 × 2 m (H × W × D) netted cage. Beetles emerging from the logs were collected daily, and their size, sex, and the date of collection were recorded and each beetle was kept in a plastic container (1–4 L) with a small hole for ventilation until each in-cage-release experiment. A short Japanese black pine twig was used as food and replaced with a fresh one every two to three days.

Three unmated females were released into one box, and three unmated males were released into the other. After 72 h, the number of feeding scars on each of the five black pine seedlings and the number of nematodes isolated from each feeding scar were investigated. Following previous reports [8,9], mature and immature beetles were discriminated based on the number of days after emergence. The experiments were repeated nine times for sexually immature beetles (males aged 5–10 days, females aged 5–10 days) and sexually mature beetles (males aged 14–31 days, females aged 19–30 days), respectively. Furthermore, following each release test, each tested beetle was individually dissected, and nematodes remaining in the beetle body were isolated using the Baermann funnel method and counted.

2. Examination of the Conditions under which Asymptomatic Carrier Trees Occur: Experimental Inoculation of Pine Seedlings Planted in the Field with a Small Number of Nematodesl

Some trials of the cage experiments showed that the number of nematodes invading through a single *Monochamus* beetle feeding scar was generally low. Therefore, in the second series of inoculation experiments, we set the inoculation dose to a lower level of 50 nematodes per site (50 ± 8.3). Two experiments were conducted: The first experiment aimed to determine how host pines develop disease when infected with a small number of nematodes in different seasons and to examine the distribution and quantity of nematodes in trees that survive as asymptomatic carrier trees. The second experiment investigated the relationship between the number of beetle feeding scars and the mortality rate of pine saplings.

2.1. Differences in Disease Development and Nematode Distribution in the Host Tree According to Infection Time

Previous inoculation experiments have experienced that late inoculation often results in delayed onset of symptoms until the following year [14]. Additionally, it is known that physiological changes in pines seem to occur in late stages of the pine wilt season (late August to September), affecting the infection of pathogenic nematodes [15]. To confirm this, 50 nematodes were inoculated on the one-year-old branches of 80 seedlings of three-year-old black pine in the early pine wilt season (1 July 2018) and on 65 seedlings in the late pine wilt season (25 August 2018). Subsequently, 20 seedlings designated from each inoculation timing group were separated from others for periodical observation for disease progression, recording resin exudation status as physiological symptoms and changes in external symptoms through photography from 29 June just prior to nematode inoculation to 30 July of the following year for the early inoculation group and, similarly, from 22 August to 30 July of the following year for the late inoculation group. Although designated control plots were not established due to operational constraints, dozens of untreated healthy seedlings growing in the same nursery were continuously monitored throughout the experiment. These untreated seedlings served as an informal control group, allowing comparison of symptoms with those of the nematode-inoculated seedlings.

Furthermore, to determine in which part, for how long, and how many nematodes survived, a total of 11 times after inoculation on 1 July, 2018 until April of the following year for the early inoculation group and 7 times after inoculation on 25 August, 2018 until April of the following year for the late inoculation group, five randomly selected inoculated seedlings from each inoculation timing group were collected periodically, using a random number table. From each of the collected seedlings, five tissue samples were taken: (1) the inoculation site, (2) another branch of the same whorl of the inoculated branch, (3) the upper part, (4) the lower part of the main stem, and (5) the roots. Each sample was shredded and soaked in a small Baermann apparatus for 24 h, and the isolated nematodes were counted under a stereomicroscope. After nematode isolation, each pine tissue was air-dried, weighed, and the number of nematodes was evaluated per gram of dry weight.

2.2. Relationship Between the Number of Beetle Feeding Scars and Mortality Rate of Pine Saplings: Effect of Multipoint Infection

When beetles feed on healthy pines, they create feeding scars in multiple places while walking on branches. In the cage-release experiments, it was observed that, on average, each beetle left about four feeding marks on the seedlings during a three-day observation period. Even with a low number of nematodes per site, an increase in the number of feeding sites and total nematodes infecting a single seedling may increase damage to the host pine and raise the probability of death. To investigate this, 50 saplings of five-year old black pine planted in the field were inoculated with 50 nematodes per site, varying the number of inoculated sites (number of nematodes per sapling) from 1 site (50) to 2 sites (100), 3 sites (150), 5 sites (250), and 7 sites (350), on 21 July 2017. Ten saplings were used for each number of inoculation sites (number of inoculated nematodes). After inoculation, resin exudation (symptom) of each sapling was observed regularly.

In experiments 2.1 and 2.2, the inoculum used for inoculation was the highly pathogenic *B. xylophilus*, Ka-4 strain cultured on *Botrytis cinerea* (grey mold) colonies. The pathogenicity of the nematodes used for inoculation was confirmed before each experiment by inoculating them into five healthy black pine trees (at a rate of 1000 nematodes per seedling and 4500 nematodes per sapling), resulting in nearly complete tree mortality.

In all experiments, the progression of symptoms was assessed based on visual changes observed and judged from photo recordings, while physiological symptoms were assessed based on the amount of resin exuding from pinholes made in the main stem of seedlings two hours after pinning, which is a modification of Oda's method [12] for small seedlings.

3. Results

3.1. In-Cage Insect-Release Experiments

3.1.1. Number of Nematodes Invading Feeding Scars Left on Healthy Trees Surrounding Breeding Target Trees

In experiments with sexually immature beetles, no significant difference was observed in the number of feeding scars left on black pine seedlings between the healthy trees placed around and the weakened trees placed in the center. Among the 36 healthy trees tested, feeding scars were observed on 28 trees (77.8%), totaling 224 scars. However, examination of 145 feeding scars revealed only 127 nematodes detected from a total of 17 scars on eight healthy trees combined (Figure 3 top). Despite the mean number of nematodes retained by the 54 sexually immature beetles used in this experiment being low, at 686 with a standard deviation of 2248, it is noteworthy that two individuals retained over 10,000 nematodes each and three beetles retained between 1000 and 10,000 nematodes. This suggests that nematodes retained by beetles during their sexually immature period very rarely fall into feeding scars, and the number of nematodes invading pine tissues is limited to a very small extent (Table 1).

Table 1. Number of Feeding Scars and Nematodes per Surrounding Healthy Seedling Examined.

Number of Feeding Scars per Seedling Examined		Number of Nematodes Recovered per Seedling Examined	
Matured beetle	**Immature beetle**	**Matured beetle**	**Immature beetle**
5.44 ± 4.05	**4.02 ± 3.20**	**34.56 ± 155.15**	**3.53 ± 12.18**

However, sexually mature beetles exhibited a notable difference in behavior compared to sexually immature beetles. They left significantly more feeding scars on the asymptomatic carrier trees placed in the center of the cage than on the healthy trees placed around them (Tukey's multiple comparison test, $p < 0.05$). However, they also left feeding scars evenly on the healthy trees placed in all four directions. In other words, sexually mature beetles were observed not only to be attracted to weakened (asymptomatic carrier) trees for reproductive activities, where they mated and laid eggs, but also to continue vigorous feeding activity during this reproductive period moving between these oviposition target seedlings and surrounding healthy seedlings [13].

Feeding scars, which are traces of their feeding activities, were observed on 32 (89%) of 36 healthy trees tested in the nine repeat trials. The total number of feeding scars on these 32 trees was 231, which was nearly the same as that observed in the case of immature individuals (t-test, df = 69, $p = 0.148$). Examination of nematodes invading 196 of these feeding scars revealed nematodes detected from 20 scars (Figure 3 bottom). The average number of nematode infections per seedling was approximately 35, which was ten times higher than the average infection rate of approximately 3.5 nematodes per seedling by immature beetles. However, the standard deviation was large in this case, so it cannot be immediately concluded that mature beetles infect more nematodes (t-test, df = 35, $p = 0.240$).

As shown above, t-tests were conducted to compare the mean number of feeding scars per seedling and the mean number of nematodes infected per seedling between immature and mature beetles. However, no significant differences were detected ($p > 0.05$) because large variances and overlapping distributions were observed in the data. Therefore, we limited our analysis to summary and basic significance testing, which we judged sufficient for this comparison.

Due to the experimental design—where feeding behavior was measured as the cumulative outcome of three females and three males on four healthy trees (and one asymptomatic carrier tree) per replicate—it was not possible to calculate individual-level variation in beetle behavior.

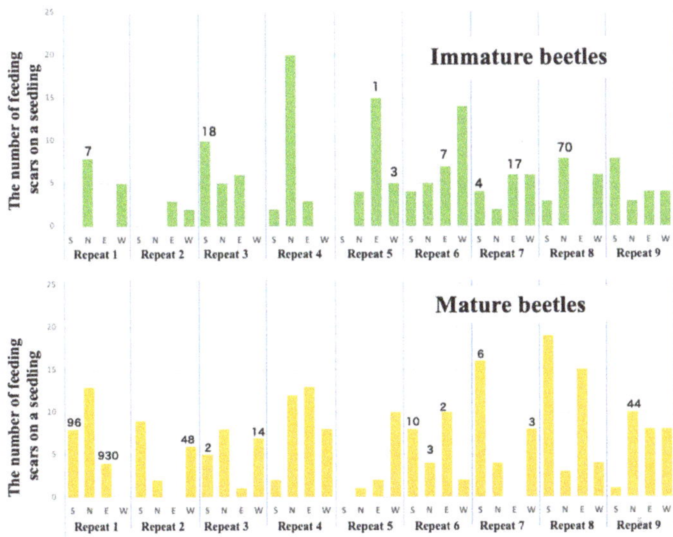

Figure 3. The number of feeding scars and infecting nematodes per seedling. Top: immature beetles, bottom: mature beetles. The numbers above some of the bars indicate the total number of nematodes that invaded from the scars on each seedling. The abbreviations N, S, E, W on the x-axis indicate directional orientation of each seedling.

3.1.2. The Number of Nematodes Remaining in the Beetle Bodies After the Release Test

Furthermore, when comparing the number of nematodes remaining in the beetle bodies after the release test between 53 mature individuals and 54 immature individuals, it was found that mature individuals retained approximately twice as many nematodes (1526.0 ± 3918.6 *) as immature individuals (686.3 ± 2248.0 *). However, due to the large standard deviation, the difference was not significant (t-test, df = 102, $p = 0.211$). * denotes average ± standard deviation.

Nevertheless, from this result, it can be concluded that, at least, contrary to the previously believed notion, the nematodes retained by sexually mature individuals are not significantly fewer than those retained by immature individuals.

Furthermore, the cage-release tests revealed that the number of nematodes infecting from one feeding scar of the beetle is often below several tens, indicating that the infection of thousands to tens of thousands of nematodes from a single wound, as has been commonly used in many inoculation experiments, very rarely occurs.

3.2. Experimental Inoculation of Pine Seedlings Planted in the Field with a Small Number of Nematodes

3.2.1. Differences in Disease Development and Nematode Distribution in the Host Tree According to Infection Time

When a small number (50 nematodes per tree) of *Bx* nematodes were inoculated in early pine wilt season (1 July 2018), resin abnormality occurred by the fourth week, reaching 80% of the inoculated seedlings by the eighth week (22 August). In addition, external symptoms appeared in 35% of the inoculated seedlings in the tenth week after inoculation, presenting as whole or partial discoloration, and this proportion rose to 60%, of which 35% were totally dead and 25% partially dead, by the 13th week. However, the remaining 40% seedlings remained asymptomatic until the end of July the following year.

Among the seedlings that exhibited resin abnormality, half withered by the beginning of the following season, while the remaining half survived without withering until the end of the observation (30 July 2019) (Figure 4).

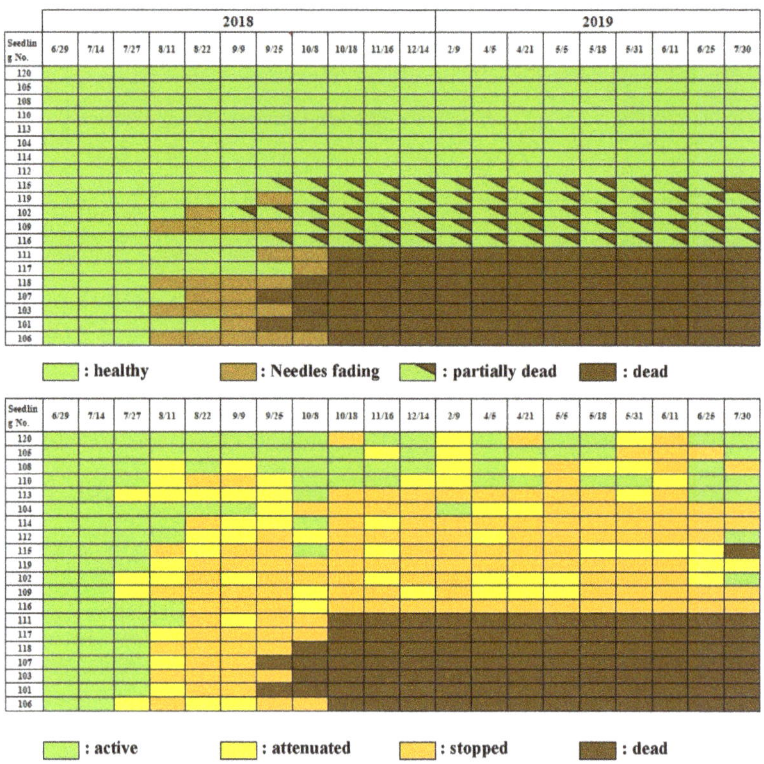

Figure 4. Progress of external disease symptoms (**top**) and resin exudation decline (**bottom**) in black pine seedlings inoculated with small number of nematodes early in the pine wilt season.

On the other hand, when a small number of *Bx* nematodes (50 individuals per seedling) were inoculated on 25 August 2018, late in the season, seedlings showing resin abnormality appeared four weeks after inoculation. By the eighth week (15 October), this proportion reached 50% of the inoculated individuals. However, external symptoms such as discoloration or fading did not appear at all until 9 February of the following year, when discoloration appeared on one branch of one seedling, while the other seedlings maintained a healthy appearance until the end of the experiment, except for two. Among the two showing abnormalities, one was the seedling with branch discoloration mentioned above, and the other withered in the tenth month after inoculation. Thus, when the inoculation timing was late, the decrease in resin exudation, an internal symptom, was similar to that of early inoculation timing. However, the leaf color of most seedlings remained healthy until the next pine wilt season (Figure 5).

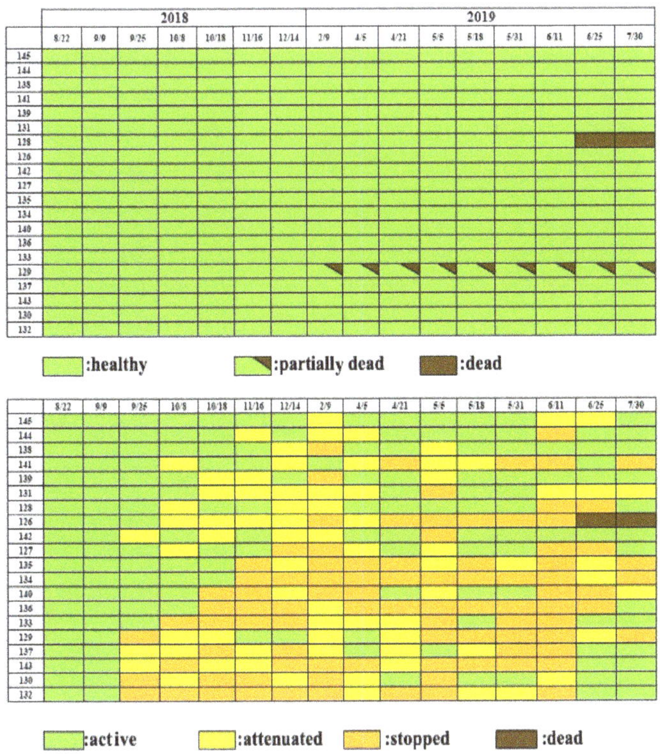

Figure 5. Progress of external disease symptoms (**top**) and resin exudation decline (**bottom**) in black pine seedlings inoculated late in the pine wilt season.

The chart formats used in Figures 4 and 5 were designed to visualize symptom progression and resin response in individual seedlings over time. In addition, Figure 6 summarizes group-level trends in disease progression as a time-series line graph.

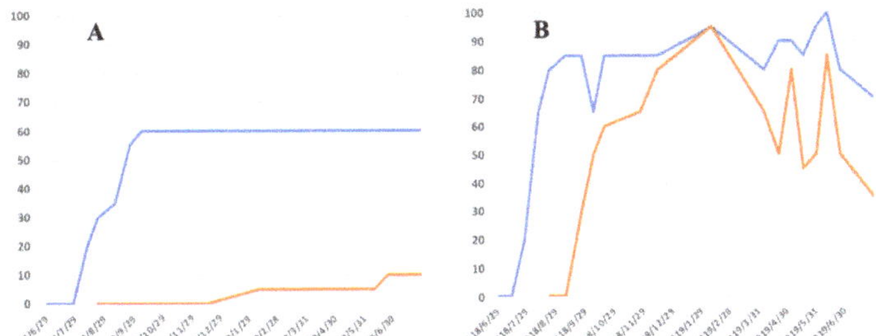

Figure 6. Comparison of disease progression between early and late inoculation groups. (**A**) Percentage of dead seedlings over time. (**B**) Percentage of seedlings showing reduced or stopped resin secretion over time. ———: Early inoculation group; ———: Late inoculation group.

As described above, when a small number (50 individuals) of nematodes were inoculated in early pine wilting season (1 July) compared to late inoculation (25 August), physiological abnormalities progressed similarly, but the expression of external symptoms differed significantly. How did the distribution trend of the inoculated nematodes within the host tree differ between the two inoculation times? In the case of inoculation in July, nematodes not only spread within the seedlings three to four weeks after inoculation, but instances of successful proliferation were also observed, leading to an increase in weakened (denoted as W in the figure) and dead seedlings (denoted as D) accordingly (Figure 7a).

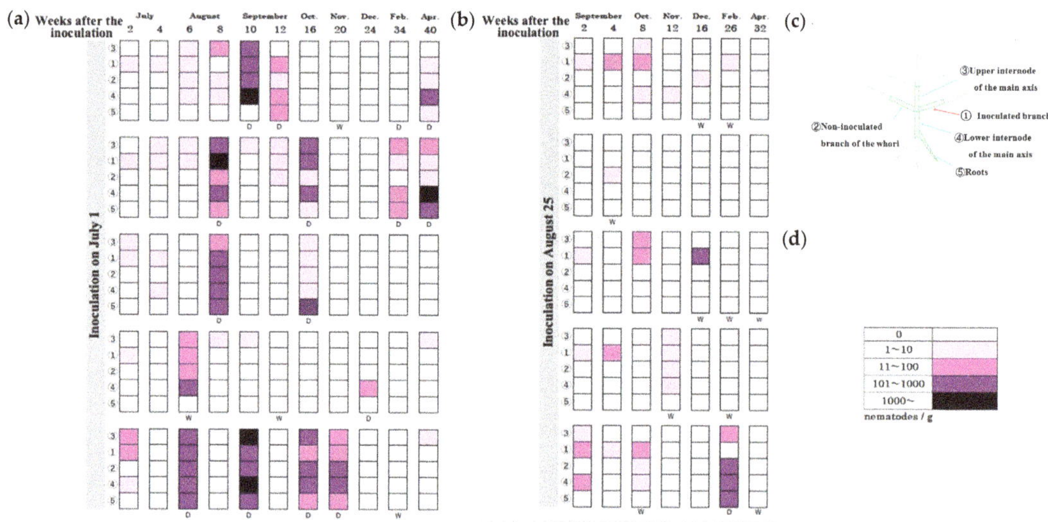

Figure 7. The distribution trend of the inoculated nematodes within the host seedling when inoculated in the early season, 1 July (**a**) and the late season, 25 August (**b**). As illustrated in the figure (**c**), tissue samples were taken from five points of the seedlings, including ① the inoculation site, ② other branches at the same branch level as the inoculated branch, ③ the upper part of the main stem, ④ the lower part of the main stem, and ⑤ the roots. Nematode density per gram of dried tissue was expressed as five levels of color shade (**d**).

On the other hand, when inoculated late, it appears that the proliferation of nematodes within the seedlings is suppressed, perhaps due to changes in the physiological status of the seedlings in response to seasonal climatic changes (Figure 7b). In both cases, nematode proliferation within the seedlings is reduced after November, possibly due to low temperatures, and the number of seedlings from which nematodes cannot be isolated increases. However, this does not mean that nematodes completely disappear from the seedlings. Some seedlings with a healthy appearance still yielded a small number of nematodes even in the following February and April.

3.2.2. Relationship Between the Number of Beetle Feeding Scars and Mortality Rate of Pine Saplings: Effect of Multipoint Infection

As the number of inoculation points per sapling increased, individuals exhibiting abnormal resin exudation appeared earlier and a higher proportion of individuals eventually died (Figure 8). When examining the relationship between the number of inoculation points and the mortality rate at five months (143 days) post-inoculation, a strong positive correlation (r^2 = 0.949) was observed (Figure 9). This experimental result suggests that even if the number of nematodes per feeding scar is small, pine saplings become more susceptible to wilting when multiple feeding scars occur.

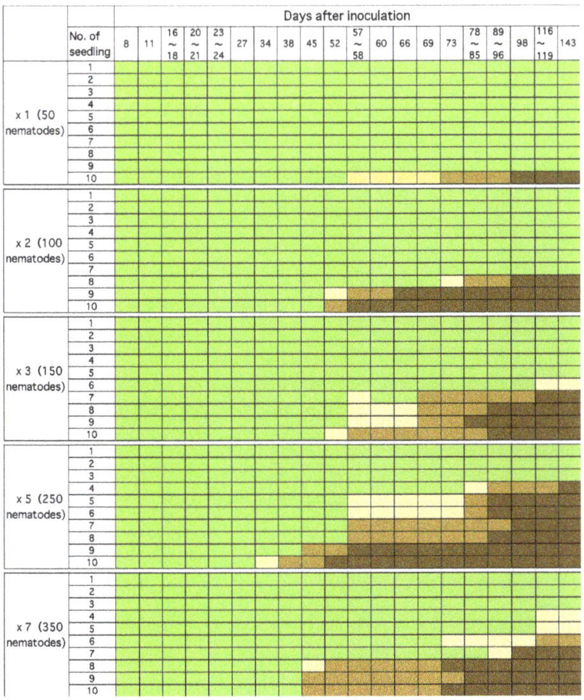

Figure 8. The relationship between the number of inoculation points (inoculated nematodes) and the mortality rate. Green, beige, light brown, and dark brown represent healthy, resin decline, resin stop, and withering, respectively.

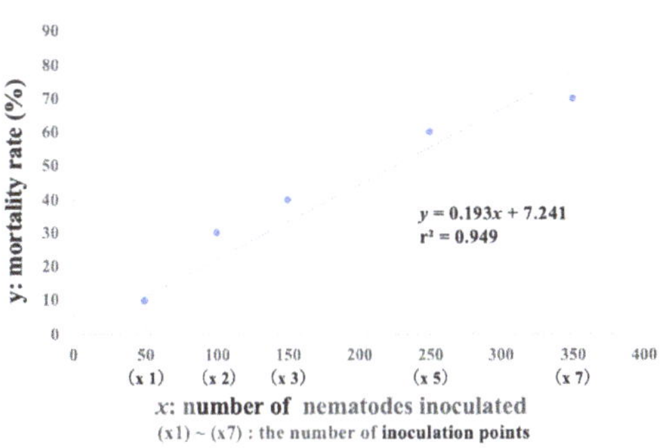

Figure 9. Regression analysis of the relationship between the number of inoculation points (nematodes inoculated) and the mortality rate.

4. Discussion

Our study challenges conventional assumptions about the infection cycle of pine wilt disease (PWD). In the cage-release experiment, sexually mature beetles (males aged 14–31 days, females aged 19–30 days) exhibited distinct feeding behavior compared to sexually immature beetles (males aged 5–10 days, females aged 5–10 days). They created significantly more feeding scars on asymptomatic carrier trees placed at the center of the cage than on healthy trees positioned around them (Tukey's multiple comparison test, $p < 0.05$). However, they also left feeding scars evenly across all healthy trees. This suggests that, contrary to previous understanding, sexually mature beetles are not exclusively attracted to weakened or dead trees for reproduction. Instead, they continue vigorous feeding during their reproductive period, moving between oviposition sites and surrounding healthy seedlings [13].

Previously, it was believed that sexually immature *Monochamus* beetles transmitted most of the nematodes they carried through feeding scars on healthy pines, causing primary infections (Figure 1). However, our cage-release experiments revealed that nematode transmission by immature beetles was minimal; very few nematodes entered feeding scars, and only a limited number successfully invaded pine tissues (Table 1). This aligns with earlier reports showing that pine trees fed on by immature beetles within the first 10 days post-emergence exhibited no mortality, whereas trees fed on by sexually mature beetles 2 to 7 weeks post-emergence showed high mortality [16]. Additionally, previous studies indicate that nematodes use intestinal storage lipid reserves as an internal clock, exiting the beetle only after a certain amount of lipid is consumed [17].

On the other hand, it has long been assumed that sexually mature beetles arriving at weakened trees for reproduction carry few nematodes, as most nematodes were thought to have already left the beetle's body during its earlier feeding activity on healthy pines. However, our experiments challenge this assumption. In the cage-release study, sexually mature beetles were individually maintained on young pine branches before reaching sexual maturity (~10 days for males, ~20 days for females). Even after feeding on pine seedlings for three days during the experiment, dissection revealed that males still retained an average of 850 nematodes and females an average of 1900. This suggests that sexually mature beetles actually harbor far more nematodes than previously believed when they arrive at weakened trees for reproduction. The results of our cage-release experiment, as shown in Table 1, suggest that sexually mature beetles transmit nematodes to healthy trees surrounding weakened trees (latent carrier trees) as sexually mature beetles are attracted to the weakened trees for reproductive activity.

Our findings are supported by studies on the timing of nematode transmission. Togashi [18] and Naves et al. [19] found that peak nematode transmission occurred 20–35 days and 15–42 days after beetle emergence, respectively, coinciding with sexual maturation and attraction to declining trees. Once these beetles begin reproduction, they continue feeding on both the weakened trees they inhabit and nearby healthy trees, thereby establishing new infection routes. However, the role of sexually mature beetles in transmitting nematodes to healthy trees has been largely overlooked.

In our experiments, sexually mature beetles frequently fed on healthy trees surrounding weakened ones (32 out of 36 healthy trees tested), creating 231 feeding scars—similar in number to those made by immature beetles. When we examined 196 of these scars, 20 contained nematodes, with an average of 35 nematodes per seedling. Importantly, even a small number of invading nematodes can have significant consequences. As demonstrated in Experiment 2b, an increase in feeding scars, rather than the size of individual scars, leads to greater nematode entry, accelerating disease progression. Interestingly, previous research found no correlation between feeding scar size and the number of transmitted nematodes [20], reinforcing the idea that total feeding site number is the primary determinant of transmission efficiency.

The results of our nematode inoculation experiment further support the role of latent infections. In the early-season inoculation experiment (1 July 2018), where a small number (50 nematodes) was introduced, disease progression was slow. Among 20 monitored black pine seedlings, 40% were completely dead, and 20% were partially dead by July of the following year (Figure 4). Partial wilting is often overlooked in control efforts, yet studies on *Pinus koraiensis* forests revealed that even after wilted branches were removed, 7 out of 13 trees eventually died [11]. These findings highlight the importance of monitoring not only symptomatic trees but also asymptomatic carrier trees and those with partial wilting.

Notably, late-season inoculations resulted in fewer cases of severe wilting, with most trees remaining asymptomatic until the following year or later. This suggests that certain infection conditions contribute to the formation of latent carrier trees. One key factor is the low number of infecting nematodes. Some infected trees survive without visible symptoms yet harbor small nematode populations internally. Increased temperatures and drought conditions [21] can weaken host resistance, triggering nematode proliferation, disease progression, and the eventual creation of breeding sites for sawyer beetles. Another factor is lower temperatures during late-summer or early-autumn infection, which may suppress nematode activity and reproduction and thereby enhance host resistance keeping in healthy looking trees the following year. In essence conditions that slow disease development contribute to asymptomatic infections. Even when trees at the same site are infected simultaneously, individual variations in these factors can determine whether a tree dies or becomes a latent carrier [22].

The ease of nematode invasion from beetle feeding scars is a critical factor in host susceptibility. This property varies by pine species and seasonal conditions [15]. In susceptible black pines, only about 10% of nematodes at feeding scars successfully invade host tissues [23]. Since *B. xylophilus* reproduces sexually, very low infection densities could reduce mating opportunities, leading to population collapse. However, nematode sex pheromones facilitate encounters between males and females [24], enhancing population persistence. If reproduction fails, the nematodes disappear and the tree survives. If successful, the nematode population remains at low density until host resistance declines, triggering rapid proliferation and wilting. This was evident in our experiments, where nematodes were recovered from symptom-free seedlings the following April.

As mentioned above, however, it is often difficult to detect low density of nematodes in plant tissue by the Baermann funnel method. Molecular tools such as PCR-based methods [25,26] and the LAMP method [27] that are more sensitive and applicable to field samples would be needed to improve nematode detection sensitivity in future studies of asymptomatic carriers. These molecular approaches offer the advantage of high-throughput screening without requiring taxonomic expertise, making them suitable for rapid and large-scale diagnostics. Nevertheless, due to their relatively high cost, such methods are not practical for long-term monitoring studies like ours, which require repeated sampling of a large number of trees over time.

Long-term latent infections have also been observed in other pine species. Bergdahl and Halik [28] found that in Vermont, USA, despite inoculating 30,000 nematodes, many *Pinus sylvestris* (Scots pine) trees survived for 7–11 years as asymptomatic carriers, with nematodes persisting for 2–11 years. The cooler climate and higher resistance of *P. sylvestris* compared to *P. thunbergii* may have contributed to this prolonged latent phase [29], but the study nonetheless illustrates how nematodes can persist long term in asymptomatic carriers.

After dead trees are removed from a forest, sawyer beetles may migrate and feed on remaining healthy trees. However, these trees take time to develop disease symptoms and reduced resin exudation, which makes them suitable for egg-laying. Research suggests that a one-month gap exists between beetle emergence and the onset of resin-abnormal trees [30]. Because *Monochamus* beetles lack suitable oviposition sites for ~20 days (males) and ~10 days (females) after emergence, latent infections from previous years provide critical breeding sites, ensuring population continuity. This means that mature beetles contribute significantly to outdoor nematode transmission by introducing small numbers of nematodes into multiple feeding sites on healthy trees surrounding weakened hosts (Figure 10). This overlooked process underscores the importance of asymptomatic carriers in sustaining PWD, nullifying control efforts and facilitating disease spread.

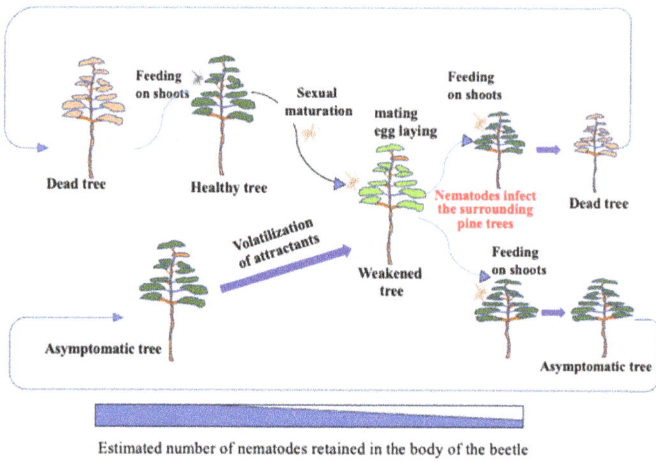

Figure 10. New understanding of the PWD infection chain. Immature beetles moving among the pines are shown in gray, while mature beetles are shown in light brown. The explanation shown in red text is most important phase of disease expansion.

Recent advancements in PWD detection methods have improved monitoring [31]. Aerial detection using remote sensing and UAVs (Unmanned Aerial Vehicles) identifies symptomatic trees at a large scale but remains ineffective for detecting asymptomatic carriers [32–35]. Direct detection methods, such as volatile compound analysis [36,37], DNA-based techniques [25,38,39], and pH measurement of wood samples [40], offer high accuracy but are costly and impractical for large-scale surveys.

Traditional resin exudation measurement remains the most practical approach for detecting infected trees in the field [12,41].

To explore the progression of nematode distribution within the seedlings, we employed a time-consuming method, but many seedlings with no visible symptoms yielded no nematodes upon isolation. This difficulty arises from the methodological constraints of the Baermann funnel technique, which makes it challenging to completely isolate low-density nematodes within plant tissues. Therefore, around November, when the number of isolates decreased, we collected samples from visually healthy seedlings. We divided these samples into two, with one part immediately undergoing nematode isolation and the other left at room temperature for five days to promote nematode proliferation before attempting isolation. As a result, among five seedlings where nematodes were not initially detected, nematodes were successfully isolated from three seedlings after the second attempt.

5. Conclusions

Finally, our study highlights the importance of revisiting traditional views on the pine wilt infection cycle. Asymptomatic carrier trees play a critical role as hidden drivers of pine wilt disease spread—not by directly harboring high nematode loads but by attracting sexually mature *Monochamus* beetles that subsequently disperse and infect pathogenic nematodes to surrounding healthy trees. Recognizing this behavioral and ecological role of mature beetles is essential for effective PWD management.

Author Contributions: Conceptualization, K.F.; methodology, K.F.; formal analysis, K.F. and H.I.; investigation, H.I.; resources, H.I.; data curation, H.I.; writing—original draft preparation, K.F.; writing—review and editing, K.F.; visualization, K.F.; supervision, K.F.; project administration, K.F. All authors have read and agreed to the published version of the manuscript.

Funding: This research received no external funding.

Data Availability Statement: The data published in this study will be shared by the co-authors and are available upon request from the corresponding author.

Acknowledgments: We thank Yuko Takeuchi of the Graduate School of Agricultural Sciences, Kyoto University, for her technical support and for providing us with much scientific information on pine wilt disease.

Conflicts of Interest: The authors declare no conflicts of interest.

Abbreviations

The following abbreviations are used in this manuscript: **PWD**, pine wilt disease; **UAV**, Unmanned Aerial Vehicle.

References

1. JAPAN Forestry Agency. *Annual Report on Forest and Forestry in Japan*; JAPAN Forestry Agency: Tokyo, Japan, 2022. (In Japanese with English Summary).
2. Zhao, B.G. Pine wilt disease in China. In *Pine Wilt Disease*; Springer: Tokyo, Japan, 2008; pp. 18–25.
3. Shin, S.C. Pine wilt disease in Korea. In *Pine Wilt Disease*; Springer: Tokyo, Japan, 2008; pp. 26–32.
4. Mota, M.; Braasch, H.; Bravo, M.A.; Penas, A.C.; Burgermeister, W.; Metge, K.; Sousa, E. First report of *Bursaphelenchus xylophilus* in Portugal and in Europe. *Nematology* **1999**, *1*, 727–734. [CrossRef]
5. Abelleira, A.; Picoaga, A.; Mansilla, J.P.; Aguin, O. Detection of *Bursaphelenchus xylophilus*, causal agent of pine wilt disease on *Pinus pinaster* in Northwestern Spain. *Plant Dis.* **2011**, *95*, 776. [CrossRef] [PubMed]
6. Arbuzova, E.N.; Karagyan, G.H.; Kozyreva, N.I.; Shchukovskaya, A.G.; Ghrejyan, T.L.; Kalashian, M.Y.; Akopyan, K.V. First finding of *Bursaphelenchus xylophilus* in pine plantations of the Republic of Armenia. *J. Nematol.* **2024**, *57*, 20250004. [CrossRef]
7. Katsuyama, N.; Sakurai, H.; Tabata, K.; Takeda, T. Effect of age of post-feeding twig on the ovarian development of Japanese pine sawyer, *Monochamus alternatus*. *Res. Bull. Fac. Agric. Gifu Univ.* **1989**, *54*, 81–89.
8. Nobuchi, A. Fertilization and oviposition of *Monochamus alternatus* Hope. *Trans. Ann. Meet. Jpn. For. Soc.* **1976**, *87*, 247–248. (In Japanese)
9. Enda, N.; Nobuchi, A. Studies on the pine bark and wood boring beetles: Maturation of the ovary and its parasitic nematodes. *Trans. Ann. Meet. Jpn. For. Soc.* **1970**, *81*, 274–276. (In Japanese)
10. Togashi, K. Spatial pattern of pine wilt disease caused by *Bursaphelenchus xylophilus* (Nematoda: Aphelenchoididae) within a *Pinus thunbergia* stand. *Res. Popul. Ecol.* **1991**, *33*, 245–256. [CrossRef]
11. Futai, K. Role of asymptomatic carrier trees in epidemic spread of pine wilt disease. *J. For. Res.* **2003**, *8*, 253–260. [CrossRef]
12. Oda, K. Target trees of pine wilt disease and its diagnostic method. *For. Pests* **1967**, *16*, 263–266. (In Japanese)
13. Ishiguro, H.; Futai, K. When coming to a wilting pine tree, *Monochamus alternatus* causes pinewood nematode infection on the surrounding healthy trees. *Tree For. Health* **2022**, *26*, 59–64, (In Japanese with English Summary).
14. Arihara, T.; Saito, K. Mortality of pine trees in Fukushima Prefecture. Effects of an inoculation of older pine trees with pinewood nematodes in early autumn. *Trans. Meet. Jpn. For. Soc.* **1985**, *96*, 467–469. (In Japanese)
15. Futai, K. Host specific aggregation and invasion of *Bursaphelenchus xylophilus* (Nematod: Aphelenchoididae) and *B. mucronatus*. *Mem. Coll. Agric. Kyoto Univ.* **1985**, *126*, 35–43.
16. Mineo, K.; Kontani, S. The departure of pinewood nematodes from *Monochaamus* beetles and transfer to pine trees. *Trans. 81st Ann. Meet. Jpn. For. Soc.* **1975**, *86*, 307–308.
17. Stamps, W.T.; Linit, M.J. Neutral storage lipid and exit behavior of *Bursaphelenchus xylophilus* fourth-stage dispersal juveniles from their beetle vectors. *J. Nematol.* **1998**, *30*, 255–261.
18. Togashi, K. Transmission Curves of *Bursaphelenchus xylophilus* (Nematoda: Aphelenchoidi-dae) from its vector, *Monochamus alternatus* (Coleoptera: Cerambycidae), to pine trees with reference to population performance. *Appl. Entomol. Zool.* **1985**, *20*, 246–251. [CrossRef]
19. Naves, P.M.; Camacho, S.; Sousa, E.; Quartau, J. Transmission of the pine. wood nematode *Bursaphelenchus xylophilus* through feeding activity of *Monochamus galloprovincialis* (Col., Cerambycidae). *J. Appl. Entomol* **2007**, *131*, 21–25. [CrossRef]
20. Linit, M.J. Nematode-Vector Relationships in the Pine Wilt Disease System. *J. Nematol.* **1988**, *20*, 227–235.
21. Estorninho, M.; Chozas, S.; Mendes, A.; Colwell, F.; Abrantes, I.; Fonseca, L.; Fernandes, P.; Costa, C.; Máguas, C.; Correia, O.; et al. Differential Impact of the Pinewood Nematode on *Pinus* Species Under Drought Conditions. *Front. Plant Sci.* **2022**, *13*, 1–12. [CrossRef]
22. Donald, P.; Stamps, W.T.; Linit, M.; Todd, T.C. Pine Wilt. *Plant Health Instr.* **2016**, *16*. [CrossRef]

23. Futai, K. Host resistances shown at the time of pine wood nematode invasion. *Mem. Coll. Agric. Kyoto Univ.* **1985**, *126*, 45–53.
24. Kiyohara, T. Sexual attraction in Bursaphelenchus xylophilus. *Jpn. J. Nematol.* **1982**, *11*, 7–12.
25. Takeuchi, Y.; Kanzaki, N.; Futai, K. A nested PCR-based method for detecting the pine wood nematode, *Bursaphelenchus xylophilus*, from pine wood. *Nematology* **2005**, *7*, 775–782.
26. Leal, I.; Green, M.; Allen, E.; Humble, L.; Rott, M. Application of a real-time PCR method for the detection of pine. wood nematode, *Bursaphelenchus xylophilus*, in wood samples from lodgepole pine. *Nematology* **2007**, *9*, 351–362. [CrossRef]
27. Kikuchi, T.; Aikawa, T.; Oeda, Y.; Karim, N.; Kanzaki, N. A rapid and precise diagnostic method for detecting the pinewood nematode *Bursaphelenchs xylophilus* by Loop-mediated isothermal amplification. *Phytopathology* **2009**, *99*, 1365–1369. [CrossRef]
28. Bergdahl, D.R.; Halik, S. Inoculated *Pinus sylvestris* serve as long-term hosts for *Bursaphelenchus xylophilus*. In *Sustainability of Pine Forests in Relation to Pine Wilt and Decline*; Shokado: Kyoto, Japan, 1999; pp. 73–78. ISBN 4-87974-999-0.
29. Futai, K.; Furuno, T. The variety of resistances among Pine-species to Pine wood nematode, *Bursaphelenchus lignicolus*. *Bull. Kyoto Univ. For.* **1979**, *51*, 23–36, (In Japanese with English Summary).
30. Kishi, Y. *The Pine Wood Nematode and the Japanese Pine Sawyer*; Thomas Co., Ltd.: Tokyo, Japan, 1995; 302p.
31. Li, M.; Li, H.; Ding, X.; Wang, L.; Wang, L.; Chen, F. The detection of pine wilt disease: A literature review. *Int. J. Mol. Sci.* **2022**, *23*, 10797. [CrossRef]
32. Ren, D.; Peng, Y.; Sun, H.; Yu, M.; Yu, J.; Liu, Z. A Global multi-scale channel adaptation network for pine wilt disease tree detection on UAV imagery by circle sampling. *Drones* **2022**, *6*, 353. [CrossRef]
33. Wu, W.; Zhang, Z.; Zheng, L.; Han, C.; Wang, X.; Xu, J.; Wang, X. Research progress on the early monitoring of pine wilt disease using hyperspectral techniques. *Sensors* **2020**, *20*, 3729. [CrossRef]
34. Zhang, Z.; Zhu, L. A review on unmanned aerial vehicle remote sensing: Platforms, sensors, data processing methods, and applications. *Drones* **2023**, *7*, 398. [CrossRef]
35. Zhang, B.; Ye, H.; Lu, W.; Huang, W.; Wu, B.; Hao, Z.; Sun, H. A Spatiotemporal Change. Detection Method for Monitoring Pine Wilt Disease in a Complex Landscape Using High-Resolution Remote Sensing Imagery. *Remote Sens.* **2021**, *13*, 2083. [CrossRef]
36. Takeuchi, Y.; Kanzaki, N.; Futai, K. Volatile compounds in pine stands suffering from pine wilt disease: Qualitative and quantitative evaluation. *Nematology* **2006**, *8*, 869–879.
37. Hwang, H.-S.; Han, J.-Y.; Choi, Y.-E. Enhanced Emission of Monoterpene 3-Carene in *Pinus densiflora* infected by Pine Wood Nematode and Characterization of 3-Carene Synthase. *Forests* **2021**, *12*, 514. [CrossRef]
38. Bogale, M.; Baniya, A.; Di Gennaro, P. Nematode Identification Techniques and Recent Advances. *Plants* **2020**, *9*, 1260. [CrossRef] [PubMed]
39. Lee, J.-P.; Sekhon, S.S.; Kim, J.H.; Kim, S.C.; Cho, B.-K.; Ahn, J.-Y.; Kim, Y.-H. The pine wood nematode *Bursaphelenchus xylophilus* and molecular diagnostic methods. *Mol. Cell. Toxicol.* **2021**, *17*, 1–13. [CrossRef]
40. Yuyan, W.; Chaoran, S.; Haiyan, L.; Zhihong, G. A study on the techniques of the rapid quarantine detection for diseased wood caused by pine wood nematode. *Sci. Silvae Sin.* **2000**, *36*, 59–62, (In Chinese with English Summary).
41. Bonifacio, L.; Sousa, E. Early detection methods for pine wood nematode infections of Maritime pine in Portugal. *Silva Lusit. N* **2011**, *19*, 49–60.

biology

MDPI

Article

Habitat Loss in the IUCN Extent: Climate Change-Induced Threat on the Red Goral (*Naemorhedus baileyi*) in the Temperate Mountains of South Asia

Imon Abedin [1], Tanoy Mukherjee [1], Joynal Abedin [2], Hyun-Woo Kim [3,4,5,*] and Shantanu Kundu [6,7,*]

1. Agricultural and Ecological Research Unit, Indian Statistical Institute, Kolkata 700108, India; imon.jabedin@gmail.com (I.A.); mukherjeetanoy@gmail.com (T.M.)
2. Dibru-Saikhowa Conservation Society, Tinsukia 786147, India; banashree.eco@gmail.com
3. Department of Marine Biology, Pukyong National University, Busan 48513, Republic of Korea
4. Marine Integrated Biomedical Technology Center, National Key Research Institutes in Universities, Pukyong National University, Busan 48513, Republic of Korea
5. Department of Biology, Faculty of Science and Technology, Airlangga University, Surabaya 60115, Indonesia
6. Institute of Fisheries Science, College of Fisheries Science, Pukyong National University, Busan 48513, Republic of Korea
7. International Graduate Program of Fisheries Science, Pukyong National University, Busan 48513, Republic of Korea

* Correspondence: kimhw@pknu.ac.kr (H.-W.K.); shantanu1984@pknu.ac.kr or shantanu1984@gmail.com (S.K.)

Simple Summary: Global warming has had a dramatic impact on many mammalian species, leading to declines and even extinctions, particularly in high-altitude regions of the Himalayas. The Red Goral, a cliff-dwelling ungulate with a restricted range in India, Myanmar, and China, is especially vulnerable due to its small population and limited habitat. Recent habitat modeling indicates that only a small portion of its extent is currently suitable, with even less protected within conservation areas. The projections for the future are concerning and reveal significant declines in suitable habitats and increased fragmentation. The key factors affecting the species distribution were found to be precipitation and elevation. The study highlights the importance of maintaining and improving connectivity between fragmented habitats and identifies crucial protected areas for conservation efforts. The recommendations stress the need for strategic management, enhanced international cooperation, and addressing traditional hunting practices to better protect the species. Overall, the findings underscore the urgent need for effective conservation strategies to ensure the long-term survival of the Red Goral in its natural environment.

Abstract: Climate change has severely impacted many species, causing rapid declines or extinctions within their essential ecological niches. This deterioration is expected to worsen, particularly in remote high-altitude regions like the Himalayas, which are home to diverse flora and fauna, including many mountainous ungulates. Unfortunately, many of these species lack adaptive strategies to cope with novel climatic conditions. The Red Goral (*Naemorhedus baileyi*) is a cliff-dwelling species classified as "Vulnerable" by the IUCN due to its small population and restricted range extent. This species has the most restricted range of all goral species, residing in the temperate mountains of northeastern India, northern Myanmar, and China. Given its restricted range and small population, this species is highly threatened by climate change and habitat disruptions, making habitat mapping and modeling crucial for effective conservation. This study employs an ensemble approach (BRT, GLM, MARS, and MaxEnt) in species distribution modeling to assess the distribution, habitat suitability, and connectivity of this species, addressing critical gaps in its understanding. The findings reveal deeply concerning trends, as the model identified only 21,363 km^2 (13.01%) of the total IUCN extent as suitable habitat under current conditions. This limited extent is alarming, as it leaves the species with very little refuge to thrive. Furthermore, this situation is compounded by the fact that only around 22.29% of this identified suitable habitat falls within protected areas (PAs), further constraining the species' ability to survive in a protected landscape. The future projections paint even degraded scenarios, with a predicted decline of over 34% and excessive fragmentation in suitable habitat extent. In addition, the present study identifies precipitation seasonality and elevation as the primary contributing predictors to the distribution of this species. Furthermore, the study identifies nine designated transboundary PAs within the IUCN extent of the Red Goral and the connectivity among them to highlight the crucial role in supporting the species' survival over time. Moreover, the Dibang Wildlife Sanctuary (DWLS) and Hkakaborazi National Park are revealed as the PAs with the largest extent of suitable habitat in the present scenario. Furthermore, the highest mean connectivity was found between DWLS and Mehao Wildlife Sanctuary (0.0583), while the lowest connectivity was observed between Kamlang Wildlife Sanctuary and Namdapha National Park (0.0172). The study also suggests strategic management planning that is a vital foundation for future research and conservation initiatives, aiming to ensure the long-term survival of the species in its natural habitat.

Keywords: artiodactyla; ecological niche; global warming; international conservation; threatened species; transboundary PAs

1. Introduction

In recent years, global warming has significantly impacted numerous vertebrate species, including mammals, leading to their rapid decline or even extinction within their ecological niches [1,2]. These ecological niches are crucial for species survival and their interactions with environmental conditions, promoting viable populations and their persistence over time [3]. The rapid decline of these ecological niches is expected to worsen as the pace and magnitude of global environmental warming continue to accelerate [4]. This is more concerning for mountainous and temperate regions because these areas are highly sensitive to temperature changes, which can lead to rapid and significant ecological disruptions as even small increases in temperature can cause accelerated snowmelt, shifts in vegetation zones, and habitat loss, creating more immediate and severe impacts compared to more stable lowland environments [5,6]. Mountain mammals cope with thermic stress and decreasing suitable areas by shifting to higher altitudes, altering their activity patterns, seeking cooler microhabitats, adjusting their diet, and modifying reproductive strategies, while conservation efforts focus on expanding protected areas and improving habitat management to support their adaptation [7,8]. Therefore, this situation is particularly pronounced in remote and highly elevated mountainous regions, such as the Himalayas and adjacent ranges in South Asia [1]. The continental drift theory suggests that the Indian and Eurasian plates collided around 50 million years ago, forming the Himalayas, also known as "the abode of snow" in Sanskrit, and creating a complex geological wedge with three tectonic domains from north to south [9,10]. The Himalayan range is divided into three sections: the western, central, and eastern ranges, each home to unique flora and fauna, including many mountainous ungulates. Unfortunately, many of these species lack adaptive strategies to cope with new climatic conditions [11–13]. The eastern Himalayan range, in particular, suffers from limited research and adaptive conservation strategies due to its remoteness, challenging terrain, and inadequate infrastructure. This leaves many species understudied, necessitating the identification of suitable areas where large ungulates with narrow niches can find refuge under future climatic conditions to improve conservation efforts [14,15]. This strategic approach is of utmost importance for the Red Goral (*Naemorhedus baileyi*), a cliff-dwelling Bovidae species (order Artiodactyla) classified as "Vulnerable" by the IUCN Red List due to its small and declining population [16]. Morphologically, this species is a vibrant foxy-red mammal characterized by its long, soft, shaggy hair, with a thin, dark stripe running along its back from head to tail [17]. The legs of this species exhibit the same rich red hue as the rest of its body, while the undersides are a lighter buff color. The Red Goral has the most restricted range of all goral species, confined to the temperate mountains of northeast India (Arunachal Pradesh), northern Myanmar, and China (southeast Tibet and Yunnan). This is the only goral species in this range and inhabits higher elevations than most gorals, found between altitudes of 2000 and 4500 m [16].

Since this species has an extremely restricted range and a very small population size, its vulnerability to climatic shifts and habitat alterations is markedly increased, making it particularly susceptible to changes in environmental conditions and disruptions to its habitat [18]. Therefore, habitat mapping and modeling corridors across the species' distribution are essential for prioritizing conservation strategies [19]. These efforts will demonstrate the suitability and viability of habitat patches for wildlife occupancy and movement between them, thereby facilitating natural gene flow and enhancing long-term survival [20,21]. Furthermore, identifying the key factors driving shifts in species' distribution ranges due to climatic changes is crucial, as understanding these drivers will offer insight into the environmental alterations that impact habitat preferences and movement patterns, ultimately guiding the development of effective conservation and management strategies [22,23].

In recent years, there have been significant advances and widespread adoption of species distribution models (SDMs) for assessing habitat suitability [24,25]. This methodology facilitates the mapping of species distribution patterns and the quantitative evaluation of various environmental factors [25–28]. Consequently, the ensemble model has emerged as a robust tool for estimating habitat suitability across different species [29]. This approach integrates multiple modeling algorithms to predict species distributions across geographic areas, leveraging the strengths of various models to capture diverse aspects and underlying processes influencing distribution [30]. The integrated approach combines various modeling methods, each capturing different aspects and underlying processes that influence species distribution. This strategy seeks to balance the strengths and limitations of individual models, leading to improved prediction accuracy and reliability. Furthermore, understanding the response of key driving variables to climate change is essential for identifying suitable habitats, which is critical for developing effective conservation strategies and landscape-level management plans [31,32]. Moreover, studying the IUCN extent is justified as it involves a systematic process that includes mapping and analyzing the geographical distribution of a species using occurrence records, adjusting for geographical barriers, and verifying with expert knowledge to accurately represent the species' natural distribution for conservation planning and risk assessment.

Therefore, this study utilizes an ensemble approach in SDM within the IUCN range of the Red Goral, *N. baileyi*, to (i) determine the suitable extent under current and future climate change scenarios; (ii) analyze the patterns of habitat fragmentation resulting from climate change; (iii) identify biological corridors within the range; and (iv) assess the suitability and connectivity of transboundary protected areas (PAs). This study is vital for enhancing conservation efforts for this species that lacks extensive research and faces vulnerability in the ecological context. The identification of suitable habitats under current and future climate scenarios will reveal the areas in which this species can thrive within the IUCN extent and the habitat shifts with changing conditions to guide conservationists in India and adjoining countries. Moreover, understanding patterns of habitat fragmentation and connectivity will aid in safeguarding these corridors that will enable movement for this species and ensure transboundary migration and genetic connectivity within the known range. Furthermore, evaluating the suitability and connectivity of PAs will ensure coordinated conservation efforts across borders, creating an international network of safe havens and supporting broader conservation initiatives to save this threatened ungulate species in the wild.

2. Materials and Methods

2.1. Study Area and Species Occurrence Data

The IUCN range of the Red Goral encompasses three countries: India, China, and Myanmar (Figure 1). In India, this species is recorded in specific regions of eastern Arunachal Pradesh, particularly in the districts of Upper Siang, Dibang Valley, Lower

Dibang Valley, Anjaw, Lohit, and Changlang, including the Namdapha National Park (NNP) and the community forests near Vijaynagar, which borders with Myanmar. Additionally, there have been unconfirmed sightings near the River Kameng in western Arunachal Pradesh. In Myanmar, the Red Goral is found in the northern mountains of Kachin State, with a confirmed presence in Putao District's Hponkanrazi Wildlife Sanctuary (HpWLS) and Hkakaborazi National Park (HkNP). Furthermore, China has the largest distribution area for this species, with populations confirmed in the Tibet Autonomous Region (Xizang) and Yunnan. In Tibet, the primary distribution area is Nyingchi City prefecture, while in Yunnan, the species is reported in Nùjiāng Lisu Autonomous Prefecture, specifically in Gongshan Derung and Nu Autonomous County [16]. Therefore, to achieve a comprehensive understanding of the distribution of the Red Goral, the entire IUCN extent was designated as the study area (Figure 1). This was conducted to ensure the analysis encompassed the full geographical range of the species and provided an accurate picture of its habitat distribution across different countries and regions aimed to identify the suitable habitats and their variations due to climate change within the IUCN known extent. Additionally, examining the IUCN extent is warranted because it employs a systematic process that entails mapping and analyzing a species' geographical distribution using occurrence records, adjusting for geographical barriers, and verifying with expert knowledge to accurately reflect the species' natural range for effective conservation planning and risk assessment.

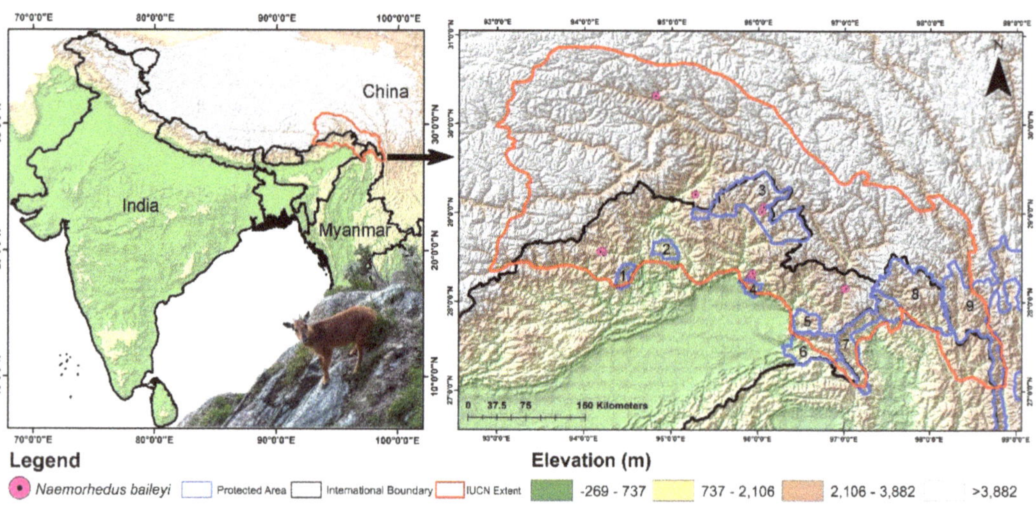

Figure 1. Map showing the study area for the present study along with the IUCN extent of Red Goral (*N. baileyi*). The figure also highlights the location points acquired from primary and secondary sources used for training the model. The photograph of the Red Goral was taken by Mr. Ravi Mekola in Dibang Valley, Arunachal Pradesh, India. Protected areas are represented by blue lines: 1. YardiRabe Supse Wildlife Sanctuary; 2. Mouling National Park; 3. Dibang Wildlife Sanctuary; 4. Mehao Wildlife Sanctuary; 5. Kamlang Wildlife Sanctuary; 6. Namdapha National Park; 7. Hponkanrazi Wildlife Sanctuary; 8. Hkakaborazi National Park; 9. Three Parallel Rivers of Yunnan Protected Areas.

The study utilized 14 location points sourced from the GeoCAT website, which compiles data from GBIF and iNaturalist, accessed on 25 June 2024 [33]. Furthermore, to ensure the reliability of the dataset and accurately represent the ecological areas of interest, records of museum specimens or captive individuals were excluded. Additionally, five location points were gathered during field visits to the temperate mountainous regions of Arunachal Pradesh, including sightings at Mayudia Pass (n = 1), the 65 km area in Dibang Valley

(n = 2), Walong (near the cliff around 4 km from the airbase) (n = 1), and near Kaho in Anjaw district (n = 1). Moreover, 12 additional location points were obtained through direct communications with tourist guides and local people and hunters in Dibang Valley (near Anini), NNP, and Anjaw (near Kibitoo area) districts. The spatial correlation between occurrences was conducted at a resolution of 1 km^2 in SDM toolbox v2.4. This specific resolution was chosen to align with the size of one pixel in the raster data, thereby reducing the risk of overfitting the model and ensuring a more accurate analysis. Consequently, the final model was generated using a total of 31 location points for the Red Goral within its IUCN extent.

2.2. Selection of Predictors for the Ensemble Model

The study employed various predictor sets, including bioclimatic, topographic, and habitat variables, to model the habitat of the Red Goral. The bioclimatic variables (n = 19) were sourced from the WorldClim website (https://www.worldclim.org/, accessed on 25 June 2024), while topographic variables, such as elevation, slope, and aspect, were obtained from the Diva-GIS website (http://srtm.csi.cgiar.org/srtmdata/, accessed on 25 June 2024) at a spatial resolution of 90 m [34]. The habitat variable, i.e., temperate forest (identified as of major importance by the IUCN), was extracted from the land cover classes in the Copernicus dataset and converted into continuous raster datasets using the Euclidean Distance function in ArcGIS [35]. This process aimed to evaluate the significance of this forest type and the species' response to its proximity, as many species show a strong preference for areas near suitable forest types [36]. All variables were then resampled to 30 arcseconds (~1 km^2) using the Spatial Analyst Extension in ArcGIS 10.6. The spatial multicollinearity testing was conducted using the SAHM (Software for Assisted Habitat Modelling) package in VisTrails software V.2, with covariates showing a correlation (r) greater than 0.8 being excluded from the analysis [37,38]. Furthermore, after addressing the correlation among the variables, a total of 8 variables were retained for the final model for both species (Figure S1).

2.3. SDM Utilizing Ensemble Approach

The assessment of distribution models involved employing multiple modeling algorithms through an ensemble approach to formulate the final distribution model for both species. Therefore, four distinct algorithms—Maximum Entropy (MaxEnt), Boosted Regression Tree (BRT), Generalized Linear Model (GLM), and Multivariate Adaptive Regression Splines (MARS)—were utilized [27,39,40] (Table S1). These algorithms were executed using the SAHM package in VisTrails software [37,41]. The execution produced probability surfaces ranging from "0" (least suitability) to "1" (highest suitability), and binary maps were generated using the minimum training presence as the threshold. The ensemble count map was constructed on a scale from "0" to "4", where each pixel denoted the number of model agreements, with a value of "4" indicating unanimous agreement across all four models, facilitating habitat configuration analysis. Additionally, to assess and compare model performance, various evaluation metrics, including AUC, True Skill Statistic (TSS), Cohen's Kappa, Proportion Correctly Classified (PCC), specificity, and sensitivity, were calculated for both the training data and cross-validation sets (n = 10) [42–45].

Furthermore, future projections were used to project potential climate change scenarios across two distinct shared socio-economic pathways (SSP)—namely SSP245 and SSP585—spanning the periods 2041–2060 and 2061–2080. The SSPs are scenarios used in climate change research to explore future socioeconomic conditions and their implications for greenhouse gas emissions and climate impacts. SSP245 represents a future where moderate efforts are made to mitigate emissions and adapt to climate change, assuming moderate population growth, technological development, and a balanced approach to environmental and social policies. Here, greenhouse gas emissions increase gradually over the 21st century, stabilizing towards the end of the century with international cooperation on climate policies, albeit with challenges in implementation [46]. Conversely, SSP585 depicts a future with high greenhouse gas emissions and limited adaptation efforts, assuming rapid population growth, high energy demand, and minimal environmental regulation, leading to continued emission increases throughout the century. This scenario reflects a world with little international cooperation on climate policies and insufficient societal efforts to mitigate emissions [47]. The study utilized the Hadley Centre Global Environment Model in Global Coupled Configuration 3.1 (HadGEM3-GC31 LL), the sixth Coupled Model Intercomparison Project (CMIP6) [48]. The selection of this General Circulation Model (GCM) was based on its recognized performance in South and Southeast Asia and its ability to capture temporal fluctuations and excel in representing temperature distribution, as evidenced by previous research [49,50]. For the present study, non-climatic raster data, including the habitat variable and topographic variables, remained constant due to the difficulty of projecting their change pattern in the future [51,52]. This deliberate decision aimed at isolating the impact of climate change on the objective of the study [53]. Additionally, to facilitate the development of an effective conservation action plan, a comprehensive assessment of habitat suitability was conducted on the transboundary protected areas (PAs), given the distinct legal framework for each country. This qualitative assessment of suitable habitats across different transboundary PAs in their distribution range was conducted using the zonal statistics function in ArcGIS v.10.6 [36,53–55].

2.4. Assessment of Habitat Shape Geometry and Connectivity

For the assessment of the qualitative and geometric characteristics of suitable patches in both current and projected future scenarios, various class-level metrics were employed using FRAGSTATS version 4.2.1 [56,57]. This specialized software for landscape ecology, urban planning, and environmental management analyzes spatial patterns in landscapes and ecosystems, providing a suite of metrics and indices to quantify and understand landscape structure and composition. The study encompassed metrics such as the number of patches (NP), aggregate index (AI), patch density (PD), largest patch index (LPI), edge density (ED), total edge (TE), and landscape shape index (LSI). Furthermore, metrics like NP, PD, ED, TE, and LPI provide detailed information about the geometry of the patches, including their size, edge characteristics, and density within a region. In contrast, LSI focuses on the shape complexity of the patches, indicating how convoluted or irregular they are, whereas AI measures the degree of proximity or clustering among patches, reflecting how aggregated or dispersed they are within the landscape. These metrics hold biological significance, shedding light on habitat ecological processes and offering valuable insights into the impacts of changes in suitable areas on landscape dynamics [58,59]. This methodology facilitates a deeper understanding of landscape characteristics and enables comprehensive analysis across the distribution range of the species. Consequently, these metrics were utilized to evaluate habitat features and levels of fragmentation in the modeled area across various scenarios, including present conditions and future climate change projections [60,61].

Furthermore, given the significance of enhancing habitat connectivity as a crucial conservation strategy for species preservation and gene flow, assessing the biological connectivity between habitat patches was imperative [36]. Therefore, to achieve this objective, the circuit model, commonly employed in designing animal corridors, was utilized [62]. The circuitscape toolbox for ArcGIS 10.6 facilitated the simulation of these corridors, where location points were used as nodes, with the conductance surface derived from the probability output generated by the ensemble model [60,63]. This toolbox facilitates the simulation of ecological corridors by modeling species movement or flow across diverse landscapes. In this process, pairwise source/ground mode settings were employed, where the probability maps generated from the model were utilized as the conductance raster. The location points were designated as the focal node raster in the pairwise setup module. The resulting output was generated as current maps, which were then used for further detailed assessment and analysis of connectivity. This corridor simulation was conducted for the present and all the future climatic scenarios.

3. Results

3.1. Ensemble Habitat Modeling

The models exhibited an Area Under the Curve (AUC) range of 0.659 to 0.815 during training and between 0.637 and 0.719 in cross-validation for the Red Goral (Figures 2, S2 and S3, Table 1). Notably, the ΔAUC value, which measures the difference between training and cross-validation AUCs, was the smallest for the MARS model, recording a value of 0.094. In contrast, the GLM demonstrated the highest ΔAUC values, reaching 0.18 across the replicate runs. These findings collectively underscore the sensitivity of the data employed for model fitting across all models. The evaluation metrics, including True Skill Statistic (TSS), Proportion Correctly Classified (PCC), Cohen's Kappa, sensitivity, and specificity, further affirmed the high-quality performance of the models in both the training and cross-validation phases. Among the four selected models, MaxEnt utilized all provided variables during the replicate runs, whereas the BRT model opted for the fewest variables, selecting only two out of the eight provided for this species (Figure 2, Table 1).

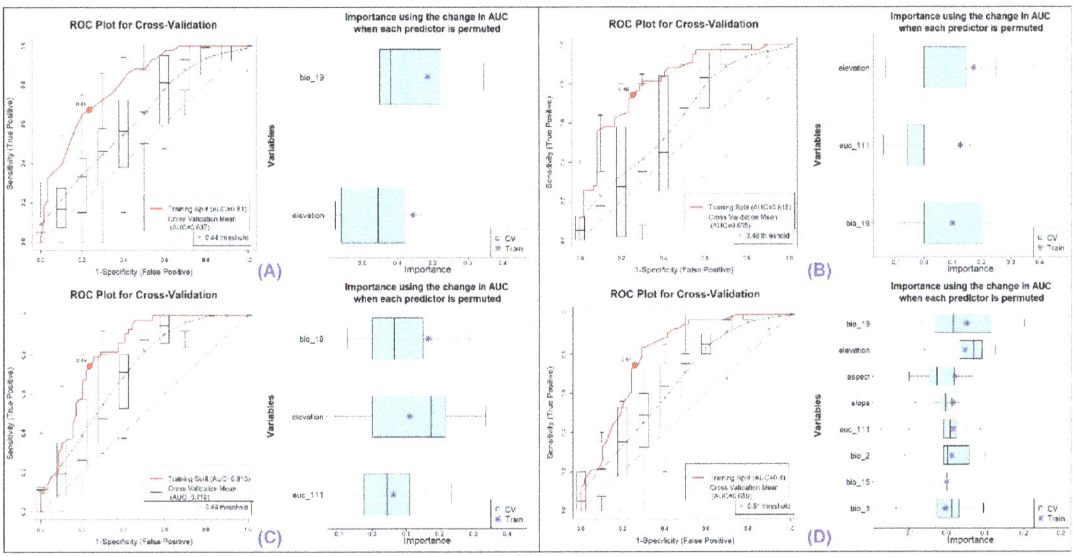

Figure 2. Model evaluation plot showing the average training ROC of both training and cross-validation (CV) and the predictors chosen by the model for the replicate runs under four models of Red Goral: (**A**) showing ROC plot of Boosted Regression Tree (BRT), (**B**) Generalized Linear Model (GLM), (**C**) Multivariate Adaptive Regression Spines (MARS), and (**D**) Maximum Entropy (MaxEnt).

Table 1. Model fit metrics for each of the participating modeling methods and for the final ensemble model for estimation of habitat suitability of *N. baileyi*. A total of four model algorithms i.e., Maximum Entropy (MaxEnt), Boosted Regression Tree (BRT), Generalized Linear Model (GLM), and Multivariate Adaptive Regression Splines (MARS). AUC: Area under Curve, ΔAUC: Change in Area under Curve (Training–Cross Validation), PCC: Proportion Correctly Classified, TSS: True Skill Statistic.

Species	Model	Dataset	AUC	ΔAUC	PCC	TSS	Kappa	Specificity	Sensitivity
Naemorhedus baileyi	BRT	Train	0.81	0.173	72.5	0.437	0.437	0.763	0.674
		CV	0.637		64.7	0.275	0.273	0.68	0.595
	GLM	Train	0.815	0.18	74.5	0.49	0.484	0.746	0.744
		CV	0.635		59.2	0.157	0.165	0.657	0.5

Table 1. *Cont.*

Species	Model	Dataset	AUC	ΔAUC	PCC	TSS	Kappa	Specificity	Sensitivity
Naemorhedus baileyi	MARS	Train	0.813	0.094	75.5	0.507	0.502	0.763	0.744
		CV	0.719		65.3	0.288	0.284	0.683	0.605
	MaxEnt	Train	0.8	0.141	74.3	0.486	0.48	0.741	0.744
		CV	0.939		88.1	0.746	0.697	0.887	0.859

3.2. *Effective Habitat Suitability Predictors*

The ensemble model for the Red Goral revealed that on average (μ) across the four models, the primary contributor to habitat suitability was the bioclimatic variable precipitation of the coldest quarter (bio_19), which accounted for 35.87% of the prediction power (Table 2, Figure S4). Furthermore, the topographic variable elevation was identified as the second most significant predictor, contributing 33.69%, whereas the habitat variable temperate forests (euc_111) emerged as the third highest predictor, accounting for 14.63% of the model ensemble run. In contrast, among the eight selected variables, the bioclimatic variable isothermality (bio_3) was the least influential in predicting the distribution of the Red Goral.

Table 2. The mean percentage contribution of each covariate generated from the final ensemble model for *N. baileyi*.

Variable	Abbreviation	BRT	GLM	MARS	MaxEnt	μ (Mean)	μ (Mean) %
Aspect	aspect	0.000	0.000	0.000	0.023	0.006	1.62
Precipitation Seasonality	bio_15	0.000	0.000	0.000	0.011	0.003	0.76
Precipitation of Coldest Quarter	bio_19	0.186	0.100	0.167	0.052	0.126	35.87
Mean Diurnal Range (Mean of Monthly (Max Temp Min Temp)	bio_2	0.000	0.000	0.000	0.014	0.003	0.98
Isothermality	bio_3	0.000	0.000	0.000	0.003	0.001	0.23
Elevation	elevation	0.144	0.172	0.111	0.047	0.119	33.69
Temperate Forest	euc_111	0.000	0.127	0.062	0.017	0.052	14.63
Slope	slope	0.000	0.000	0.000	0.172	0.043	12.21

3.3. *Suitable Habitat Extent in the IUCN Extent and Protected Areas*

The estimated extent of occurrence for this species is reported to be 164,150 km^2 within its distribution range, according to the IUCN. However, the findings of the ensemble model are concerning, revealing that only 21,363 km^2, or a mere 13.01% of this area, represents suitable habitat under current conditions (Figure 3). The projections for future climate change scenarios present even more alarming trends, indicating a substantial reduction in suitable habitat that could reach up to 45% due to climatic shifts. Specifically, under the SSP245 scenario, the model predicts habitat declines of 34.82% for the period 2041–2060 and 41.55% for the period 2061–2080 (Figure 4, Table S2). Furthermore, the high emission scenario SSP585 forecasts even more severe reductions, with suitable habitat expected to decrease by 39.68% during 2041–2060 and by 45.74% during 2061–2080, compared to the present scenario, respectively (Figure 5, Table S2).

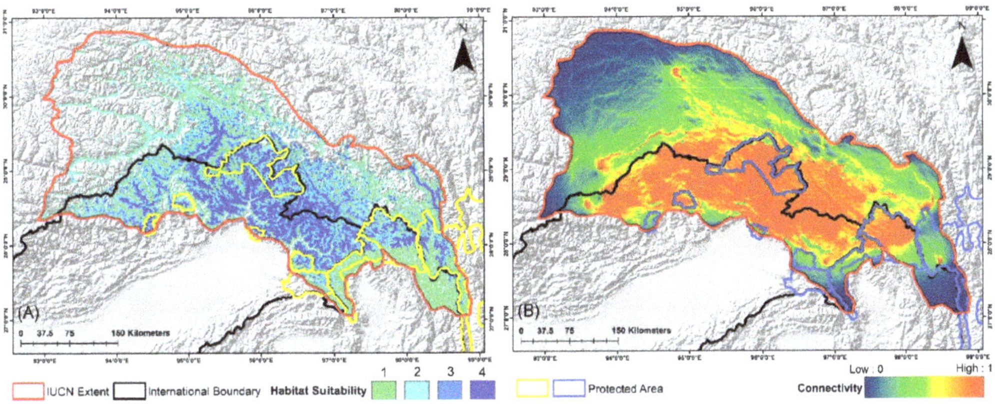

Figure 3. (**A**) This figure shows the present suitable habitats for *N. baileyi* in the study area. The four classes (1–4) defined in the map show the four model arguments used in the present study. Class "0" of habitat suitability is not indicated in the map as it represents no suitability and zero model agreement. (**B**) Map representing the habitat connectivity in the IUCN extent in the present scenario.

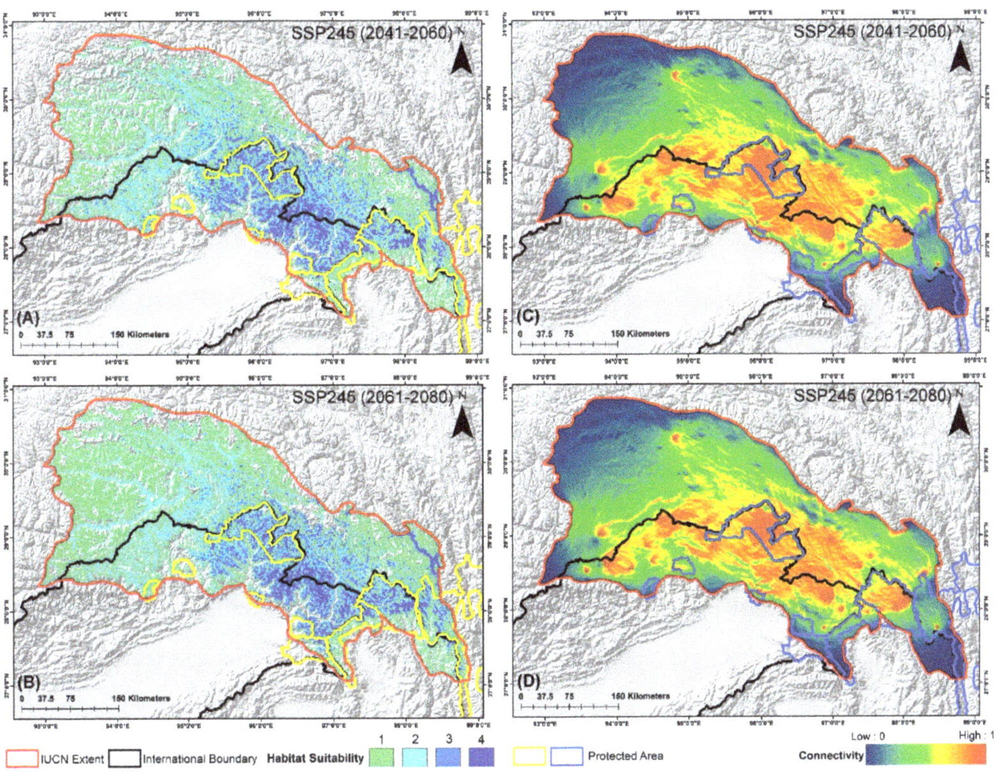

Figure 4. Maps representing the two-time frames of the SSP245 scenario for *N. baileyi*: (**A**,**B**) determine the habitat suitable in the IUCN extent, whereas (**C**,**D**) determine the connectivity in the landscape in these scenarios.

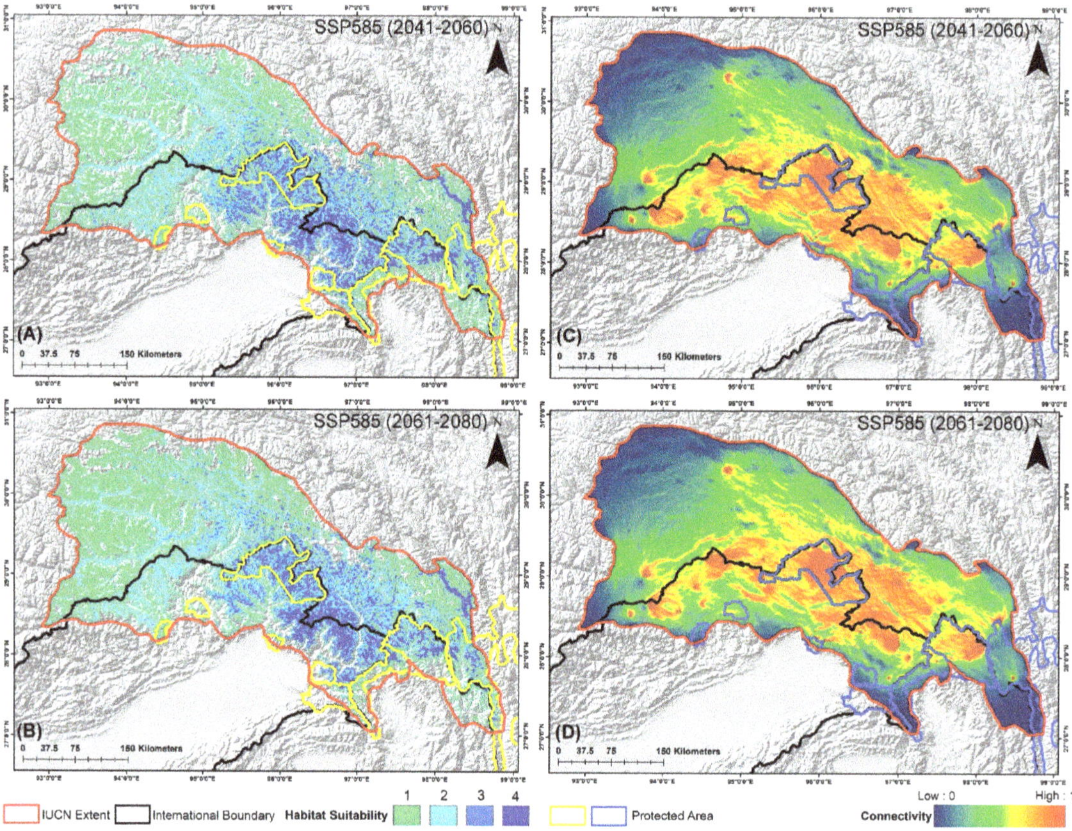

Figure 5. Maps representing the two-time frames of the SSP585 scenario for *N. baileyi*: (**A**,**B**) determine the habitat suitable in the IUCN extent, whereas (**C**,**D**) determine the connectivity in the landscape in these scenarios.

Additionally, nine transboundary PAs fall within the IUCN distribution range of this species across India, China, and Myanmar (Table 3). Despite its greater extent, only 4762 km^2 of suitable habitat is included within the PAs, accounting for just 22.29% of the habitat identified by the model. Among them, Dibang Wildlife Sanctuary (DWLS) in India has the largest extent of suitable habitat at 1852 km^2, while Mehao Wildlife Sanctuary (MWLS) in India has the smallest extent (71 km^2) (Table 3). However, the assessment of mean habitat suitability yields different results as Yardi-Rabe Supse Wildlife Sanctuary (YRSWLS) in India has the highest mean habitat suitability score of 2.497, whereas Mouling National Park (MNP) has the lowest at 1.445 under current conditions (Table 4). Furthermore, the future scenarios predict a rapid decrease in both area extent and mean habitat suitability. Specifically, YRSWLS and MNP are projected to experience the most significant declines, exceeding 89% of area extent in both future scenarios for the 2041–2060 and 2061–2080 timeframes. Similarly, DWLS is projected to see a 35% decline in area extent under the SSP245 (2041–2060) scenario. In contrast, HkNP in Myanmar is expected to experience the least decline among all PAs, ranging from 9% to 15%. Interestingly, HpWLS is the only PA that shows an increase in area extent by over 4%, reaching 59.53% under the SSP245 (2061–2080) scenario. Furthermore, the SSP585 (2061–2080) scenario predicts the most severe declines compared to the other scenarios. Moreover, in terms of mean habitat suitability, MNP is expected to suffer the greatest decline, with an 87% decrease under the

SSP585 (2061–2080) scenario. In contrast, DWLS is the only PA projected to see an increase in mean habitat suitability, up to 8%, due to climatic shifts. These projections underscore the pronounced declines and shifts caused by climate change, although the impacts are relatively less severe in the SSP245 (2041–2060) and SSP585 (2041–2060) scenarios compared to the 2061–2080 timeframe, which marks the onset of more significant climate-induced changes (Tables 3 and 4).

Table 3. The suitable habitat extent (in km^2) for *N. baileyi* in both current and future scenarios within protected areas (descending order) in the IUCN extent. The gain is represented by "+", whereas loss is represented by "−". WLS: wildlife sanctuary, NP: national park, PA: protected area, and GR: growth rate.

Country	Protected Areas	Present	SSP 245 (2041–2060)	GR from Present (%)	SSP 245 (2061–2080)	GR from Present (%)	SSP 585 (2041–2060)	GR from Present (%)	SSP 585 (2061–2080)	GR from Present (%)
India	Dibang WLS	1852	1443	−22.084	1195	−35.475	1252	−32.397	1304	−29.590
Myanmar	Hkakaborazi NP	1191	1081	−9.236	1050	−11.839	1129	−5.206	1004	−15.701
China	Three Parallel Rivers of Yunnan PA	825	607	−26.424	514	−37.697	700	−15.152	592	−28.242
Myanmar	Hponkanrazi WLS	257	269	+4.669	410	+59.533	331	+28.794	281	+9.339
India	Kamlang WLS	210	120	−42.857	116	−44.762	109	−48.095	79	−62.381
India	Yardi-Rabe Supse WLS	134	14	−89.552	2	−98.507	4	−97.015	2	−98.507
India	Namdapha NP	126	120	−4.762	115	−8.730	112	−11.111	93	−26.190
India	Mouling NP	96	1	−98.958	0	−100.000	0	−100.000	0	−100.000
India	Mehao WLS	71	31	−56.338	27	−61.972	27	−61.972	10	−85.915

Table 4. The mean habitat suitability for *N. baileyi* in both current and future scenarios within protected areas (descending order) in the IUCN extent. The gain is represented by "+", whereas loss is represented by "−". WLS: wildlife sanctuary, NP: national park, PA: protected area, and GR: growth rate.

Country	Protected Areas	Present	SSP 245 (2041–2060)	GR from Present (%)	SSP 245 (2061–2080)	GR from Present (%)	SSP 585 (2041–2060)	GR from Present (%)	SSP 585 (2061–2080)	GR from Present (%)
India	Yardi-Rabe Supse WLS	2.497	1.081	−56.705	0.976	−60.911	1.147	−54.075	0.755	−69.763
India	Dibang WLS	2.361	2.225	−5.759	2.212	−6.315	2.231	−5.498	2.161	−8.461
India	Kamlang WLS	2.059	1.400	−31.996	1.322	−35.800	1.315	−36.113	1.011	−50.912
India	Mehao WLS	2.028	1.284	−36.682	0.905	−55.374	1.000	−50.701	0.616	−69.626
Myanmar	Hkakaborazi NP	2.028	1.735	−14.425	1.886	−6.996	1.796	−11.430	1.628	−19.716
India	Namdapha NP	1.905	1.356	−28.781	1.253	−34.230	1.431	−24.858	0.956	−49.811
China	Three Parallel Rivers of Yunnan PA	1.784	1.484	−16.823	1.659	−7.032	1.556	−12.787	1.339	−24.975
Myanmar	Hponkanrazi WLS	1.710	1.201	−29.752	1.282	−25.051	1.226	−28.320	0.841	−50.836
India	Mouling NP	1.445	0.246	−82.977	0.244	−83.079	0.396	−72.579	0.180	−87.564

3.4. Habitat Fragmentation and Biological Corridors

The future habitat loss induced by climate change has caused significant fragmentation in the suitable habitat patches for this species. The results reveal severe fragmentation of viable patches, as indicated by the NP metric, which increases by 26.39% in SSP245 (2041–2060) and 21.38% in SSP585 (2041–2060) compared to the present (Table 5). However, this rate slows to 17.75% and 13.55% in SSP245 (2061–2080) and SSP585 (2061–2080), likely due to greater overall habitat loss. Furthermore, the heightened fragmentation is reflected in increases of PD and ED by over 1658.93% and 23.14%, respectively, across all future scenarios, indicating increased density of the smaller patches with greater edge areas. Moreover, the LPI declines by over 38.02%, showing a substantial reduction in the size of viable patches, supported by the TE metric, which also declines between 17.91% and 25.53%, revealing the shrinkage of patches. Additionally, the decrease in both patch size and the number of patches explains the decline in the LSI by up to 6.77%, indicating simpler patch geometry due to reduced sizes as supported by the AI value, which decreases by more than 16%, revealing that patches are now farther apart. Overall, climate-induced shifts result in the fragmentation of currently viable patches into numerous smaller, more isolated patches, highlighting extensive habitat fragmentation (Table 5).

Table 5. Assessment of habitat shape geometry of *N. baileyi* in present and future scenarios. NP: no. of patches, PD: patch density, LPI: largest patch index, ED: edge density, TE: total edge, LSI: landscape shape index, AI: aggregation index.

Scenario	NP	PD	LPI	TE	ED	LSI	AI
Present	1239	96,210,504.14	16.135	236.352	1,728,689.791	50.416	65.878
SSP 245 (2041–2060)	1566	1,757,433,743	14.473	194.032	2,177,512.031	51.386	56.932
SSP 245 (2061–2080)	1459	1,825,941,129	11.782	177.648	2,223,267.922	49.567	56.037
SSP 585 (2041–2060)	1504	1,823,684,619	13.651	180.944	2,194,047.804	49.601	56.620
SSP 585 (2061–2080)	1407	1,692,277,346	10.854	176.992	2,128,781.464	48.518	57.933

The assessment of biological corridors revealed the highest connectivity between DWLS and MWLS, with a mean connectivity value of 0.0583, followed by MNP and DWLS, with a mean connectivity value of 0.0538 in the present scenario (Figure 2, Table 6). Conversely, the lowest connectivity was found between Kamlang Wildlife Sanctuary (KWLS) and NNP, with a mean connectivity value of 0.0172, despite their close proximity to the landscape. Furthermore, the climate change projections indicate a decline in connectivity among the PAs in all future scenarios. Specifically, in the SSP245 scenario, the most significant decline of 25.60% from the present scenario was observed between YRSWLS and MNP during the 2041–2060 period. This decline was further compounded in the SSP245 (2061–2080) scenario, exceeding 31% (Figure 4). A similar trend of decline between YRSWLS and MNP was also observed in the SSP585 scenarios (Figure 5). However, three connectivity corridors between the PAs in this landscape showed relatively less decline in future projections. These include the corridor between HkNP and Three Parallel Rivers of Yunnan Protected Areas (TPRYPA) (decline by 1.2%), HpWLS and HkNP (decline up to 10%), and NNP and HpWLS (decline between 3.97% and 12.67%) (Table 6). Overall, the highest decline in connectivity was observed in the SSP585 (2061–2080) scenario, indicating that future climate change poses a significant threat to connectivity and, consequently, the conservation of the Red Goral.

Table 6. Assessment of corridor connectivity of *N. baileyi* among the protected areas in both the current and future climate change scenarios. The gain is represented by "+", whereas loss is represented by "−". YRSWLS: Yardi-Rabe Supse Wildlife Sanctuary, MNP: Mouling National Park, DWLS: Dibang Wildlife Sanctuary, MWLS: Mehao Wildlife Sanctuary, KWLS: Kameng Wildlife Sanctuary, NNP: Namdapha National Park, HpWLS: Hponkanrazi Wildlife Sanctuary, HkNP: Hkakaborazi National Park, TPRYPA: Three Parallel Rivers of Yunnan Protected Areas, GR: growth rate.

Corridors	Present	SSP 245 (2041–2060)	GR from Present (%)	SSP 245 (2061–2080)	GR from Present (%)	SSP 585 (2041–2060)	GR from Present (%)	SSP 585 (2061–2080)	GR from Present (%)
YRSWLS_MNP	0.0426	0.0317	−25.60	0.0292	−31.35	0.0310	−27.15	0.0255	−40.11
MNP_DWLS	0.0538	0.0481	−10.57	0.0460	−14.54	0.0477	−11.39	0.0430	−20.02

Table 6. *Cont.*

Corridors	Present	SSP 245 (2041–2060)	GR from Present (%)	SSP 245 (2061–2080)	GR from Present (%)	SSP 585 (2041–2060)	GR from Present (%)	SSP 585 (2061–2080)	GR from Present (%)
DWLS_MWLS	0.0583	0.0532	−8.81	0.0505	−13.40	0.0516	−11.51	0.0468	−19.86
DWLS_KWLS	0.0446	0.0401	−10.12	0.0383	−14.10	0.0390	−12.60	0.0348	−22.09
KWLS_NNP	0.0172	0.0151	−12.05	0.0147	−14.74	0.0147	−14.29	0.0128	−25.84
NNP_HpWLS	0.0332	0.0316	−4.88	0.0319	−3.97	0.0316	−4.84	0.0290	−12.67
HpWLS_HkNP	0.0482	0.0461	−4.45	0.0466	−3.32	0.0461	−4.35	0.0434	−10.05
HkNP_TPRYPA	0.0493	0.0490	−0.49	0.0487	−1.22	0.0490	−0.55	0.0494	0.29

4. Discussion

Recent research has highlighted troubling trends in the global decline of mammalian species and populations, underscoring the severity of the situation [64]. This widespread decrease in mammalian diversity can largely be attributed to changes in land cover and environmental crises driven by climate change, particularly in South Asia [65]. Although numerous studies have been conducted worldwide to address these issues in various mammalian species, this work stands out as the first comprehensive analysis of the distribution, habitat suitability, and connectivity of viable patches for the Red Goral. The intricate details provided by this study not only fill a critical gap in understanding this particular species but also offer valuable insights that could guide future conservation efforts.

The findings reveal deeply concerning trends for the Red Goral, as the model identified only 21,363 km^2, or a mere 13.01% of the total area, as suitable habitat within the IUCN extent under current conditions (Figure 3, Table S2). This limited, suitable habitat is alarming, as it leaves the species with very little space to thrive. Furthermore, this situation is compounded by the fact that only around 22.29% of this identified suitable habitat falls within the PAs, further constraining the species' ability to survive in a protected landscape. The future projections paint an even grimmer picture, with a predicted decline of over 34% in suitable habitat extent. Given the restricted range of the species, this additional reduction due to climate change significantly worsens the situation. These findings are consistent with previous studies that concluded red gorals might face more severe challenges in the Tibetan Plateau adjoining the eastern Himalayas from the pronounced impacts of climate change [66,67]. Furthermore, the gorals are particularly vulnerable to habitat loss and environmental changes because of their limited movement range in restricted extent [68]. The alarming results underscore the necessity of enhancing research efforts focused on habitat modeling and identifying key predictors for species conservation. This type of research is especially critical for lesser-known ungulate species with very restricted ranges, such as the Red Goral [23]. Additionally, these research initiatives can provide valuable insights and inform conservation strategies to mitigate the adverse effects of habitat loss and climate change, ultimately aiding in the preservation of these vulnerable species.

Therefore, the present study identifies precipitation seasonality (bio_15) as the primary contributing predictor to the distribution of the Red Goral, accounting for 35.87% of the overall model (Table 2). This bioclimatic variable is closely followed by elevation, which is the second highest contributor, aligning with the observation that the species is predominantly found in the high-altitude mountains of the eastern Himalayas [16]. This finding corroborates previous studies conducted on various ungulate species, which have similarly highlighted the critical roles of precipitation and temperature as the top bioclimatic variables influencing their distribution [1,32]. Specifically, it was established that precipitation was the most influential variable in the distribution of ungulate species. This aligns with the results of the present study, where precipitation seasonality (bio_15) and precipitation of the coldest quarter (bio_19) are identified as significantly influential variables. However, the model in this study selected a different set of precipitation and temperature predictors compared to previous research, which could be attributed to the unique geographical context of the eastern Himalayas rather than the western ranges. Furthermore, elevation emerged as another key factor in predicting the habitat of the

Red Goral, accounting for 33.69% of the overall model, suggesting that these animals may prefer higher-elevated temperate forests [10,69,70]. Therefore, the temperate forest variable (euc_111) also emerged as one of the most significant predictors for this species, contributing 14.63% to the prediction of the model.

The implications of these findings are particularly critical in the context of climate change in the future, as the distribution of the Red Goral is projected to shift towards the northeastern direction, corroborating the previous studies [20,71]. This northward shift in the Eastern Himalayan and Tibetan regions is clearly exhibited in the present study, emphasizing the need for proactive conservation efforts. Moreover, as the climate continues to change, it will be essential to focus conservation strategies on PAs that are likely to experience these shifts. Furthermore, ensuring that these PAs are well-prepared to accommodate the changing distribution of the Red Goral will be crucial for the species' long-term survival. The study identifies nine designated transboundary PAs within the IUCN range of the Red Goral, highlighting their crucial role in supporting the species' survival over time (Figures 3–5). Among these, the DWLS in Arunachal Pradesh, India, is revealed as the PA with the largest extent of suitable habitat in the present scenario (Table 3). Additionally, HkNP in Myanmar also possesses a significant area of suitable habitat. These two PAs maintain considerable suitable habitat patches in future projections despite suffering some losses due to climatic shifts. This trend can be attributed to the anticipated northeastern shift in the distribution of Red Goral caused by climate change aligning with the geographic locations of these PAs and mitigating their losses to some extent. Moreover, the TPRYPA in China is projected to experience significant habitat loss. It is important to note that this PA is extensive, and only a small portion of it falls within the IUCN range of the Red Goral, which may explain discrepancies between the suitable habitat extent and mean habitat suitability results. The mean habitat suitability averages the quality of various habitats within the studied area, reflecting the overall potential for the species to thrive based on different environmental factors [72]. As a result, YRSWLS in India has the highest mean habitat suitability score (2.497) despite having a much smaller suitable extent (134 km^2) than the other areas, which may be attributed to its relatively smaller size. This difference highlights the importance of considering both suitable extent areas and mean habitat suitability to gain a comprehensive understanding of the landscapes. It is notable that DWLS is unique in ranking high in both suitable extent and mean habitat suitability, making it a priority conservation area for the Red Goral. However, the entire distribution within the PAs is projected to be affected by climatic shifts, leading to a decline in both suitable habitat extent and habitat suitability scores. This trend results in significant fragmentation, especially in PAs like YRSWLS and MNP, where future scenarios predict a dramatic decline of over 90% in suitable areas. Additionally, the threat of irreversible biodiversity decline is particularly severe in these landscapes, which are already fragmented due to ongoing habitat loss from development activities. These activities reduce patch sizes and increase distances between patches, disrupting connectivity and exacerbating edge effects. The increased habitat fragmentation is linked to higher rates of species decline due to additional habitat loss [73]. The present study corroborates these findings, demonstrating that extensive reductions in suitable habitat have led to significant spatial alterations and fragmentation, adversely affecting the Red Goral and other ungulate species.

The distribution extent of the Red Goral has undergone extensive fragmentation due to climatic shifts (Table 5). This severe fragmentation of viable patches is evidenced by a notable increase in the NP metric by 26.39% in SSP245 (2041–2060) and 21.38% in SSP585 (2041–2060) when compared to the present scenario. These increases reflect a significant disruption in the continuity of the species' habitat, indicating that climate change is causing the habitat to break apart into smaller, more isolated patches. However, this rate of increase in fragmentation reduces to 17.75% in SSP245 (2061–2080) and 13.55% in SSP585 (2061–2080). This deceleration in the rate of fragmentation may be attributed to the complete loss of some viable patches, which reduces the total number of fragments, indicating a critical threshold of habitat destruction. The increase in NP is also linked to the disintegration of suitable habitats, which has led to higher PD and ED in future scenarios while decreasing the LPI. This indicates that the fragmented patches have decreased in size and become more restricted for the species. Furthermore, the decrease in AI exacerbates the situation, as it signifies that these small, disintegrated patches will be far from each other, thus hindering the movement of the species. This fragmentation is particularly concerning because it could restrict gene flow among different populations of the Red Goral, potentially increasing inbreeding rates as these isolated patches are less likely to support viable populations over the long term [19,74]. Therefore, the identification and protection of corridor connectivity among fragmented habitats and PAs are crucial for the conservation of this species, as ensuring such connectivity will help maintain genetic diversity and facilitate the natural movement of the species, mitigating some of the adverse effects of habitat fragmentation induced by climate change.

The study assessed the corridors among the PAs within the Red Goral's IUCN extent, identifying eight significant connectivity pathways (Table 6). It revealed that the highest mean connectivity (0.0583) was found between DWLS and MWLS in the present scenario, followed closely by the corridor between MNP and DWLS, with a mean connectivity of 0.0538. Conversely, the lowest connectivity was observed between KWLS and NNP, with a mean suitability score of 0.0172. This is particularly intriguing given their close proximity, highlighting the necessity of landscape-level connectivity. Furthermore, important transboundary corridors, such as the one between HkNP and the TPRYPA, with a mean connectivity score of 0.0493, and the corridor between NNP and HpWLS, with a mean connectivity score of 0.0332, were identified, which necessitates transboundary cooperation and collaboration. However, future scenarios project a decline in crucial corridors due to climatic shifts in the landscape level for this species. Notably, a significant decrease of over 25% in mean connectivity is anticipated between YRSWLS and MNP, whereas the corridor between HkNP and TPRYPA is projected to experience the least decline in mean connectivity, up to 1%. This minimal decline witnessed between HkNP and TPRYPA is likely due to their close proximity, underscoring the importance of maintaining continuous transboundary landscapes. The findings underscore the paramount importance of implementing conservation strategies to maintain and augment connectivity between the PAs to ensure the long-term viability of the Red Goral and other species that are dependent on these temperate mountain habitats.

Thus, the PAs identified by this study require enhanced protection through implementation by governmental agencies; specifically, DWLS and HkNP, having the highest suitable extent, need immediate attention for conservation efforts. Additionally, considering the current sociopolitical situation in Myanmar, wildlife conservation efforts are severely jeopardized [75]. As a result, the study strongly urges international communities to support conservation efforts in these regions by providing aid to indigenous people and non-governmental organizations. In the Indian context, the construction of new hydropower dams in Arunachal Pradesh, particularly in Dibang Valley, has led to the clearing of many hectares of forest land. The study suggests conducting impartial environmental impact assessments (EIAs) before initiating any construction and insists that construction agencies adhere to EIA regulations while supporting local conservation efforts. Furthermore, traditional practices like trophy hunting and bushmeat hunting, embedded in tribal cultures due to historically high forest cover and low human presence, must be addressed. The recent development in these areas calls for awareness programs to reduce hunting practices and encourage engagement in conservation work supported by the growing tourism industry. Moreover, the TPRYPA in China requires a landscape-level assessment to identify habitats for this species that are not included in the IUCN extent. Furthermore, it is crucial to monitor and prevent the trade of body parts of Red Goral and other ungulates for medicine in the international market through these significant wildlife trafficking routes. Moreover, it is necessary to encourage the establishment of more captive breeding facilities for both breeding and research, especially since there is currently only one such facility at the Shanghai Zoo in China [76]. In order to effectively safeguard transboundary corridors, it is essential to maintain their continuity. When disruptions occur, it is important to implement measures, such as constructing underpasses or overpasses, to ensure better habitat connectivity. Achieving this goal requires collaboration and cooperation among the three countries involved. Moreover, forming community reserves with support from both international and national agencies is necessary for the long-term preservation of these vital corridors. Furthermore, a reassessment by the IUCN is also needed, potentially considering the species under the "Endangered" category, as this reclassification would help enhance protection and attention towards the species, which occupies a very restricted range and is significantly affected by climate change. Therefore, implementing these strategic measures will substantially enhance the conservation efforts for the Red Goral and other ungulate species.

5. Conclusions

In conclusion, this study offers a crucial, comprehensive analysis of distribution, habitat suitability, and connectivity, addressing significant gaps in understanding the vulnerable Red Goral. The findings reveal a troubling scenario with a very limited area identified as a suitable habitat under current conditions, and projections indicate a further decline due to climate change. This situation is exacerbated by the limited protection within designated areas, with only a small portion of suitable habitats falling under PAs. The identification of key PAs, such as DWLS in India and HkNP in Myanmar, as priority regions for conservation efforts is particularly significant for this species. Additionally, the study emphasizes the importance of maintaining and enhancing connectivity among fragmented habitats and transboundary corridors to mitigate the adverse effects of habitat fragmentation and climate change. However, a more exhaustive field survey on the species outside of the known IUCN extent will help to expand the training area for further SDM research to understand the climate change shift of the species on a broader scale across this temperate mountainous region. The recommendations for immediate conservation actions, including international support, stricter adherence to EIA, and addressing traditional hunting practices, are essential steps towards safeguarding the Red Goral. The insights provided a vital foundation for future research and conservation initiatives, aiming to ensure the long-term survival of this ungulate species in its natural habitat.

Supplementary Materials: The following supporting information can be downloaded at: https://www.mdpi.com/article/10.3390/biology13090667/s1, Figure S1. Figure showing the correlation between the covariates chosen for the final model for *N. baileyi*; Figure S2. Evaluation matrix performance across model runs for *N. baileyi*. Brown—represents the correlation coefficient among the four different models; Yellow—represents the proportion of deviance explained; Green—represents the proportion of correctly classified; Blue—represents Area under Curve (AUC); and Pink—represents True Skill Statistic. Figure S3. Confusion matrixes, model calibration plots, and residual plots for *N. baileyi*. Row 1 represents spatial pattern of residuals, where the color ramp indicates the magnitude of deviance and size represents the quantity. Row 2 represents the model calibration plot across all four different models for cross-validation split. Row 3 represents the confusion matrix for all four models, plotted by observed vs. predicted, where the color ramp from the lowest value of 0% (white) to 100% (red) indicates the quantification of particular pair types. Column a. represents plots for BRT, Column b. represents plots for GLM, Column c. represents plots for MARS, and Column d. represents plots for MaxEnt. Figure S4. The response curves of the covariates selected by each of the participating ensemble models for *N. baileyi*. (A) BRT, (B) GLM, (C) MARS, and (D) MaxEnt. Table S1. The details of the models used in the SAHM (Software for Assisted Habitat Modelling) package in VisTrails software; Table S2. The total suitable habitat extent of *N. baileyi* in the present and future climate change scenarios within its IUCN extent.

Author Contributions: Conceptualization: I.A. and S.K.; methodology: I.A. and T.M.; software: I.A. and T.M.; validation: J.A. and H.-W.K.; formal analysis: I.A., T.M. and S.K.; investigation: T.M. and J.A.; resources: H.-W.K. and S.K.; data curation: I.A., T.M. and J.A.; writing—original draft: I.A. and S.K.; writing—review and editing: T.M. and H.-W.K.; visualization: J.A. and S.K.; supervision: T.M. and S.K.; project administration: H.-W.K. and S.K.; funding acquisition: H.-W.K. and S.K. All authors have read and agreed to the published version of the manuscript.

Funding: This research was supported by the Pukyong National University 2023 Research Grant (202315370001).

Institutional Review Board Statement: Not applicable.

Informed Consent Statement: Not applicable.

Data Availability Statement: Data used for the analysis were sourced from open-access resources.

Acknowledgments: The authors wish to express their sincere gratitude to Ravi Mekola for providing the photograph of this elusive Red Goral featured in this study. They also extend their heartfelt thanks to Reema Abedin, Barnita Dey Kundu, Sabana Sargam Rahman, Rezina Mihu, and field guides (Dipul Duarah, Souravjyoti Boruah, and Drama Mekola) for their valuable information and support throughout the research. T.M. thanks the Department of Science & Technology, Government of India (Sanction No: DST/INSPIRE/04/2021/001149).

Conflicts of Interest: On behalf of all co-authors, it is certified that the authors do not have any conflicts of interest with respect to the publication of this manuscript, and all co-authors approved the final manuscript.

References

1. Ahmad, S.; Yang, L.; Khan, T.U.; Wanghe, K.; Li, M.; Luan, X. Using an ensemble modelling approach to predict the potential distribution of Himalayan gray goral (*Naemorhedus goral bedfordi*) in Pakistan. *Glob. Ecol. Conserv.* **2020**, *21*, e00845. [CrossRef]
2. Crooks, K.R.; Burdett, C.L.; Theobald, D.M.; King, S.R.B.; Di Marco, M.; Rondinini, C.; Boitani, L. Quantification of habitat fragmentation reveals extinction risk in terrestrial mammals. *Proc. Natl. Acad. Sci. USA* **2017**, *114*, 7635–7640. [CrossRef]
3. Santangeli, A.; Mammola, S.; Lehikoinen, A.; Rajasärkkä, A.; Lindén, A.; Saastamoinen, M. The effects of protected areas on the ecological niches of birds and mammals. *Sci. Rep.* **2022**, *12*, 11601. [CrossRef]
4. Meehl, G.A.; Covey, C.; Delworth, T.; Latif, M.; McAvaney, B.; Mitchell, J.F.B.; Stouffer, R.J.; Taylor, K.E. The WCRP CMIP3 Multimodel Dataset: A New Era in Climate Change Research. *Bull. Am. Meteorol. Soc.* **2007**, *88*, 1383–1394. [CrossRef]
5. Knight, J. Scientists' warning of the impacts of climate change on mountains. *PeerJ* **2022**, *10*, e14253. [CrossRef] [PubMed]
6. Chan, W.P.; Lenoir, J.; Mai, G.S.; Kuo, H.C.; Chen, I.C.; Shen, S.F. Climate velocities and species tracking in global mountain regions. *Nature* **2024**, *629*, 114–120. [CrossRef] [PubMed]
7. Brivio, F.; Apollonio, M.; Anderwald, P.; Filli, F.; Bassano, B.; Bertolucci, C.; Grignolio, S. Seeking temporal refugia to heat stress: Increasing nocturnal activity despite predation risk. *Proc. R. Soc. B* **2024**, *291*, 20231587. [CrossRef]
8. Anderwald, P.; Campell Andri, S.; Palme, R. Reflections of ecological differences? Stress responses of sympatric Alpine chamois and red deer to weather, forage quality, and human disturbance. *Ecol. Evol.* **2021**, *11*, 15740–15753. [CrossRef]
9. Chatterjee, S.; Goswami, A.; Scotese, C.R. The longest voyage: Tectonic, magmatic, and paleoclimatic evolution of the Indian plate during its northward flight from Gondwana to Asia. *Gondwana Res.* **2013**, *23*, 238–267. [CrossRef]
10. Haq, S.M.; Waheed, M.; Ahmad, R.; Bussmann, R.W.; Arshad, F.; Khan, A.M.; Casini, R.; Alataway, A.; Dewidar, A.Z.; Elansary, H.O. Climate Change and Human Activities, the Significant Dynamic Drivers of Himalayan Goral Distribution (*Naemorhedus goral*). *Biology* **2023**, *12*, 610. [CrossRef]
11. Huntley, B.; Berry, P.M.; Cramer, W.; McDonald, A.P. Special Paper: Modelling Present and Potential Future Ranges of Some European Higher Plants Using Climate Response Surfaces. *J. Biogeogr.* **1995**, *22*, 967. [CrossRef]
12. Pearson, R.G.; Dawson, T.P. Predicting the Impacts of Climate Change on the Distribution of Species: Are Bioclimate Envelope Models Useful? *Glob. Ecol. Biogeogr.* **2003**, *12*, 361–371. [CrossRef]
13. Angert, A.L.; LaDeau, S.L.; Ostfeld, R.S. Climate change and species interactions: Ways forward. *Ann. N. Y. Acad. Sci.* **2013**, *1297*, 1–7. [CrossRef]
14. Zhou, W.; Wang, M.; Gao, K.; Gao, H.; Wei, F.; Nie, Y. Behavioural thermoregulation by montane ungulates under climate warming. *Divers. Distrib.* **2022**, *28*, 2229–2238. [CrossRef]
15. Lehikoinen, P.; Tiusanen, M.; Santangeli, A.; Rajasärkkä, A.; Jaatinen, K.; Valkama, J.; Virkkala, R.; Lehikoinen, A. Increasing protected area coverage mitigates climate-driven community changes. *Biol. Conserv.* **2021**, *253*, 108892. [CrossRef]
16. Nijhawan, S. *Naemorhedus baileyi* (Amended Version of 2020 Assessment). The IUCN Red List of Threatened Species 2020, e.T14294A179947455. Available online: https://www.iucnredlist.org/species/14294/179947455 (accessed on 30 June 2024).
17. Hayman, R.W. The Red Goral of the North-East Frontier Region. *Proc. Zool. Soc. Lond.* **1961**, *136*, 317–323. [CrossRef]

18. Wiens, J.J. Climate-Related Local Extinctions Are Already Widespread among Plant and Animal Species. *PLoS Biol.* **2016**, *14*, e2001104. [CrossRef] [PubMed]
19. Niyogi, R.; Sarkar, M.S.; Hazra, P.; Rahman, M.; Banerjee, S.; John, R. Habitat Connectivity for the Conservation of Small Ungulates in a Human-Dominated Landscape. *ISPRS Int. J. Geo-Inf.* **2021**, *10*, 180. [CrossRef]
20. Khadka, K.K.; James, D.A. Modeling and Mapping the Current and Future Climatic-Niche of Endangered Himalayan Musk Deer. *Ecol. Inform.* **2017**, *40*, 1–7. [CrossRef]
21. Bao, S.; Yang, F. Identification of Potential Habitats and Adjustment of Protected Area Boundaries for Large Wild Herbivores in the Yellow-River-Source National Park, China. *Land* **2024**, *13*, 186. [CrossRef]
22. Ebrahimi, E.; Sayahnia, R.; Ranjbaran, Y.; Vaissi, S.; Ahmadzadeh, F. Dynamics of threatened mammalian distribution in Iran's protected areas under climate change. *Mamm. Biol.* **2021**, *101*, 759–774. [CrossRef]
23. Feng, B.; Xiao, Y.; Hu, L.; Yang, X.; Dong, X.; Zhang, J.; Yang, Z.; Qi, D.; Zhou, C.; Bai, W. Predicted Climate Change Impacts on Distribution and Habitat Structure of Forest Ungulates in Southwest China. *Ecosyst. Health Sustain.* **2024**, *10*, 0173. [CrossRef]
24. Ye, X.; Yu, X.; Yu, C.; Tayibazhaer, A.; Xu, F.; Skidmore, A.K.; Wang, T. Impacts of Future Climate and Land Cover Changes on Threatened Mammals in the Semi-Arid Chinese Altai Mountains. *Sci. Total Environ.* **2018**, *612*, 775–787. [CrossRef]
25. Loiseau, N.; Mouquet, N.; Casajus, N.; Grenié, M.; Guéguen, M.; Maitner, B.; Mouillot, D.; Ostling, A.; Renaud, J.; Tucker, C.; et al. Global Distribution and Conservation Status of Ecologically Rare Mammal and Bird Species. *Nat. Commun.* **2020**, *11*, 5071. [CrossRef] [PubMed]
26. Mohammadi, A.; Almasieh, K.; Vaissi, S. Ungulates Conservation in the Face of Human Development: Mining and Roads' Influences on Habitat and Connectivity in Iran's Central Plateau. *Ecol. Inform.* **2024**, *81*, 102656. [CrossRef]
27. Elith, J.; Leathwick, J.R. Species Distribution Models: Ecological Explanation and Prediction Across Space and Time. *Annu. Rev. Ecol. Evol. Syst.* **2009**, *40*, 677–697. [CrossRef]
28. Di Febbraro, M.; Bosso, L.; Fasola, M.; Santicchia, F.; Aloise, G.; Lioy, S.; Tricarico, E.; Ruggieri, L.; Bovero, S.; Mori, E.; et al. Different facets of the same niche: Integrating citizen science and scientific survey data to predict biological invasion risk under multiple global change drivers. *Glob. Chang. Biol.* **2023**, *29*, 5509–5523. [CrossRef]
29. Hao, T.; Elith, J.; Lahoz-Monfort, J.J.; Guillera-Arroita, G. Testing Whether Ensemble Modelling Is Advantageous for Maximising Predictive Performance of Species Distribution Models. *Ecography* **2020**, *43*, 549–558. [CrossRef]
30. Jamali, F.; Amininasab, S.M.; Taleshi, H.; Madadi, H. Using an ensemble modeling to predict the potential distribution and habitat suitability of caracal (*Caracal caracal*) in southwestern Iran. *Glob. Ecol. Conserv.* **2024**, *52*, e02968. [CrossRef]
31. Zhang, J.; Jiang, F.; Li, G.; Qin, W.; Wu, T.; Xu, F.; Hou, Y.; Song, P.; Cai, Z.; Zhang, T. The Four Antelope Species on the Qinghai-Tibet Plateau Face Habitat Loss and Redistribution to Higher Latitudes under Climate Change. *Ecol. Indic.* **2021**, *123*, 107337. [CrossRef]
32. Li, Z.; Khattak, R.H.; Han, X.; Zhang, N.; Wu, J.; Liu, Z.; Teng, L. Distribution Update of Water Deer (*Hydropotes inermis*) and Prediction of Their Potential Distribution in Northeast China. *Sci. Rep.* **2023**, *13*, 5610. [CrossRef] [PubMed]
33. Bachman, S.; Moat, J.; Hill, A.W.; de la Torre, J.; Scott, B. Supporting red list threat assessments with GeoCAT: Geospatial conservation assessment tool. *ZooKeys* **2011**, *150*, 117–126. [CrossRef] [PubMed]
34. Su, J.; Aryal, A.; Hegab, I.M.; Shrestha, U.B.; Coogan, S.C.P.; Sathyakumar, S.; Dalannast, M.; Dou, Z.; Suo, Y.; Dabu, X.; et al. Decreasing Brown Bear (*Ursus arctos*) Habitat Due to Climate Change in Central Asia and the Asian Highlands. *Ecol. Evol.* **2018**, *8*, 11887–11899. [CrossRef] [PubMed]
35. Buchhorn, M.; Bertels, L.; Smets, B.; De Roo, B.; Lesiv, M.; Tsendbazar, N.E.; Masiliunas, D.; Li, L. *Copernicus Global Land Operations "Vegetation and Energy": Algorithm Theoretical Basis Document—Moderate Dynamic Land Cover 100 m*, version 3; Copernicus Global Land Service: Barcelona, Spain, 2020. [CrossRef]
36. Mukherjee, T.; Sharma, L.K.; Kumar, V.; Sharief, A.; Dutta, R.; Kumar, M.; Joshi, B.D.; Thakur, M.; Venkatraman, C.; Chandra, K. Adaptive Spatial Planning of Protected Area Network for Conserving the Himalayan Brown Bear. *Sci. Total Environ.* **2021**, *754*, 142416. [CrossRef]
37. Morisette, J.T.; Jarnevich, C.S.; Holcombe, T.R.; Talbert, C.B.; Ignizio, D.; Talbert, M.K.; Silva, C.; Koop, D.; Swanson, A.; Young, N.E. VisTrails SAHM: Visualization and Workflow Management for Species Habitat Modeling. *Ecography* **2013**, *36*, 129–135. [CrossRef]
38. Warren, D.L.; Glor, R.E.; Turelli, M. ENMTools: A Toolbox for Comparative Studies of Environmental Niche Models. *Ecography* **2010**, *33*, 607–611. [CrossRef]
39. Guisan, A.; Zimmermann, N.E.; Elith, J.; Graham, C.H.; Phillips, S.; Peterson, A.T. What matters for predicting the occurrences of trees: Techniques, data, or species' characteristics? *Ecol. Monogr.* **2007**, *77*, 615–630. [CrossRef]
40. Miller, J. Species Distribution Modeling. *Geogr. Compass* **2010**, *4*, 490–509. [CrossRef]
41. Talbert, C.B.; Talbert, M.K. User Manual for SAHM Package for VisTrails. 2012. Available online: https://pubs.usgs.gov/publication/70118102 (accessed on 25 June 2024).
42. Cohen, J. Weighted kappa: Nominal scale agreement provision for scaled disagreement or partial credit. *Psychol. Bull.* **1968**, *70*, 213–220. [CrossRef]
43. Allouche, O.; Tsoar, A.; Kadmon, R. Assessing the accuracy of species distribution models: Prevalence, kappa and the true skill statistic (TSS). *J. Appl. Ecol.* **2006**, *43*, 1223–1232. [CrossRef]

44. Phillips, S.J.; Elith, J. POC Plots: Calibrating Species Distribution Models with Presence-Only Data. *Ecology* **2010**, *91*, 2476–2484. [CrossRef] [PubMed]
45. Jiménez-Valverde, A.; Acevedo, P.; Barbosa, A.M.; Lobo, J.M.; Real, R. Discrimination Capacity in Species Distribution Models Depends on the Representativeness of the Environmental Domain. *Glob. Ecol. Biogeogr.* **2013**, *22*, 508–516. [CrossRef]
46. O'Neill, B.C.; Kriegler, E.; Riahi, K.; Ebi, K.L.; Hallegatte, S.; Carter, T.R.; Mathur, R.; van Vuuren, D.P. A New Scenario Framework for Climate Change Research: The Concept of Shared Socioeconomic Pathways. *Clim. Chang.* **2014**, *122*, 387–400. [CrossRef]
47. Riahi, K.; van Vuuren, D.P.; Kriegler, E.; Edmonds, J.; O'Neill, B.C.; Fujimori, S.; Bauer, N.; Calvin, K.; Dellink, R.; Fricko, O.; et al. The Shared Socioeconomic Pathways and Their Energy, Land Use, and Greenhouse Gas Emissions Implications: An Overview. *Glob. Environ. Chang.* **2017**, *42*, 153–168. [CrossRef]
48. Andrews, M.B.; Ridley, J.K.; Wood, R.A.; Andrews, T.; Blockley, E.W.; Booth, B.; Burke, E.; Dittus, A.J.; Florek, P.; Gray, L.J.; et al. Historical Simulations with HadGEM3-GC3.1 for CMIP6. *J. Adv. Model. Earth Syst.* **2020**, *12*, e2019MS001995. [CrossRef]
49. Desmet, Q.; Ngo-Duc, T. A novel method for ranking CMIP6 global climate models over the southeast Asian region. *Int. J. Clim.* **2022**, *42*, 97–117. [CrossRef]
50. Norgate, M.; Tiwari, P.R.; Das, S.; Kumar, D. On the Heat Waves over India and Their Future Projections under Different SSP Scenarios from CMIP6 Models. *Int. J. Clim.* **2024**, *44*, 973–995. [CrossRef]
51. Allen, B.J.; Hill, D.J.; Burke, A.M.; Clark, M.; Marchant, R.; Stringer, L.C.; Williams, D.R.; Lyon, C. Projected future climatic forcing on the global distribution of vegetation types. *Philos. Trans. R. Soc. B Biol. Sci.* **2024**, *379*, 20230011. [CrossRef]
52. Atsawawaranunt, K.; Whibley, A.; Cain, K.E.; Major, R.E.; Santure, A.W. Projecting the current and potential future distribution of New Zealand's invasive sturnids. *Biol. Invasions* **2024**, *26*, 1345–1366. [CrossRef]
53. Abedin, I.; Mukherjee, T.; Kim, A.R.; Kim, H.-W.; Kang, H.-E.; Kundu, S. Distribution Model Reveals Rapid Decline in Habitat Extent for Endangered Hispid Hare: Implications for Wildlife Management and Conservation Planning in Future Climate Change Scenarios. *Biology* **2024**, *13*, 198. [CrossRef]
54. Liang, G.; Niu, H.; Li, Y. A multi-species approach for protected areas ecological network construction based on landscape connectivity. *Glob. Ecol. Conserv.* **2023**, *46*, e02569. [CrossRef]
55. Abedin, I.; Mukherjee, T.; Kim, A.R.; Lee, S.R.; Kim, H.W.; Kundu, S. Fragile futures: Evaluating habitat and climate change response of hog badgers (Mustelidae: Arctonyx) in the conservation landscape of mainland Asia. *Ecol. Evol.* **2024**, *14*, e70160. [CrossRef] [PubMed]
56. Zungu, M.M.; Maseko, M.S.T.; Kalle, R.; Ramesh, T.; Downs, C.T. Effects of landscape context on mammal richness in the urban forest mosaic of EThekwini Municipality, Durban, South Africa. *Glob. Ecol. Conserv.* **2020**, *21*, e00878. [CrossRef]
57. McGarigal, K.; Marks, B.J. FRAGSTATS: Spatial Pattern Analysis Program for Quantifying Landscape Structure. In *General Technical Report—PNW-GTR-351*; U.S. Department of Agriculture, Forest Service, Pacific Northwest Research Station: Portland, OR, USA, 1995; 122p. [CrossRef]
58. Midha, N.; Mathur, P.K. Assessment of Forest Fragmentation in the Conservation Priority Dudhwa Landscape, India Using FRAGSTATS Computed Class Level Metrics. *J. Indian Soc. Remote Sens.* **2010**, *38*, 487–500. [CrossRef]
59. Barwicka, S.; Milecka, M.; Chmielewski, S.; Olszewska-Guizzo, A.; Masoudi, M.; Szczepańska, M. The Use of Selected Landscape Metrics to Evaluate the Transformation of the Rural Landscape as a Result of the Development of the Mining Function—A Case Study of the Puchaczów Commune. *Sustainability* **2021**, *13*, 12279. [CrossRef]
60. Abedin, I.; Mukherjee, T.; Kang, H.E.; Yoon, T.H.; Kim, H.W.; Kundu, S. Unraveling the unknown: Adaptive spatial planning to enhance climate resilience for the endangered Swamp Grass-babbler (*Laticilla cinerascens*) with habitat connectivity and complexity approach. *Heliyon* **2024**, *10*, e30273. [CrossRef]
61. Kundu, S.; Mukherjee, T.; Kamalakannan, M.; Barhadiya, G.; Ghosh, C.; Kim, H.-W. Matrilineal Phylogeny and Habitat Suitability of the Endangered Spotted Pond Turtle (*Geoclemys hamiltonii*; Testudines: Geoemydidae): A Two-Dimensional Approach to Forecasting Future Conservation Consequences. *PeerJ* **2023**, *11*, e15975. [CrossRef]
62. Wang, F.; McShea, W.J.; Wang, D.; Li, S.; Zhao, Q.; Wang, H.; Lu, Z. Evaluating Landscape Options for Corridor Restoration between Giant Panda Reserves. *PLoS ONE* **2014**, *9*, e105086. [CrossRef]
63. McRae, B.H.; Dickson, B.G.; Keitt, T.H.; Shah, V.B. Using Circuit Theory to Model Connectivity in Ecology, Evolution, and Conservation. *Ecology* **2008**, *89*, 2712–2724. [CrossRef]
64. Ceballos, G.; Ehrlich, P.R.; Soberón, J.; Salazar, I.; Fay, J.P. Global mammal conservation: What must we manage? *Science* **2005**, *309*, 603–607. [CrossRef]
65. Newbold, T.; Hudson, L.N.; Hill, S.L.L.; Contu, S.; Lysenko, I.; Senior, R.A.; Börger, L.; Bennett, D.J.; Choimes, A.; Collen, B.; et al. Global Effects of Land Use on Local Terrestrial Biodiversity. *Nature* **2015**, *520*, 45–50. [CrossRef] [PubMed]
66. Luo, Z.; Jiang, Z.; Tang, S. Impacts of Climate Change on Distributions and Diversity of Ungulates on the Tibetan Plateau. *Ecol. Appl.* **2015**, *25*, 24–38. [CrossRef] [PubMed]
67. Bhasin, A.; Ghosal, S.; Raina, P.; Hore, U. Climate change impacts on high altitude wildlife distribution: Predicting range shifts for four ungulates in Changthang, eastern Ladakh. *Ecol. Front.* **2024**, *44*, 365–380. [CrossRef]
68. Suggitt, A.J.; Wheatley, C.J.; Aucott, P.; Beale, C.M.; Fox, R.; Hill, J.K.; Isaac, N.J.B.; Martay, B.; Southall, H.; Thomas, C.D.; et al. Linking Climate Warming and Land Conversion to Species' Range Changes across Great Britain. *Nat. Commun.* **2023**, *14*, 6759. [CrossRef]
69. Cavallini, P. Survey of the goral *Nemorhaedus goral* (Hardwicke) in Himachal Pradesh. *J. Bombay Nat. Hist. Soc.* **1992**, *89*, 302–307.

70. Hirzel, A.H.; Le Lay, G. Habitat Suitability Modelling and Niche Theory. *J. Appl. Ecol.* **2008**, *45*, 1372–1381. [CrossRef]
71. Parmesan, C.; Yohe, G. A Globally Coherent Fingerprint of Climate Change Impacts across Natural Systems. *Nature* **2003**, *421*, 37–42. [CrossRef]
72. Elith, J.; Phillips, S.J.; Hastie, T.; Dudík, M.; Chee, Y.E.; Yates, C.J. A statistical explanation of MaxEnt for ecologists. *Divers. Distrib.* **2011**, *17*, 43–57. [CrossRef]
73. Taubert, F.; Fischer, R.; Groeneveld, J.; Lehmann, S.; Müller, M.S.; Rödig, E.; Wiegand, T.; Huth, A. Global Patterns of Tropical Forest Fragmentation. *Nature* **2018**, *554*, 519–522. [CrossRef]
74. Jangtarwan, K.; Kamsongkram, P.; Subpayakom, N.; Sillapaprayoon, S.; Muangmai, N.; Kongphoemph, A.; Wongsodchuen, A.; Intapan, S.; Chamchumroon, W.; Safoowong, M.; et al. Predictive Genetic Plan for a Captive Population of the Chinese Goral (*Naemorhedus griseus*) and Prescriptive Action for Ex Situ and In Situ Conservation Management in Thailand. *PLoS ONE* **2020**, *15*, e0234064. [CrossRef]
75. IUCN. Nature Conservation in Times of Conflict: Myanmar. 2013. Available online: https://www.iucn.nl/en/story/nature-conservation-in-times-of-conflict-myanmar/ (accessed on 11 July 2024).
76. Yuan, Y.; Huang, K.; Liu, Q. Population Status and Genetic Analysis of Captive Red Goral (*Naemorhedus baileyi*) in Shanghai Zoo, China. *Folia Zool.* **2019**, *68*, 285–293. [CrossRef]

biology

MDPI

Article

Testing the Resilience, Physiological Plasticity and Mechanisms Underlying Upper Temperature Limits of Antarctic Marine Ectotherms

Simon A. Morley [1,*], Amanda E. Bates [2], Melody S. Clark [1], Elaine Fitzcharles [1], Rebecca Smith [1], Rose E. Stainthorp [1,3] and Lloyd S. Peck [1]

[1] British Antarctic Survey, Natural Environment Research Council, Cambridge CB3 0ET, UK; mscl@bas.ac.uk (M.S.C.); emfi@bas.ac.uk (E.F.); rebsmi@bas.ac.uk (R.S.); rosestainthorp@gmail.com (R.E.S.); lspe@bas.ac.uk (L.S.P.)

[2] Department of Biology, University of Victoria, P.O. Box 1700, Victoria, BC V8W 2Y2, Canada; amandabates@uvic.ca

[3] National Oceanography Centre, University of Southampton, Southampton SO14 3ZH, UK

* Correspondence: smor@bas.ac.uk

Simple Summary: Antarctic marine invertebrates live in the constant of the Southern Ocean and are characterised by sensitivity to small increases in temperature. We conducted a series of aquarium experiments that tested this ability and found species-specific responses to warming. We found that some species were able to survive for many months at up to 10 °C, a temperature which is up to 4 °C warmer than previously recorded. We found that the survivors of three species had adjusted their biological systems (acclimated) and were better able to survive additional rapid warming, but one anemone species did not elevate its upper temperature limit, even though it survived for 270 days at 6 °C. There were also species-specific effects of increasing oxygen concentration on long-term survival to elevated temperatures, with extended, no change, or reduced survival duration all found in different species. Thermal sensitivity is clearly the product of multiple ecological and physiological capacities, and this diversity of response needs further investigation and interpretation to improve our ability to predict future patterns of biodiversity.

Abstract: Antarctic marine ectotherms live in the constant cold and are characterised by limited resilience to elevated temperature. Here we tested three of the central paradigms underlying this resilience. Firstly, we assessed the ability of eight species, from seven classes representing a range of functional groups, to survive, for 100 to 303 days, at temperatures 0 to 4 °C above previously calculated long-term temperature limits. Survivors were then tested for acclimation responses to acute warming and acclimatisation, in the field, was tested in the seastar *Odontaster validus* collected in different years, seasons and locations within Antarctica. Finally, we tested the importance of oxygen limitation in controlling upper thermal limits. We found that four of 11 species studied were able to survive for more than 245 days (245–303 days) at higher than previously recorded temperatures, between 6 and 10 °C. Only survivors of the anemone *Urticinopsis antarctica* did not acclimate CT_{max} and there was no evidence of acclimatisation in *O. validus*. We found species-specific effects of mild hyperoxia (30% oxygen) on survival duration, which was extended (two species), not changed (four species) or reduced (one species), re-enforcing that oxygen limitation is not universal in dictating thermal survival thresholds. Thermal sensitivity is clearly the product of multiple ecological and physiological capacities, and this diversity of response needs further investigation and interpretation to improve our ability to predict future patterns of biodiversity.

Keywords: Antarctic; Southern Ocean; resilience; physiology; acclimation; acclimatisation; climate change; warming

1. Introduction

Understanding the mechanisms underlying species' resilience to environmental perturbations is key to predictions of future patterns of biodiversity. Physiological plasticity is a key element of species' capacity to survive in fluctuating environments [1]. This plasticity consists of biochemical "resistance" pathways that can be switched on or off (e.g., anaerobic metabolism [2]; heat shock proteins [3]), but also the capacity to adjust biochemical pathways in response to altered environmental conditions [4]. Acclimation, the phenotypic response to a single stressor, often in response to laboratory-induced stress, and acclimatisation, the response to multiple stressors, often under field conditions [5], can occur over scales from hours to months. Physiological plasticity is an important component of species' capacity, providing resilience to both short-term acute heating (e.g., heatwaves [6]) and long-term change (e.g., climate change [7]). For acclimation to be beneficial, it requires predictable "zeitgebers", i.e., cues that are predictive of future conditions, ensuring that the cost of any physiological remodelling is outweighed by the benefit [8]. The benefit of physiological acclimation has been argued [9], as has its capacity to buffer against global warming [10,11].

Physiological responses vary over geographic and temporal scales, with many dimensions known to govern the evolution of thermal tolerance [1]. At one extreme for marine life, Antarctic marine ectotherms have evolved in the constant cold of the Southern Ocean for tens of millions of years and are predicted to be sensitive to warming, but also have limited physiological plasticity [12,13]. Antarctic marine ectotherms have atypical thermal responses, including the expression of heat shock proteins, an almost ubiquitous protection of proteins to acute warming (HSP [14]), failure to select preferred temperatures [15] and in some species, lack of escape response to dangerously high temperatures [15,16]. This combined evidence supports that universal protective responses to heat regimes experienced in the Antarctic could be missing in these taxa. Recent studies have shown that the responses of Antarctic marine ectotherms are much more complicated. For example, longer-term incubations can lead to acclimation of some Antarctic marine ectotherms [11] but not others [11,17], and cellular responses, including HSP expression, vary markedly between species and treatment regimens [18]. In the cold of the Southern Ocean all physiological processes take longer, and acclimation in Antarctic marine ectotherms can take many months and can occur more slowly in invertebrates than fish [11].

Mixed responses indicate that the mechanisms underlying thermal sensitivity are complex. Oxygen and capacity limitation of thermal tolerance (OCLTT [2]) is one of the mechanisms whose applicability has been questioned in the Antarctic [1], but also more broadly with studies finding mixed responses both inside and outside the Southern Ocean [19–21].

Here we synthesise a series of experiments conducted from 2006 to 2015 that investigated both the acclimation and acclimatisation capacity of selected Antarctic marine ectotherms. Previous estimates of upper thermal limits for the long-term survival of Antarctic marine ectotherms were extrapolated to temperatures between 1 and 6 °C for incubations of several months [12]. To test if 6 °C represents the long-term limit for survival, species from a range of phyla were selected to represent a range of functional groups, the ophiuroids *Ophionotus victoriae* (Bell, 1902) and *Ophiura crassa* (Mortensen, 1936), the echinoid *Sterechinus neumayeri* (Meissner, 1900), the asteroid *Odontaster Validus* (Koehler, 1906), the holothuroid *Heterocucumis steineni* (Ludwig, 1898), the anthozoa *Urticinopsis antarctica* (Verrill, 1922), and the molluscs *Margarella antarctica* (E. Lamy, 1906) and *Laternula elliptica* (P.P. King, 1832), were incubated at 6 and 8 °C for up to 10 months. Survivors were tested for whole body acclimation to see if acute thermal limits (CT_{max}—the critical thermal maximum) had been elevated through exposure to these elevated temperatures. Acclimitisation of *O. validus* was tested by comparing CT_{max} from field fresh individuals just after collection from the shallow sea around the British Antarctic Survey's Rothera Research Station in spring, summer and winter seasons, and McMurdo Sound near Antarctica New Zealand's Scott Base. The aim was to test for spatio-temporal variation in CT_{max}. To test if elevated oxygen increased survival times at elevated temperatures, species were incubated long-term under normoxic (21%) and mildly hyperoxic (30%) oxygen saturation. The arthropod *Paraceradocus miersi* (Pfeffer, 1888), the ophiuroids *Ophionotus victoriae* and *O. crassa* were incubated at 3 °C and *S. neumayeri*, *O. Validus*, and the holothuroids *Cucumaria georgiana* (Lampert, 1886) and *H. steineni* were reared at 8 °C rising to 10 °C, under both normoxia and mild hyperoxia. In this way, this study further investigated the paradigms of limited physiological resilience, plasticity and the importance of oxygen setting upper temperature limits in Antarctic marine ectotherms, searching for exceptions and potential new insights into the underlying mechanisms.

2. Materials and Methods

2.1. Collections and Animal Husbandry

Experiments were conducted between 2006 and 2015 with individuals hand collected by SCUBA divers in the austral summer, from 6–15 m depth, near Rothera Research Station, Adelaide Island (67°34′25″ S, 68°08′00″ W). Common marine ectotherms were selected from different phyla to represent a range of functional groups, for which aquarium husbandry is well established and long-term temperature limits have been estimated under normoxia [12]. All animals remained submerged throughout the transfer from the sea to the flow-through aquarium system at the station. To control for size-dependent effects on survival, we selected individuals of a similar size within each species group at the start of the experiment, and only studied fully reproductive adults.

The 3.0 °C temperature –oxygen experiment was conducted in flow through aquaria at Rothera Research Station. In Rothera, the tanks had a constant exchange of seawater that was balanced to allow temperature and oxygen treatments to be maintained while preventing any build-up of metabolic waste. In all experiments, seawater chemistry was monitored every 2–3 days using Nutrafin aquarium test kits. Ammonia, nitrite, and nitrates were maintained well below 0.4, 0.2 and 5 mg L^{-1} to prevent toxicity from metabolic by-products.

All other animals were transported back to the UK in purpose-built, recirculating, temperature-controlled transport aquaria maintained at 0.0 ± 0.3 °C with an average 12L:12D photoperiod for the six-week passage. Once in the UK, the animals were kept in a temperature-controlled aquaria at 0.1 ± 0.1 °C until the start of experiments. In the UK, holding and experiments were conducted in recirculating tanks that were fitted with biological filters (EHEIM GmbH & Co KG, Stuttgart, Germany), UV sterilisers, bubbled air, and plastic lids. Water quality was maintained with a combination of biological filtration, protein skimming, and partial seawater exchanges (approximately 5–15% every 2–3 days), which was confirmed as detailed above. Throughout transport and experimental incubations all animals were fed, to excess, on small pieces of fish or crustaceans except for the filter feeders *C. georgiana*, *H. steineni*, and *L. elliptica* which were fed on a mixture of instant concentrated *Nanochloropsis* spp. and *Tetraselmis* spp. (ZM Fish Food, Winchester, UK) and algae growing naturally in the water.

In all experiments, functional survival (CT_{max}) was monitored daily in both treatment and control animals through the following standard procedures *cf* [22]. For all but *P. miersi*, the first sign that individuals were approaching their limit was the inability to remain attached to the side of the tank. These individuals were then tested daily for the absence of movement of arms, spines, and tube feet. For *P. miersi*, we used the absence of movement of appendages (uropods or pleopods). If an individual was non-responsive it was returned to the treatment tank and checked again 24 h later. If, after 24 h, there was still no response to external stimuli then the number of days from the start of the experiment to the first day with no response was recorded as the time taken for CT_{max} to occur. To account for any effect of size on mortality, the size of each individual was recorded.

2.2. Identification of Anemone Species

A small piece of one tentacle was removed from each anemone and preserved in 96% ethanol. DNA was extracted from each tentacle using the DNeasy Blood and Tissue kit (Qiagen, Manchester, UK) according to manufacturer's instructions. The cytochrome oxidase subunit I gene (COI) mitochondrial region was amplified using 1–2 μL extracted DNA and MyTaq DNA polymerase mix (30 μL reactions; Bioline UK (now Meridien Biosceince, London, UK), with 10 nmol each of the universal COI primers for invertebrates (LCO 1490 5′-GGTCAACAAATCATAAAGATATTGG-3′; HCO 2198 5′-TAAACTTCAGGGTGACCAAAAAATCA-3′) [23]. PCR conditions were 94 °C for 5 min, five cycles of 94 °C for 1 min, 45 °C for 1.5 min, 72 °C for 1.5 min, followed by 30 cycles of 94 °C for 1 min, 50 °C for 1 min, 72 °C for 1 min and a final elongation stage of 5 min at 72 °C. The COI fragments were bi-directionally sequenced by Source Bioscience (Cambridge, UK). The species identity of each individual was analysed using Blast sequence similarity searching of INSDC (International Nucleotide Sequence Database Collaboration) (https://www.insdc.org/, accessed on 12 March 2020).

2.3. Acclimation Experiments: Temperature Incubations

Temperatures were raised at the same rate in all experiments (0.3 ± 0.1 °C d^{-1}) until the incubation temperature was reached, and subsequently monitored daily (Figure S1). *O. victoriae*, *O. crassa*, *S. neumayeri*, *O. validus*, *H. steineni*, *U. antarctica*, *M. antarctica*, and *L. elliptica* were held at temperatures of 6.0 and 8.0 °C, temperatures that were 0 to 4 °C above the long-term limits calculated from experiments with different rates of warming [12]. Control animals were kept at 0.0 °C in the main holding aquarium.

Acute thermal limits: individuals that survived beyond the duration of incubations at both 6.0 and 8.0 °C were then tested to see if acute thermal limits were elevated due to acclimation to these elevated incubation temperatures. These data were compared with acute thermal limits conducted on control individuals that had been kept in the aquarium at 0.0 °C. For the acute temperature ramping trials individuals were transferred to plastic jacketed tanks (Engineering Design and Plastics Ltd., Cambridge, UK), whose jackets were filled with 25% *v*/*v* ethanol in water solution, that was heated or cooled by LTD20G thermocirculators (Grant Instruments, Royston, UK). Temperatures were increased at 1.0 ± 0.1 °C d^{-1} until the last animal was no longer responding to the stimuli detailed above.

2.4. Field Comparisons

Acclimatisation in field animals was tested for in only one species, the common starfish *O. validus*. To test for field acclimatisation the assessment of CT_{max} of freshly collected *O. validus* was repeated between 2006 and 2015 in both summer and winter. There was also a test of individuals from Scott Base (77°50′57″ S, 166°46′06″ E) to allow for regional differences and comparison with a site with even less annual and seasonal variation than Rothera. As above, temperatures were increased at 1.0 ± 0.1 °C d^{-1} until the last animal was no longer responding to the stimuli. For analysis, field temperature from the Rothera Time Series [24] was measured by CTD (conductivity, temperature, depth) casts at 15 m depth and was averaged for the month of animal collection.

2.5. Temperature–Oxygen Incubations

In temperature–oxygen incubations, temperatures were raised from ambient at the same rate of warming as in previous trials (0.3 ± 0.1 °C d^{-1}). In the 2010/11 experiments, individuals of *P. miersi*, and the ophiuroids *O. victoriae* and *O. crassa*, were incubated at 3.0 °C (Table 1), temperatures that were 1 °C above the calculated long-term limits for *O. victoriae* and that matched those for *P. miersii*, based on the results of the acclimation experiments and calculated long-term lethal limits [12]. *S. neumayeri*, *O. validus*, *C. georgiana*, and *H. steineni* were incubated long-term at 8.0 °C. After 148 days, survival of *H. steineni*, *S. neumayeri*, and *O. validus* was greater than 95%, and to increase the chance of testing for treatment effects, the incubation temperature was raised to 10.0 °C and maintained until the end of the experiment (351 days in total).

Table 1. Summary of the conditions (± SE) for temperature–oxygen incubations and a list of the species used in each experiment with sample size per tank in parentheses. A step-change in temperature in the 2014/15 experiment is indicated in parentheses. Species are (A) *Paraceradocus miersii* (30), *Ophionotus victoriae* (30), *Ophiura crassa* (30) and (B) *Sterechinus neumayeri* (10), *Odontaster validus* (10), *Cucumaria georgiana* (10), and *Heterocucumis steineni* (10).

Year	Duration (d)	Tank	Treatment	Temperature (°C)	Oxygen (% O_2)	Oxygen (mg/L) *
(A)						
2010/11	100	1	Control	0.9 ± 0.05	21.0 ± 0.1	13.9
		2	Normoxic	3.0 ± 0.01	21.0 ± 0.1	13.2
		3	Hyperoxic	3.0 ± 0.02	29.3 ± 0.5	18.3
(B)						
2014/15	351	1	Control	0.1 ± 0.01	21.0 ± 0.1	14.2
		2	Normoxic	8.6 ± 0.5 (10.0 ± 0.01)	21.0 ± 0.3	11.4
		3	Normoxic	8.1 ± 0.2 (9.9 ± 0.01)	21.5 ± 0.5	11.8
		4	Hyperoxic	8.1 ± 0.2 (10.0 ± 0.01)	30.5 ± 0.6	16.8
		5	Hyperoxic	8.0 ± 0.2 (10.0 ± 0.01)	31.3 ± 0.7	17.2

* Mean oxygen concentration (% O_2) was converted to mg L^{-1} using standard conversion tables (PreSens Precision Sensing GmbH, Regensburg, Germany).

Oxygen levels were controlled by vigorously bubbling gas mixtures with different oxygen concentrations through the water to ensure oxygen did not fall below the required saturation and that the tanks were sufficiently mixed. In the Rothera experiments, premixed cylinders with 30% oxygen, 390 ppm of CO_2 with the balance nitrogen, were used to control hyperoxia (BOC, London, UK). In the UK experiments, cylinders of 100% oxygen and air pumps were used, with the flow of gas mixtures into each tank controlled by needle valves and adjusted as necessary to ensure that required oxygen concentrations were maintained (Table 1). Stability of the oxygen concentration in all tanks was confirmed by daily monitoring of oxygen levels during set-up of the experiment using a pre-calibrated oxygen-sensitive foil and a Fibox-3 oxygen meter. We found that the stability of dissolved oxygen concentrations was such that after set-up, monitoring was only required when a gas cylinder was changed. In each experiment, controls were set up with animals held at normoxia (21% O_2) and a temperature of 0.0 ± 0.1 °C in the respective aquaria for the duration of the experiment with no mortality.

2.6. Statistical Analysis

We used the Kaplan-Meier estimator to compare survival between treatments as it takes account of individuals that survived beyond the duration of the incubation (R package survival 3.5-8 [25]). Kaplan-Meier estimates median survival time including survivors as censored observations. Mantel-Haenszel tests were used to test for differences between temperature or oxygen treatments within each species.

For analysis of acute lethal limits after acclimation, trial (as a fixed factor) and replicate (as a random factor to account for non-independence of species) were added to account for any differences between the different experiments. To account for animal size, wet weight was added as a covariate. Lethal limits were Box-Cox transformed to normalise the residuals from ANOVA analysis (or in the case of *U. antarctica*, *H. steineni*, and *O. Validus*, acclimatisation to assess them as visually approximately normal). All data sets had equal variance (Levene's test). To test for acclimatisation, CT_{max} of *O. validus* was regressed against average water temperature during the month of collection.

3. Results

3.1. Identification of Anemone Species

COI DNA sequences identified 27 out of 32 anemones as *Urticinopsis antarctica*. Five individuals were not identified with sufficient certainty and were removed from further analysis.

3.2. Temperature Incubations

In six out of eight trials, species survived for significantly longer when incubated at 6.0 than 8.0 °C (Figure 1, Tables 2 and S1, Figure S2). The two exceptions were *M. antarctica*, whose median survival duration (13 and 17 days) was not significantly different at 6.0 and 8.0 °C ($Chi^2 = 2.4$, $p = 0.1$; Table S1), and the trial with *S. neumeyari*, when the majority of individuals survived for the duration of incubations (303 days) at both 6.0 and 8.0 °C ($Chi^2 = 3.7$, $p = 0.06$; Table S1). When considered across all trials, species for which median survival durations could be estimated at both 6.0 and 8.0 °C survived for significantly longer at the lower temperature ($Chi^2 = 24.5$, $p < 0.01$). *O. validus* also survived more than 200 days at 8.0 °C until a technical failure of the tank.

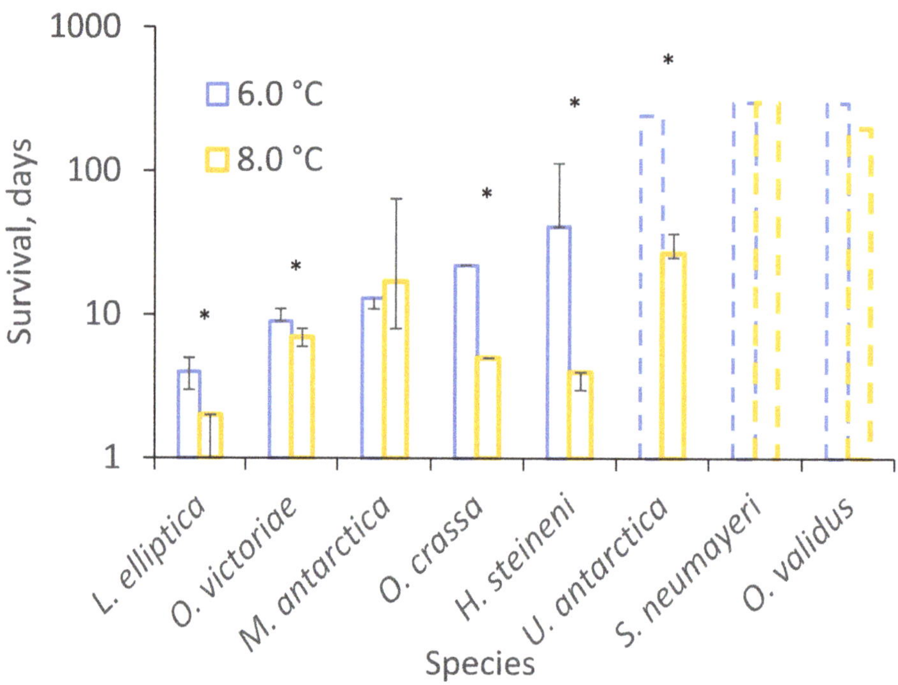

Figure 1. The Kaplan-Meier estimate of median survival duration, in days, after incubation at 6.0 and 8.0 °C. Species ordered from lowest to highest survival. Median ± interquartile range. Hashed lines indicate the duration of experiments for species for which the Kaplan-Meier estimator could not be calculated as most individuals survived until the experiment ended (censored observations), except for *O. validus* at 8.0 °C when the experiment terminated after 200 days due to a heater failure. Missing interquartile ranges indicate that the Kaplan-Meier could not estimate that value (Table S1). An asterix (*) indicates significantly different survival of that species between incubation temperatures. See Figure S1 for the stability of incubation temperatures and Figure S2 for Kaplan-Meier estimation figures.

Table 2. Survival in each incubation experiment.

Species	Incubation	Mortalities	Survivors	% Survival
S. neumayeri	6 °C	10	21	68
	8 °C	2	10	83
O. validus	6 °C	1	19	95
	8 °C	Technical failure		
O. victoriae	6 °C	27	0	0
	8 °C	23	0	0
O. crassa	6 °C	24	0	0
	8 °C	25	0	0
M. antarctica	6 °C	13	0	0
	8 °C	19	0	0
U. antarctica	6 °C	1	13	93
	8 °C	16	0	0
L. elliptica	6 °C	19	0	0
	8 °C	15	0	0
H. steineni	6 °C	21	4	16
	8 °C	25	0	0
O. validus	21%	8	11	58
	30%	12	8	40
O. crassa	21%	26	4	13
	30%	4	26	87
P. miersi	21%	15	15	50
	30%	15	15	50
O. victoriae	21%	30	0	0
	30%	30	0	0
C. georgiana	21%	20	0	0
	30%	20	0	0
S. neumayeri	21%	18	3	14
	30%	19	0	0
H. steineni	21%	20	0	0
	30%	19	1	5

3.3. Acclimation of Acute CT_{max}

U. antarctica acute CT_{max} was not significantly different when incubated at either 0 or 6 °C ($F_{(1,25)} = 0.7$, $p = 0.69$), with no effect of animal weight ($F_{(1,25)} = 0.6$, $p = 0.61$) or trial ($F_{(1.25)} = 1.0$, $p = 0.33$; Figure 2A; Table S2). *H. steineni* incubated at 6.0 °C had significantly higher CT_{max} ($F_{(1,85)} = 8.2$, $p < 0.01$) than those incubated at 0.0 °C (Figure 2B) but with no effect of animal weight ($F_{(1,85)} = 0.2$, $p = 0.66$; Table S2) or trial ($F_{(1,85)} = 0.8$, $p = 0.36$). There was, however, a significant difference between replicates ($F_{(1,85)} = 5.9$, $p = 0.02$), with the first incubation having significantly higher CT_{max} (12.7 ± 0.34 °C at 0.0 °C and 13.6 ± 0.29 °C at 6.0 °C) than the second (12.1 ± 0.31 °C at 0.0 °C and 12.8 ± 0.34 °C at 6.0 °C). *O. validus* incubated at 6.0 °C had a significantly higher CT_{max} ($F_{(1,31)} = 11.1$, $p < 0.01$) than those incubated at 0.0 °C (Figure 2C), with no effect of animal weight ($F_{(1,31)} = 3.7$, $p = 0.07$; Table S2), replicate ($F_{(1,31)} = 0.1$, $p = 0.72$) or trial ($F_{(1,31)} = 2.2$, $p = 0.15$). Acclimation of CT_{max} was also apparent for *S. neumayeri* ($F_{(1,50)} = 67.7$, $p < 0.01$), with no effect of animal weight ($F_{(1,50)} = 1.0$, $p = 0.3$) or replicate ($F_{(1,50)} = 1.5$, $p = 0.23$; Table S2). Lethal limits were higher at 6.0 °C ($T = 9.6$, $p < 0.01$) and 8.0 °C ($T = 9.4$, $p < 0.01$) than 0.0 °C (Figure 2D).

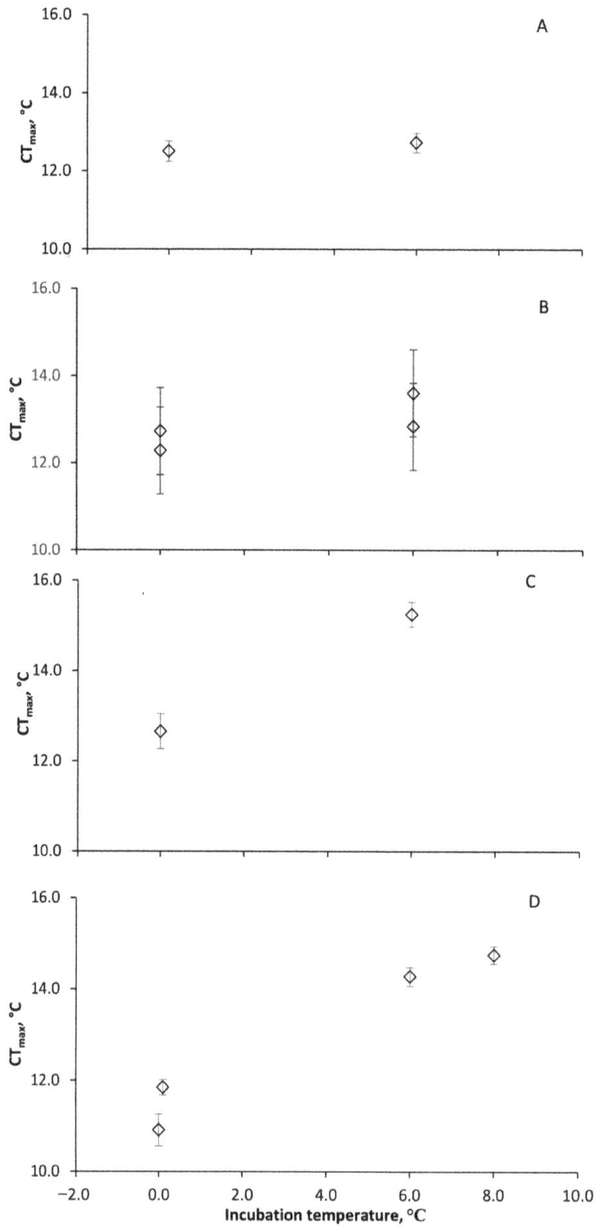

Figure 2. Lethal temperature limits in four Antarctic marine invertebrates incubated in the laboratory at 0.0, 6.0 and 8.0 °C. (**A**) *Urticinopsis antarctica*, (**B**) *Heterocucumis steineni*, (**C**) *Odontaster validus*, and (**D**) *Sterechinus neumayeri*. *U. antarctica* was the only species whose CTmax did not acclimate between control (0.0 °C) and elevated temperatures (6 and 8.0 °C). Where more than one data point is shown for a given temperature, more than one experiment was conducted at different times. *Urticinopsis antarctica*, 0 °C n = 14, 6 °C n = 13, *Heterocucumis steineni*, n = 19 to 24, *Odontaster validus*, 0 °C n = 14, 6 °C n = 19 and *Sterechinus neumayeri*, 0 °C n = 12 and 8, 6 °C n = 21, 8 °C n = 10.

There was no significant effect of average environmental (seawater) temperature during the month of collection on CT_{max} measured in *O. validus* ($F_{(1,264)} = 0.82$, $p = 0.37$; Figure 3).

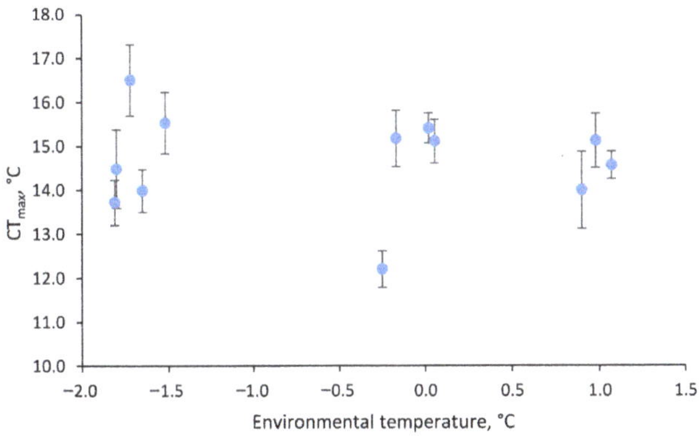

Figure 3. Upper lethal limits of *Odontaster validus* tested directly after collection from the field in different years and seasons. Mean ± 2 SE.

3.4. Temperature–Oxygen Incubations

In the first experiment all individuals of *O. victoriae* reached CT_{max} within the 99 days at 3.0 °C. The survival duration of both *O. victoriae* (median 84 versus 66.5; $Chi^2 = 47.8$, $p < 0.01$) and *O. crassa* (>100 versus 82; $Chi^2 = 41.4$, $p < 0.01$) was extended under hyperoxia (30%) compared to normoxia (21%; Figures 4 and S2; Table S3). A total of 87% of *O. crassa* reached CT_{max} within 100 days under normoxia, whereas only 13% reached CT_{max} under hyperoxia. The other difference in survival duration was *O. validus* incubated at 8.0 increased to 10.0 °C, which showed the opposite pattern, and survived significantly longer under normoxia (median > 351 days) than hyperoxia (322 days). There was no significant effect of oxygen concentration on survival of the other species but when all species were considered together, survival was longer under hyperoxia (30%) than normoxia (21%; Likelihood ratio, $Chi^2 = 15.8$, $p < 0.01$).

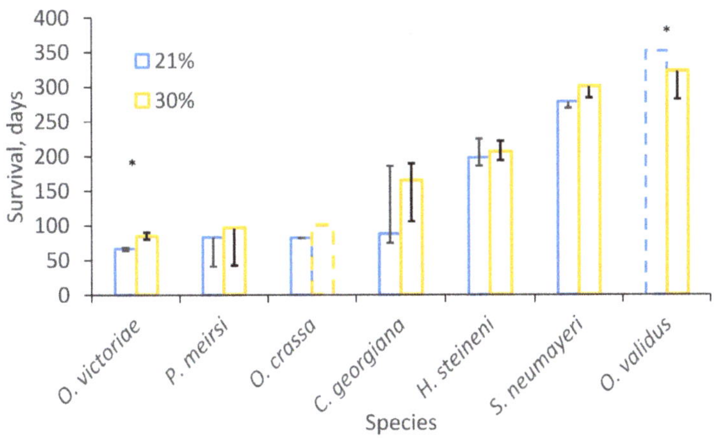

Figure 4. The Kaplan-Meier estimated median survival duration, in days (±interquartile range) that each species survived when incubated under normoxia (21%) or hyperoxia (30%). Hashed lines indicate

the duration of experiments for species for which the Kaplan-Meier estimator could not be calculated as most individuals survived until the experiment ended (censored observations). An asterix (*) indicates species with significantly different survivals under different oxygen treatments. For stability of incubation temperature see Table 1 and Figure S3 for the Kaplan-Meier estimation figures.

4. Discussion

Overall, we report mixed support for three central thermal physiology paradigms, with species-specific responses in: (1) physiological tolerance, (2) long-term acclimation, and (3) the effect of oxygen limitation on long-term survival. Previous estimates of long-term thermal tolerance of Antarctic marine ectotherms, extrapolated from CT_{max} assessments at multiple rates of warming, were between 1 and 6 °C if warming occurred over months, but dropped to a predicted average of 1.3 °C after about a year [12]. However, the current study extended known long-term incubation limits, with some individuals of *U. antarcticus* and *H. steineni* able to survive long-term at 6 °C, and *O. validus* and *S. neumayeri* at temperatures up to 10 °C (Table 1). Of all the Antarctic marine ectotherms tested to date, *O. validus* is one of the most tolerant, with the collapse of physiological processes (food processing and coelomic oxygen concentration) at 6 °C and loss of activity at 9 °C [12]. *S. neumayeri* also has a thermal tolerance towards the upper-end of the Antarctic ectotherm assemblage, with average functional thresholds in short-term warming around 8.3 ± 1.3 °C [26]. This study extended the long-term tolerance of *O. validus* and *S. neumayeri* to higher than previously reported temperatures of up to 10.0 °C, although it should be noted that individuals might have been resisting warming for long periods and might have impaired functions that would preclude, e.g., reproduction and recruitment. This is likely the case for *U. antarcticus*, which did not show evidence of acclimation, even after 245 days at 6 °C.

The current study, therefore, extends the known long-term experimental upper thermal limits of several Antarctic marine species to up to 8.2 °C, well above current maximum summer temperatures recorded in Marguerite Bay (1.8 °C [24]). The warming trend of air temperature along the Western Antarctic Peninsula since the 1970s has continued after a brief hiatus around the turn of the 21st century [27]. However, the warming signal in the Southern Ocean is less clear, with slower than average surface warming (0.02 °C per decade since 1950) [28]. The findings of the current study therefore suggest that the thermal limit of the most tolerant shallow water Antarctic marine ectotherms will not be breached for several centuries. However, thermal tolerance is species-specific, with limits for less tolerant species such as *O. victoriae* (this study and [19]), and the encrusting spirorbid worm *Protolaeospira stalagmia* [18], likely be reached in much less than 100 years, which will directly impact ecosystem biodiversity. Also, the limits for activity are often much lower than CT_{max} estimates, with species expected to suffer ecological impacts at temperatures well below CT_{max}. Species such as the scallop *Adamussium colbecki*, the limpet, *Nacella concinna*, and the bivalve mollusc, *L. elliptica*, have behavioural limits between 2 and 3 °C [28] which will clearly affect long-term survival in the wild. Understanding the variation in thermal tolerance between species, and between physiological functions, is crucial for determining the likely winners and losers in response to climate change and, therefore the impact of these changes on ecosystem function.

The responses of species to the effect of elevated oxygen on long-term survival arguing against the OCLTT being a universal mechanism controlling upper temperatures limits. While some species had clearly elevated CT_{max} under mild hyperoxia, indicating a positive effect of oxygen on whole animal CT_{max}, the response was not consistent across species and *O. validus* even showed a slight negative effect of hyperoxia on survival. As with the findings of the current study, Clark et al. [19] found multiple mechanisms underpinning species response when incubated at temperatures just below CT_{max}, with only two species showing the switch from aerobic to anaerobic metabolism [19]. Species-specific thermal tolerances have been linked to their functional traits, with the most active species having higher upper lethal limits [12]. Activity is correlated with aerobic scope, with more active species having an oxygen supply cascade that is capable of supplying more active tissues, resulting in a greater "total excess aerobic power budget" and therefore a higher capacity to resist thermal challenge [12,29]. However, where in the past Antarctic marine ectotherm species with higher activity levels have survived warming to higher temperatures [12], and this was taken as support for OCLTT, the species that acclimated to the higher temperatures in this study were all low-activity species, arguing against the importance of oxygen limitation. In the current study, the most resistant species came from a mixture of functional groups, from a mobile obligate predator through to a sessile primary consumer. It should be noted that the use of CT_{max} to test the OCLTT has been criticised and future research should include sub-lethal assays of thermal tolerance such as activity and the response of physiological pathways [30].

High concentrations of dissolved oxygen are expected to form free radicals, which could lead to tissue damage during long-term incubations [31]. If oxygen was not limiting to *O. validus* at the chosen incubation temperatures, then mild hyperoxia could have resulted in oxygen free-radical damage or additional costs for antioxidant defence mechanisms (e.g., superoxide dismutase [31]). This could potentially explain the negative effect of hyperoxia, but also provides a clear indication that oxygen was not limiting at these temperatures for this species. It is possible that a different pattern will be seen at higher temperatures.

At temperatures typical of polar oceans, the acclimation of physiological systems of marine invertebrates can take many months, with successful whole animal acclimation being recorded in three out of eight Antarctic marine invertebrate species within 2 and 9 months at 3–8 °C ([11] and current study). Acclimation in Antarctic fish was faster, with seven out of nine species acclimating within 5 to 36 days at 4 to 7 °C [32–37]. In the current study, the anemone *U. antarcticus* survived, with minimal mortality, for 270 days at 6 °C but with no whole animal adjustment of CT_{max}. Caution must be applied when interpreting the CT_{max} of survivors in thermal challenge experiments, as this will likely overestimate resilience, as only a subset of the more resilient survivors is being tested (5–95% survival). Also, the results under carefully controlled laboratory conditions may not reflect survival in the wild when sub-lethal effects will likely limit long-term population persistence, and how these reductions in sub-lethal performance will affect ecological interactions (e.g., reproduction, disease resistance, predation, and competition) is likely to be important. The slow rate of surface warming in the Southern Ocean [38] means that even a small amount of acclimation can provide species with a significant additional thermal buffer for many decades, long enough for generational plasticity and the possibility of adaptive change, even in Antarctic species [11].

It is therefore not surprising that *O. validus* did not show acclimatisation of acute CT_{max} in the field over the temperature range −1.8 to +1.1 °C. It is probable that the duration of the peak of summer temperatures (1–2 months [24]) or the maximum environmental temperature (+1.1 °C in the field) were not sufficient to trigger whole animal physiological adjustment of CT_{max}. The lack of predictable seasonal temperature change in the Southern Ocean could explain the lack of ecologically relevant acclimatisation, the observation that species do not select preferred temperatures, and why there are no observable escape response thresholds [15,16]. However, many other factors, such as tissue energy status, affect CT_{max} (e.g., [39]), and there might be greater seasonal differences in thermal tolerance in primary consumers with seasonally restricted food supplies, than in predators/scavengers which have a more constant food supply [40]. The complexity of the triggers for molecular thermal protective mechanisms has been illustrated for the HSP70 heat shock response in the Antarctic limpet, *N. concinna*. Acute laboratory thermal challenge did not elicit upregulation of HSP70 gene expression until temperatures of 15 °C, whereas limpets in the wild expressed HSP70 at foot temperatures of only 3.3 °C during a natural tidal cycle [41].

This study has highlighted species-specific variation in long-term survival at elevated temperatures, acclimation response and the influence of oxygen on CT_{max} of Antarctic marine ectotherms. The search for mechanisms underlying the evolution of physiological capacity [1,2,42] provides important frameworks against which exceptions can be identified and explanatory hypotheses constructed for future research. Thermal sensitivity is clearly the product of multiple ecological and physiological capacities, and this diversity of response needs further investigation and interpretation to improve our ability to predict future patterns of biodiversity.

5. Conclusions

Tests of three of the central paradigms underlying physiological response of Antarctic marine ectotherms to warming highlighted species-specific responses in regard to resilience, plasticity and the effect of oxygen limitation. This further highlights the complexity of thermal physiology and the need for further investigation and interpretation.

Supplementary Materials: The following supporting information can be downloaded at: https://www.mdpi.com/article/10.3390/biology13040224/s1, Figure S1. Example temperatures during ramping and incubation experiments at, A 6.0 °C and B 8.0 °C for 4 species *Heterocucumis steineni, Odontaster validus, Sterechinus neumeyeri* and *Laternula elliptica,* C for the 1st temperature-oxygen experiment. Figure S2. Kaplan Meier fits for survival of *Ophiura crassa, Ophionotus victoriae, Margarella antarctica, Laternula elliptica, Sterechnius neumayeri, Odontaster Validus, Hetercucumis steineni* and *Urticinopsis antarctica.* Figure S3. Kaplan-meier fits for survival of *Odontaster validus, Sterechinus neumayeri, Heterocucumis steineni, Cucumaria georgiana, Ophionotus victoriae, Paraceradocus miersi* and *Ophiura crassa* incubated at normoxia (21%) and hyperoxia (30%). See Table 1 and Figure 1 for incubation temperatures. Table S1. Median and interquartile range from Kaplan Meier fits in days (± Inter-Quartile Range). Nominal temperatures of 6.0 and 8.0 °C are listed that had a typical SE of ± 0.04 °C (Figure 1). Table S2. Wet mass of incubated individuals, mean, standard error and sample size. Within each species, there were no significant differences in wet mass between treatments. Table S3. Median and interquartile range from Kaplan Meier fits for *Odontaster Validus, Sterechinus neumayeri, Heterocucumis steineni, Cucumaria georgiana, Ophionotus victoriae, Paraceradocus meirsi* and *Ophiura crassa* incubated at either normoxia (21%) and hyperoxia (30%).

Author Contributions: Conceptualization, S.A.M., L.S.P., R.S. and M.S.C.; Methodology, S.A.M., L.S.P., E.F. and R.S.; Validation, S.A.M. and A.E.B.; Formal Analysis, S.A.M., R.E.S. and A.E.B.; Investigation, S.A.M., L.S.P., M.S.C., E.F. and R.S.; Resources, L.S.P., M.S.C. and R.S.; Data Curation, S.A.M.; Writing—Original Draft Preparation, S.A.M., R.E.S. and A.E.B.; Writing—Review and Editing, S.A.M., A.B, M.S.C., L.S.P., R.E.S. and E.F.; Visualization, S.A.M. and R.E.S.; Supervision, S.A.M. and A.E.B. All authors have read and agreed to the published version of the manuscript.

Funding: This study was funded by Natural Environment Research Council core funding to the British Antarctic Survey and Spitfire DTP funding to R.E.S.

Institutional Review Board Statement: All research was conducted under Antarctic Treaty permits issued by the UK Foreign, Commonwealth and Development Office to British Antarctic Survey. Experiments were conducted on marine invertebrate species not protected under the UK Animals (Scientific Procedures) Act 1986 and although ethical approval was not required the highest possible standards of animal welfare were applied.

Data Availability Statement: All data are either published with the manuscript or available on request from the lead author. Morley, S.A., Bates, A.E., Clark, M.S., Fitzcharles, E., Smith, R., Stainthorp, R.E., and Peck, L.S. (2024). Acclimation and acclimatisation of marine ectotherms collected at Rothera Research Station and Scott Base in Antarctica between 2004 and 2015 (Version 1.0) [Data set]. NERC EDS UK Polar Data Centre. https://doi.org/10.5285/60B777B4-0BD6-48C3-A301-C700854FBFA1.

Acknowledgments: The authors want to thank the marine teams and technical support at Rothera who enabled this research.

Conflicts of Interest: The authors declare no conflicts of interest.

References

1. Gaston, K.J.; Chown, S.L.; Calosi, P.; Bernardo, J.; Bilton, D.T.; Clarke, A.; Clusella-Trullas, S.; Ghalambor, C.K.; Konarzewski, M.; Peck, L.S.; et al. Macrophysiology: A conceptual reunification. *Am. Nat.* **2009**, *174*, 595–612. [CrossRef]

2. Pörtner, H.O. Climate variations and the physiological basis of temperature dependent biogeography: Systemic to molecular hierarchy of thermal tolerance in animals. *Comp. Biochem. Physiol. A* **2002**, *132*, 739–761. [CrossRef]
3. Lindquist, S. The heat shock response. *Annu. Rev. Biochem.* **1986**, *55*, 1151–1159. [CrossRef]
4. Tomanek, L. Environmental proteomics: Changes in the proteome of marine organisms in response to environmental stress, pollutants, infection, symbiosis, and development. *Ann. Rev. Mar. Sci.* **2011**, *3*, 373–399. [CrossRef]
5. Collier, R.J.; Baumgard, L.H.; Zimbelman, R.B.; Xiao, Y. Heat stress: Physiology of acclimation and adaptation. *Anim. Front.* **2019**, *9*, 12–19. [CrossRef]
6. Hobday, A.J.; Alexander, L.V.; Perkins, S.E.; Smale, D.A.; Straub, S.C.; Oliver, E.C.J.; Benthuysen, J.A.; Burrows, M.T.; Donat, M.G.; Feng, M.; et al. A hierarchical approach to defining marine heatwaves. *Prog. Oceanogr.* **2016**, *141*, 227–238. [CrossRef]
7. Stillman, J. Acclimation underlies susceptibility to climate change. *Science* **2003**, *301*, 65. [CrossRef]
8. Leroi, A.M.; Bennett, A.F.; Lenski, R.E. Temperature acclimation and competitive fitness: An experimental test of the Beneficial Acclimation Hypothesis. *Proc. Natl. Acad. Sci. USA* **1994**, *91*, 1917–1921. [CrossRef]
9. Wilson, R.S.; Franklin, C.E. Testing the beneficial acclimation hypothesis. *Trends Ecol. Evol.* **2002**, *17*, 66–70. [CrossRef]
10. Gunderson, A.R.; Stillman, J.H. Plasticity in thermal tolerance has limited potential to buffer ectotherms from global warming. *Proc. R. Soc. B* **2015**, *282*, 20150401. [CrossRef] [PubMed]
11. Morley, S.A.; Peck, L.S.; Heiser, S.; Sunday, J.; Bates, A.E. Physiological acclimation extends predicted persistence of species across the globe. *Glob. Ecol. Biodivers.* **2019**, *28*, 1018–1037. [CrossRef]
12. Peck, L.S.; Clark, M.S.; Morley, S.A.; Massey, A.; Rossetti, H. Animal temperature limits and ecological relevance: Effects of size, activity and rates of change. *Funct. Ecol.* **2009**, *23*, 248–256. [CrossRef]
13. Molina, A.N.; Pulgar, J.M.; Rezende, E.L.; Carter, M.J. Heat tolerance of marine ectotherms in a warming Antarctica. *Glob. Chang. Biol.* **2023**, *29*, 179–188. [CrossRef]
14. Clark, M.S.; Peck, L.S. Hsp70 heat shock proteins and environmental stress in Antarctic marine organisms: A mini-review. *Mar. Genom.* **2009**, *2*, 11–18. [CrossRef]
15. Bates, A.E.; Lee, R.W.; Tunnicliffe, V.; Lamare, M.D. Deep-sea hydrothermal vent animals seek cool fluids in a highly variable thermal environment. *Nat. Commun.* **2010**, *1*, 14. [CrossRef]
16. Morley, S.A.; Chu, J.W.F.; Peck, L.S.; Bates, A.E. Temperatures leading to heat escape responses in Antarctic marine ectotherms match acute thermal limits. *Front. Physiol.* **2022**, *13*, 1077376. [CrossRef]
17. Clark, M.S.; Villota Nieva, L.; Hoffman, J.I.; Davies, A.J.; Trivedi, U.H.; Turner, F.; Ashton, G.; Peck, L.S. Lack of long-term acclimation in Antarctic encrusting species suggests vulnerability to warming. *Nat. Commun.* **2019**, *10*, 3383. [CrossRef]
18. Collins, M.; Peck, L.S.; Clark, M.S. Large within, and between, species differences in marine cellular responses: Unpredictability in a changing environment. *Sci. Total Environ.* **2021**, *794*, 148594. [CrossRef]
19. Clark, M.S.; Sommer, U.; Sihra, J.K.; Thorne, A.S.; Morley, S.A.; King, M.; Viant, M.R.; Peck, L.S. Biodiversity in marine invertebrate responses to acute warming revealed by a comparative multi-omics approach. *Glob. Chang Biol.* **2017**, *23*, 318–330. [CrossRef]
20. Clark, T.D.; Sandblom, E.; Jutfelt, F. Aerobic scope measurements of fishes in an era of climate change: Respirometry, relevance and recommendations. *J. Exp. Biol.* **2013**, *216*, 2771–2782. [CrossRef]
21. Verberk, W.C.; Overgaard, J.; Ern, R.; Bayley, M.; Wang, T.; Boardman, L.; Terblanche, J.S. Does oxygen limit thermal tolerance in arthropods? A critical review of current evidence. *Comp. Biochem. Physiol. A Mol. Integr. Physiol.* **2016**, *192*, 64–78. [CrossRef]
22. Richard, J.; Morley, S.A.; Thorne, M.A.; Peck, L.S. Estimating long term survival temperatures at the assemblage level in the marine environment: Towards macrophysiology. *PLoS ONE* **2012**, *7*, e34655. [CrossRef]
23. Folmer, O.; Black, M.; Hoeh, W.; Lutz, R.; Vrijenhoek, R. DNA primers for amplification of mitochondrial cytochrome c oxidase subunit I from diverse metazoan invertebrates. *Mol. Mar. Biol. Biotech.* **1994**, *3*, 294–299.
24. Venables, H.; Meredith, M.P.; Hendry, K.R.; ten Hoopen, P.; Peat, H.; Chapman, A.; Beaumont, J.; Piper, R.; Miller, A.J.; Mann, P.; et al. Sustained, year-round oceanographic measurements from Rothera Research Station, Antarctica, 1997–2017. *Sci. Data* **2023**, *10*, 265. [CrossRef]
25. Allignol, A.; Latouche, A. CRAN Task View: Survival Analysis, Version 2023-09-10. 2023. Available online: https://CRAN.R-project.org/view=Survival (accessed on 1 January 2023).
26. De Leij, R.; Grange, L.J.; Peck, L.S. Functional thermal limits are determined by rate of warming during simulated marine heatwaves. *Mar. Ecol. Prog. Ser.* **2022**, *685*, 183–196. [CrossRef]
27. Carrasco, J.F.; Bozkurt, D.; Cordero, R.R. A review of the observed air temperature in the Antarctic Peninsula. did the warming trend come back after the early 21st hiatus? *Polar Sci.* **2021**, *28*, 100653. [CrossRef]
28. Peck, L.S.; Webb, K.E.; Bailey, D. Extreme sensitivity of biological function to temperature in Antarctic marine species. *Funct. Ecol.* **2004**, *18*, 625–630. [CrossRef]
29. Pörtner, H.O.; Bock, C.; Mark, F.C. Oxygen- and capacity-limited thermal tolerance: Bridging ecology and physiology. *J. Exp. Biol.* **2017**, *220*, 2685–2696. [CrossRef]
30. McArley, T.J.; Sandblom, E.; Herbert, N.A. Fish and hyperoxia—From cardiorespiratory and biochemical adjustments to aquaculture and ecophysiology implications. *Fish Fish.* **2020**, *22*, 324–355. [CrossRef]
31. Abele, D.; Puntarulo, S. Formation of reactive species and induction of antioxidant defence systems in polar and temperate marine invertebrates and fish. *Comp. Biochem. Physiol. A* **2004**, *138*, 405–415. [CrossRef] [PubMed]

32. Wilson, R.S.; Franklin, C.E.; Davison, W.; Kraft, P. Stenotherms at subzero temperatures: Thermal dependence of swimming performance in Antarctic fish. *J. Comp. Physiol.* **2001**, *171*, 263–269. [CrossRef] [PubMed]
33. Seebacher, F.; Davison, W.; Lowe, C.J.; Franklin, C.E. A falsification of the thermal specialization paradigm: Compensation for elevated temperatures in Antarctic fishes. *Biol. Lett.* **2005**, *1*, 151–154. [CrossRef] [PubMed]
34. Franklin, C.E.; Davison, W.; Seebacher, F. Antarctic fish can compensate for rising temperatures: Thermal acclimation of cardiac performance in *Pagothenia borchgrevinki*. *J. Exp. Biol.* **2007**, *210*, 3068–3074. [CrossRef] [PubMed]
35. Robinson, E.; Davison, W. The Antarctic notothenioid fish *Pagothenia borchgrevinki* is thermally flexible: Acclimation changes oxygen consumption. *Polar Biol.* **2008**, *31*, 317–326. [CrossRef]
36. Bilyk, K.T.; Devries, A.L. Heat tolerance and its plasticity in Antarctic fishes. *Comp. Biochem. Physiol. A Mol. Integr. Physiol.* **2011**, *158*, 382–390.
37. Strobel, A.; Bennecke, S.; Leo, E.; Mintenbeck, K.; Pörtner, H.O.; Mark, F.C. Metabolic shifts in the Antarctic fish *Notothenia rossii* in response to rising temperature and PCO_2. *Front. Zool.* **2012**, *9*, 28. [CrossRef] [PubMed]
38. Cai, W.; Gao, L.; Luo, Y.; Li, X.; Zheng, X.; Zhang, X.; Cheng, X.; Jia, F.; Purich, A.; Santoso, A.; et al. Southern Ocean warming and its climatic impacts. *Sci. Bull.* **2023**, *68*, 946–960. [CrossRef] [PubMed]
39. Brokordt, K.B.; Himmelman, J.H.; Guderley, H.E. Effect of reproduction on escape responses and muscle metabolic capacities in the scallop Chlamys islandica Müller. *J. Exp. Mar. Biol. Ecol.* **2000**, *251*, 205–225. [CrossRef] [PubMed]
40. Obermüller, B.O.; Morley, S.A.; Clark, M.S.; Barnes, D.K.A.; Peck, L.S. Antarctic intertidal limpet ecophysiology: A winter-summer comparison. *J. Exp. Mar. Biol. Ecol.* **2011**, *403*, 39–45. [CrossRef]
41. Clark, M.S.; Geissler, P.; Waller, C.; Fraser, K.P.P.; Barnes, D.K.A.; Peck, L.S. Low heat shock thresholds in wild Antarctic inter-tidal limpets (Nacella concinna). *Cell Stress Chaperones* **2008**, *13*, 51–58. [CrossRef]
42. Brown, J.H.; Gilooly, J.F.; Allen, A.P.; Savage, V.M.; West, G.B. Towards a metabolic theory of ecology. *Ecology* **2004**, *85*, 1771–1789. [CrossRef]

biology MDPI

Article

The Mechanism by Which Umbrella-Shaped Ratchet Trichomes on the *Elaeagnus angustifolia* Leaf Surface Collect Water and Reflect Light

Zhanlin Bei [1,2], Xin Zhang [2] and Xingjun Tian [1,*]

[1] School of Life Sciences, Nanjing University, Nanjing 210023, China; realpal00147@163.com
[2] School of Biological Science and Engineering, North Minzu University, Yinchuan 750021, China; x_zhang@nmu.edu.cn
* Correspondence: tianxj@nju.edu.cn

Simple Summary: Plants in arid areas have evolved narrow linear leaves or degenerated into cone-shaped spines to reduce solar radiation and excessive water loss. We have demonstrated for the first time the widely distributed sub-tree species of *Elaeagnus angustifolia* in arid areas. *E. angustifolia* has large- and medium-sized leaves, and the surface of the leaves is covered with umbrella-shaped ratchet trichomes that can capture moisture in the air and reflect solar radiation. This study reveals that the trichomes on the *E. angustifolia* leaf surface can collect water and reflect light, which provides a reference for bionic development and research in atmospheric water harvesting, seawater desalination, energy management, microfluidic control, and daytime radiative cooling.

Abstract: Leaves are essential for plants, enabling photosynthesis and transpiration. In arid regions, water availability limits plant growth. Some plants, like *Elaeagnus angustifolia*, a sandy sub-tree species widely distributed in arid and semi-arid regions, have unique leaf structures to reduce water loss and solar radiation. Here, we describe the leaves of *Elaeagnus angustifolia* L., with special functioning trichomes. Through leaf submicroscopic structure observation, in situ water collection experiments, photosynthesis measurements, and reflection spectrum analysis, we investigated *E. angustifolia* leaves, focusing on their functioning trichomes. These trichomes capture water vapor, reflect UV and NIR light, and possess a 3D interface structure composed of 1D and 2D structures. The 1D conical structure captures water droplets, which are then gathered by the radial conical structure and guided towards the stomata through wedge-shaped grooves on the 2D umbrella structure. The trichomes also reflect sunlight, with micropapillae reflecting UV light and the umbrella structure reflecting NIR light. These mechanisms reduce leaf temperature, respiration, and water transpiration, protecting against solar radiation damage. This study provides insights into water collection and light-reflection mechanisms, revealing adaptive strategies of plants with large leaves in arid regions.

Keywords: *Elaeagnus angustifolia* leaf; umbrella-shaped ratchet trichomes; water collection; light reflection

1. Introduction

In arid regions characterized by clear skies and intense sunlight [1–3], efficient water acquisition, water conservation, and reduced solar radiation and heat absorption are critical for the survival of plants [4–6]. Plant leaves, like other organs, experience severe drought stress [7]. Consequently, leaves have developed an ingenious and effective mechanism to absorb water and reflect light in response to environmental stress. For instance, *Opuntia microdasys* leaves in desert areas possess specialized spines and glandular trichomes that minimize solar radiation and water loss during gas exchange and transpiration. Moreover, these clustered needle-like spines enable the directional collection of airborne water droplets [8], which sustains cactus survival in arid environments for several days. Another example is *Syntrichia caninervis*, a perennial plant that thrives in arid regions. Its leaves feature long white "awns" at their tips, which efficiently collect and transport water from the air, while also reflecting sunlight to minimize water evaporation. This adaptation allows *S. caninervis* to flourish in arid environments [9]. In semi-humid regions, the large palmate leaves of *Populus tomentosa* possess a non-wettability hollow white tomentose layer on their back surfaces, which reflects up to 55% of sunlight. This feature significantly reduces leaves' heat absorption and shields them from intense solar radiation [10]. However, little is known about the drought-resistant strategies of plants with large- and medium-sized leaves in arid and semi-arid regions.

Elaeagnus angustifolia L. (Elaeagnaceae) is a deciduous tree or large multilayer shrub with small size and spines. It is widely distributed in arid and semi-arid areas [11]. This species is commonly known as "Russian olives" or "oleaster" [12] and is native to southern Europe, Central Asia, and the Western Himalayas [13]. During the early 20th century, it was introduced from Eurasia to Canada, the United States, the Mediterranean coast, Southern Russia, Iran, and India [14]. The fruit of *E. angustifolia* is highly nutritious, containing proteins, sugars, vitamins, and minerals, making it a valuable natural food and medicinal herb [15]. The climate conditions suitable for *E. angustifolia* are mainly found in cold and dry warm and mid-temperate zones in winter, as well as in arid and semi-arid regions between 30° N and 50° N in China [16]. *E. angustifolia* thrives in these areas, where the annual rainfall does not exceed 300 mm [17–20]. Due to its ability to withstand severe drought, high salinity, cold, and wind stress [21], *E. angustifolia* plays a crucial role in preserving ecosystem functionality in arid areas [22].

The leaves of *E. angustifolia* are medium-sized and lanceolate in shape. They possess trichomes on both the upper and lower surfaces, exhibiting a silver-white color [23,24]. Upper sun leaves are smaller, more slender, and thicker than the lower, half-exposed shade leaves [25]. The foliage of a single *E. angustifolia* plant displays a gradient from silver-white to dark green, with peltate trichomes on the upper sun-exposed leaves, pedestalled trichomes on the medium half-sun-exposed leaves, and multicellular trichomes on the lower shade leaves [24,25]. This leaf heterogeneity likely represents a specific adaptation to the environment [25,26]. However, there is limited information available on how *E. angustifolia* leaves respond to environmental conditions. In this study, we conducted submicroscopic structure observations of *E. angustifolia* leaves, performed in situ water-collection experiments, measured photosynthesis rates, and analyzed reflection spectra in order to determine their ability to collect water and reflect sunlight with their specifically constructed trichomes. This study provides the first insights into a novel mechanism employed by *E. angustifolia* leaves to acquire water and modulate light reflection, significantly contributing to our understanding of how plants with large and medium-sized leaves respond to environmental stresses in arid regions.

2. Materials and Methods

2.1. Sample Collection and Characterization

In July 2021, fresh adult *E. angustifolia* leaves were collected from the southern edge of the Mu Us Sandy Land (37°95′70″ N, 106°42′47″ E) in China. Healthy leaves were cut into 2 × 2 mm pieces, which were gold-spayed using an ion beam sputtering system (Hitachi E-1045, Tokyo, Japan) with a thickness of 20 nm. The microstructure of the leaf surface was observed with a field-emission scanning electron microscope (ESEM; Hitachi S-4800, Japan). The maximum magnification of the SEM was 650,000 times, the acceleration voltage was 0.530 kV; the resolution was 2.2 nm at 1 kV and 1.0 nm at 15 kV. The adaxial and abaxial leaf surfaces trichome size and morphology conducted SEM image comparison through MATLAB (Version R2022a.MathWorks, Inc., Natick, MA, USA) scripts to compute the trichome similarity.

2.2. Determination of Leaf Surface Wettability

A 1 μL water droplet was applied to the leaf surface, and the contact angle (CA) was measured using a Dataphysics OCA20 CA device (Dataphysics Inc., Filderstadt, Germany) with five replicates for each leaf within 30 s [27]. The measurement was conducted at room temperature (25.00 ± 1.00 °C), and the relative humidity was kept at 80%.

2.3. Water-Collection Experiment

The trichomes were carefully fixed on a glass slide, and purified water (Milli-Q Reference, Inc., Bedford, MA, USA) was added to an ultrasonic humidifier (Yadou YC-D204, Shanghai, China) to generate fog droplets to the leaf trichomes. Water collection of the trichomes was examined under the saturation status at a flow rate of 20–30 mm s^{-1} according to Ju [8]. The trichome region was imaged using a CCD color camera (Moticam Pro 252A; Motic Instruments Inc., Richmond, BC, Canada) attached to a compound microscope (BA410; Motic) at a magnification of ×10.

2.4. Photosynthesis Determination

A portable photosynthesis system (LI-6400; LI-COR, Inc., Lincoln, NE, USA) was used to measure the photosynthetic rate, intercellular CO_2 concentration, stomatal conductance, and transpiration rate of the leaves according to Lawrence [28]. In July 2021, the southern edge (37°95′70″ N, 106°42′47″ E) of the Mu Us Sandy Land in China was utilized. Experimental results were tested statistically via paired two-tailed Student's *t*-test and expressed as mean ± SEM by OriginPro (Version 2023. OriginLab Corporation, Northampton, MA, USA). Data were considered statistically significant when $p < 0.05$.

2.5. Spectral Measurement and Statistical Analysis

The spectra of the front and back surfaces of *E. angustifolia* leaves were measured using a USB4000 miniature fiber optic spectrometer (Ocean Optics, Inc., Dunedin, FL, USA), with a spectral range of 200–850 nm. Before measurement, a standard whiteboard was used for calibration, and the viewing angle was 25°. Three replicates were performed for each sample, and the mean value was taken as the spectral reflectance of this sample. R v3.6.0 and Origin v20.0 were used to analyze and plot the different data.

3. Results

3.1. Leaf Surface Characteristics

E. angustifolia has lanceolate leaves (Figure 1a), the adaxial and abaxial surfaces of which are covered with shiny, silvery-white trichomes. There are fewer trichomes (with a coverage of <37%) on the adaxial leaf surface (Figure 1b), while trichomes cover the entire abaxial leaf surface (Figure 1c). No significant difference in the trichome size and morphology was observed between the adaxial and abaxial leaf surfaces. There are many stomata on the epidermis of the abaxial leaf surface (Figure 2e)

3.2. Structure and Characteristics of Trichomes

A scanning electron microscope (SEM) was used to further observe the individual trichomes on the *E. angustifolia* leaf surface. The trichome is in the shape of an umbrella-shaped ratchet, harboring a plurality of wedge-shaped grooved two-dimensional (2D) umbrella surfaces composed of multiple tapered conical spines (Figure 2a). There are micropapillae on the apex of the umbrella surface, and the ratchet structure is composed of multiple one-dimensional (1D) cones extending radially from the periphery of the umbrella surface (Figure 2b). The umbrella-shaped ratchet trichomes are hollow (Figure 2c), and the micropapillae on the apex of a single trichome have an apex angle (α) of 115.30 ± 9.30° (n = 20), a height (h) of 30.00 ± 9.10 μm (n = 25), a basal width of 58.20 ± 9.30 μm (n = 24), and h/w of 0.50 ± 0.20 (n = 24) (Figure 2d). A single conical spine has a length (L) of 186.27 ± 20.20 μm (n = 40), and a single conical spine has an angle (θ_1) of 7.46 ± 1.72° (n = 40); the angle between two adjacent conical spines (θ_2) is 15.48 ± 2.64° (n = 45)

(Figure 2f). The above findings indicate that a single umbrella-shaped ratchet trichome has both 1D and 2D structures.

Figure 1. Morphological characteristics of *E. angustifolia* leaves. Medium-sized *E. angustifolia* leaves are lanceolate. (**a**) Wild *E. angustifolia*; (**b**) front surface of the leaf with sparse trichomes in dark green color; (**c**) back surface of the leaves with dense trichomes in white-gray color; (**d**) SEM of leaf front; (**e**) SEM of the back of the leaf.

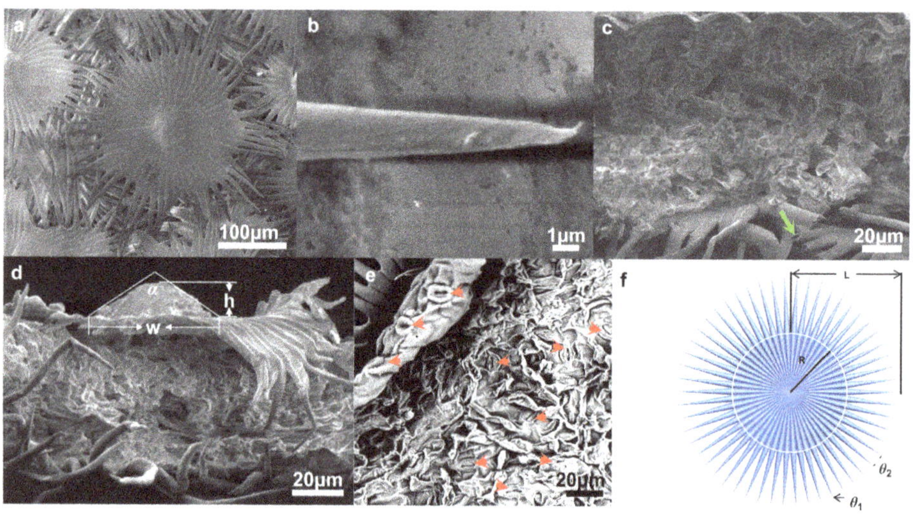

Figure 2. Submicroscopic structure of *E. angustifolia* leaves. An umbrella-shaped ratchet trichome of *E. angustifolia* leaves is composed of one-dimensional cone thorn and two-dimensional umbrella surface. (**a**) SEM image of umbrella ratchet trichome on leaf surface; (**b**) SEM image of trichome ratchet fiber; (**c**) SEM cross-section of *E. angustifolia* leaves, hollow structure of leaf trichomes (green solid arrow); (**d**) SEM of the structure of microemulsion on the top of a single trichome; (**e**) SEM image of the outer and inner layer structure on the back of the leaf with many stomata (red solid arrows); (**f**) an umbrella ratchet structure model of single trichome on the leaf surface of *E. angustifolia* composed of one-dimensional cone thorn and two-dimensional umbrella surface.

3.3. *Leaf Wettability*

The front and back surfaces of *E. angustifolia* leaves are micro-/nano-level heterogeneous rough surfaces, mainly composed of a number of umbrella-shaped ratchet trichomes. The wettability of the front and back leaf surfaces was measured. The contact angle of the adaxial leaf surface was 117.40 ± 5.40° (n = 8) (Figure 3a), and that of the abaxial leaf surface was 138.20 ± 3.70° (n = 8) (Figure 3b), suggesting that both leaf surfaces are hydrophobic interfaces with self-cleaning properties.

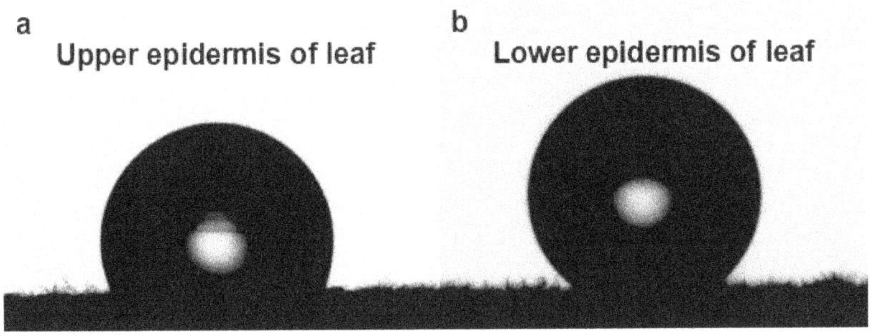

Figure 3. Leaf surface wettability of *E. angustifolia*. (**a**) The wettability of the epidermis on the leaves; (**b**) the wettability of the leaf lower epidermis.

3.4. *Multifunctional Hierarchical Water-Collection Mechanism of Trichomes*

The water-collection experiment was carried out on the trichomes of the *E. angustifolia* leaf surface, and the results showed that these trichomes could capture and directionally transport water droplets (Figures 4–6). There were many stomata (Figures 2e and 7a) on the leaf surface. Water droplets were captured by a single conical spine at the edge of the ratchet trichome (Figures 4b and 7b) and transported from the apex to the base of the conical spine through the Laplace pressure difference. Two adjacent conical spines formed an open "wedge"-shaped capillary gap for water collection (Figure 4c). The aggregated water was transported to the apex of the umbrella handle through the wedge-shaped grooves on the abaxial surface of the umbrella-shaped trichomes (Figure 7c,d). Then, the water flowed from the apex to the base of the umbrella handle, from where it flowed into the stomata on the leaf surface, entering the leaf interior (Figure 7e), and finally, the airborne water captured by the trichomes was absorbed by the leaf. Large droplets were aggregated by multiple conical spines onto the leaf surface, where they were absorbed by the stomata. Therefore, the umbrella-shaped ratchet trichomes on the *E. angustifolia* leaf surface harbor a multifunctional hierarchical water-collection mechanism (Figure 8).

3.5. *Response of Leaves to the Light*

At noon (12:00–14:00) on a cloudless day, the adaxial surface of *E. angustifolia* leaves curled or turned over with the increasing sunlight intensity, exposing the abaxial leaf surface to the sunlight (Figure 9a). From morning to noon, the photosynthetic rate of *E. angustifolia* leaves changed intermittently: the stomata gradually closed, the transpiration rate gradually decreased, and the intercellular CO_2 concentration fluctuated instantaneously (Figure 9b). The carbon dioxide concentration on the leaves was higher than in the environment, while the temperature, water, and relative humidity on the leaves were lower than in the environment (Figure 10). Both the front and back surfaces of *E. angustifolia* leaves could absorb blue (400–520 nm) and red lights (610–720 nm), while they highly reflected rays in the UV (<380 nm) and NIR regions (>700 nm) (Figure 9c). The umbrella-shaped ratchet trichomes on the *E. angustifolia* leaf surface greatly reduced the absorption of solar radiation and heat by the leaves (Figures 9d and 11).

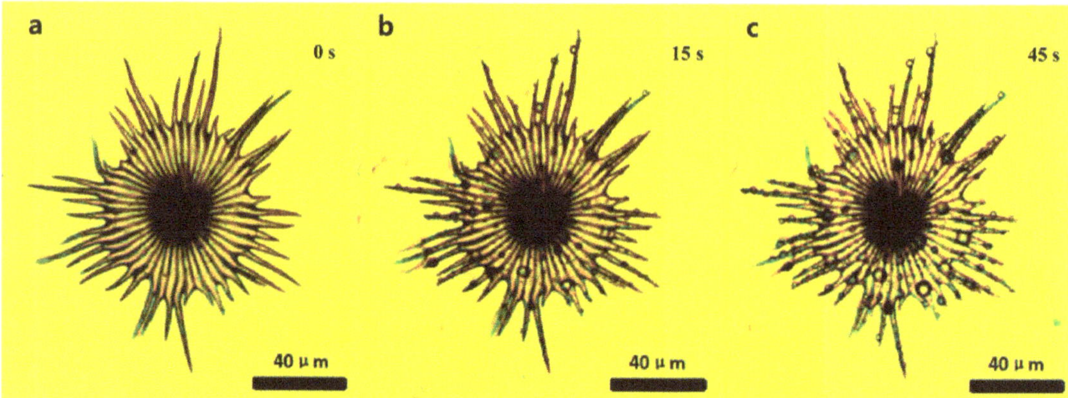

Figure 4. Water-collecting process of single trichome on the leaf surface of *E. angustifolia* (**a**→**b**→**c**). (**a**) The water-collecting of a single trichome at 0 s; (**b**) The water-collecting of a single trichome at 15 s; (**c**) The water-collecting of a single trichome at 45 s.

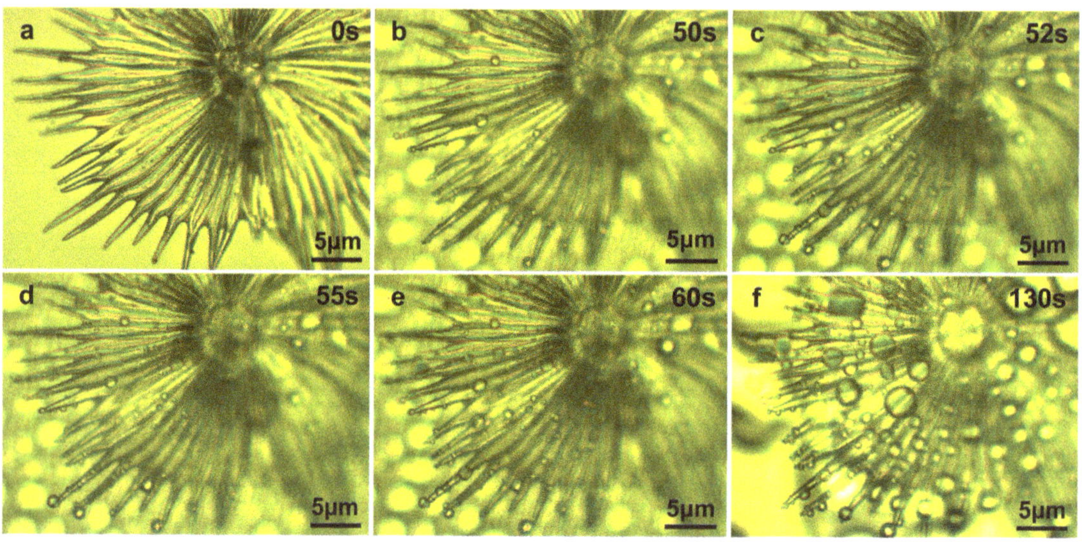

Figure 5. Water-collecting process of a single trichome with local magnification on the leaf surface of *E. angustifolia* (**a**→**b**→**c**→**d**→**e**→**f**).(**a**) The water-collecting of a single trichome with local magnification at 0 s; (**b**) The water-collecting of a single trichome with local magnification at 50 s; (**c**) The water-collecting of a single trichome with local magnification at 52 s; (**d**) The water-collecting of a single trichome with local magnification at 55 s; (**e**) The water-collecting of a single trichome with local magnification at 60 s; (**f**) The water-collecting of a single trichome with local magnification at 130 s.

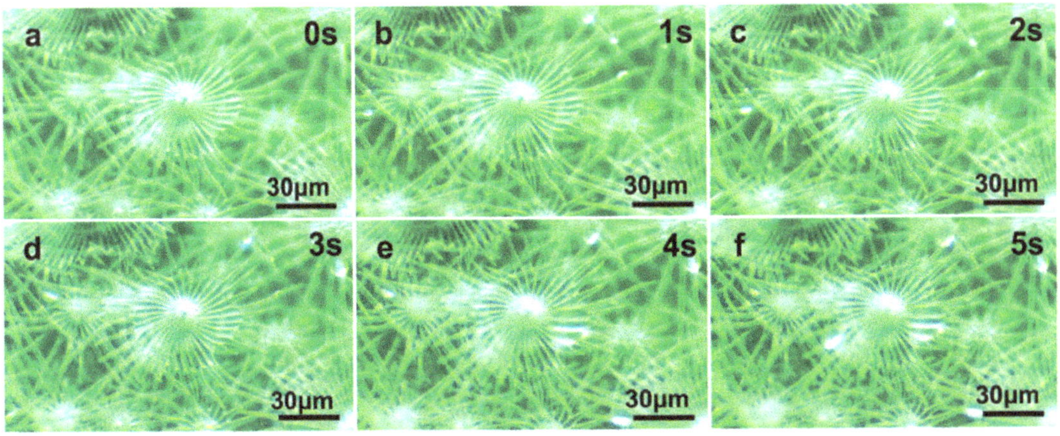

Figure 6. Water-collecting process of multiple trichomes on the leaf surface of *E. angustifolia* (**a**→**b**→**c**→**d**→**e**→**f**).(**a**) The water-collecting of multiple trichomes at 0 s; (**b**) The water-collecting of multiple trichomes at 1 s; (**c**) The water-collecting of multiple trichomes at 2 s; (**d**) The water-collecting of multiple trichomes at 3 s; (**e**) The water-collecting of multiple trichomes at 4 s; (**f**) The water-collecting ofmultiple trichomes at 5 s.

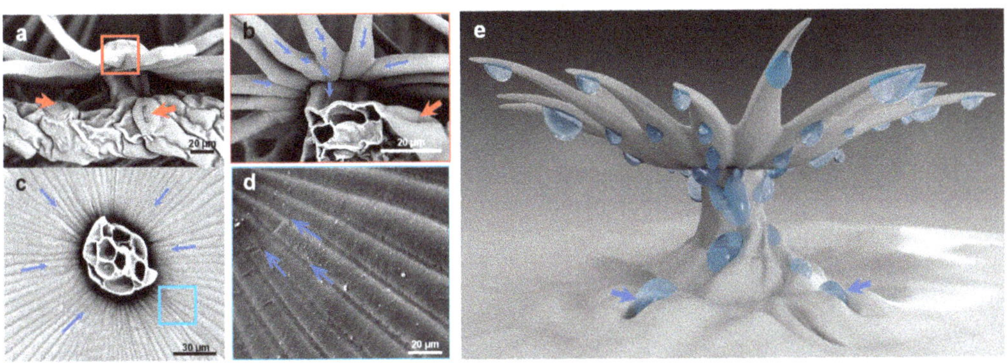

Figure 7. Water-collecting process of single trichome on leaf surface of *E. angustifolia.* (**a**) Side view of single umbrella ratchet trichome, in which the red arrow indicates the stomata on the leaf surface; (**b**) enlarged view of the trichome root of umbrella ratchet, in which the blue arrow represents the moving direction of droplets and the red arrow indicates the pores on the leaf surface; (**c**) the blue arrow in the figure represents the moving direction of the droplet; (**d**) the umbrella ratchet trichomes the groove structure on the back of the umbrella surface, and the blue arrow in the figure represents the moving direction of the droplet; (**e**) water-collection pattern of single trichome, in which the blue arrow indicates that the droplet enters the stomata on the leaf surface.

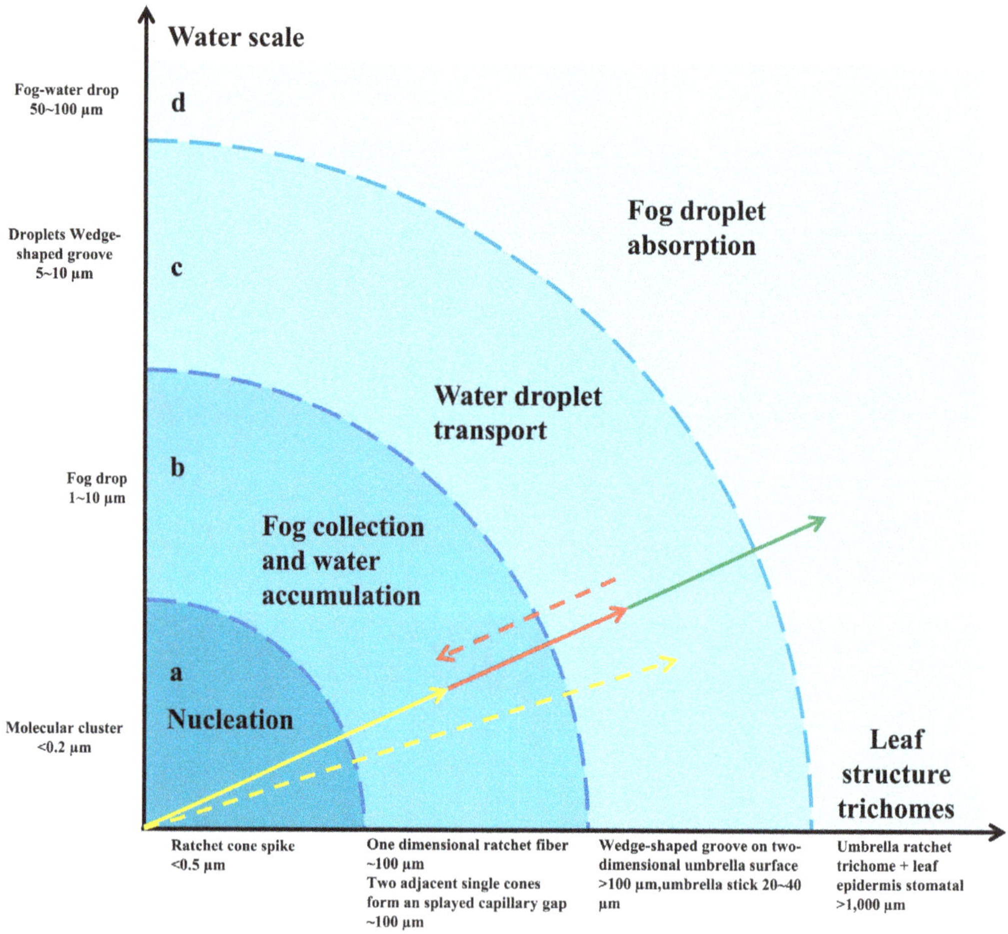

Figure 8. Multifunctional layered water-collecting mechanism of umbrella ratchet trichomes on leaf surface. The concentric shaded area shows the functional mapping between the size trichome of the structure on the umbrella ratchet trichome and the size trichome of water a–d. The arrow shows that the trichome is a catchment system composed of grading mechanisms, which together effectively capture all available water. a. Nucleation improves overall surface properties by providing a layer of water. The overall wettability is improved, providing more favorable nucleation, fog capture (yellow solid arrow), and water transport (yellow dotted arrow). b. Fog collection (one-dimensional cone prick fiber) and water accumulation (two adjacent single cones form an open octagonal capillary gap) gather large droplets (solid red arrow). c. The water droplets are transported on the back of the two-dimensional umbrella trichome (wedge-shaped groove), flow from the top of the umbrella handle to the base of the umbrella handle, enter the leaf surface pores, and prepare the next group of water droplets for the one-dimensional fiber cone to pierce the umbrella surface of the two-dimensional wedge-shaped groove (red dashed arrow). d. Large droplets are absorbed by multiple fiber cones (umbrella ratchet trichome clusters) and leaf surfaces (pores) (solid green arrows).

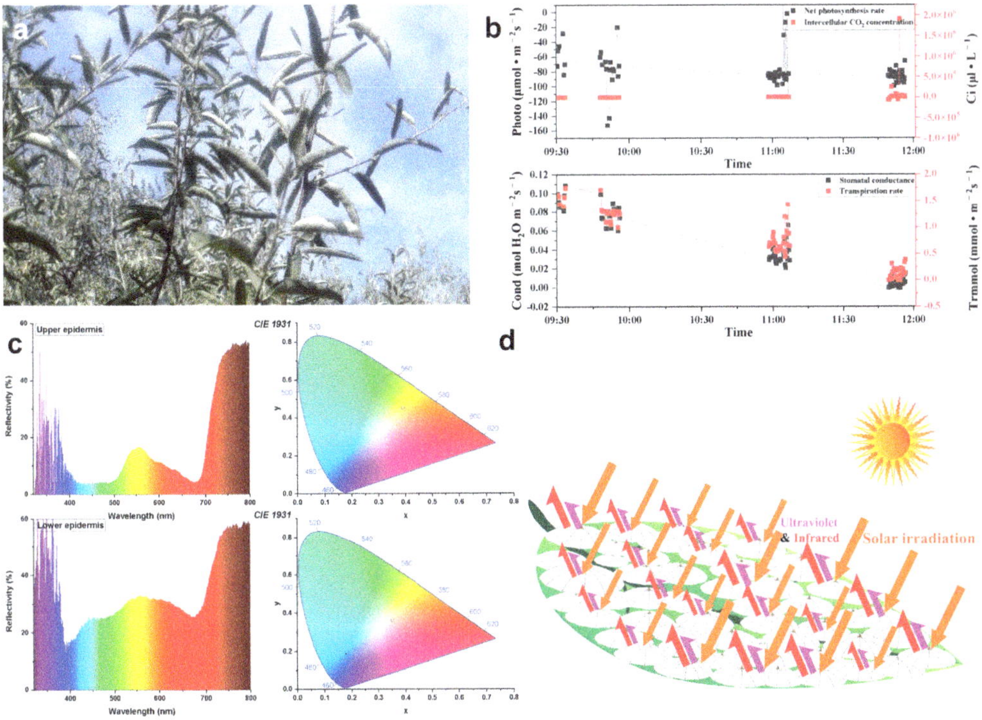

Figure 9. Response diagram of *E. angustifolia* leaves to light. When the leaves of *E. angustifolia* respond to strong sunlight, they reduce due to photosynthesis and irradiate the back of the leaves fully covered with trichomes to the sunlight so as to reduce the absorption of sunlight radiation heat by the leaves. (**a**) At noon, the leaves of *E. angustifolia* curl or turn over, exposing the back of the leaves to the sun; (**b**) changes in leaf photosynthetic rate (Net photosynthesis rate), intercellular CO_2 concentration, stomatal conductance, and transpiration rate of *E. angustifolia* leaves; (**c**) reflection spectra of the front and back of *E. angustifolia* leaves; (**d**) pattern diagram of light response of umbrella ratchet trichomes on the leaf surface of *E. angustifolia*.

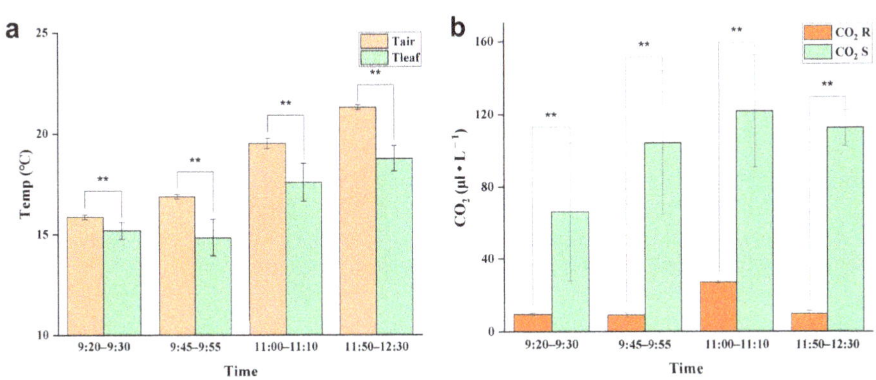

Figure 10. *Cont.*

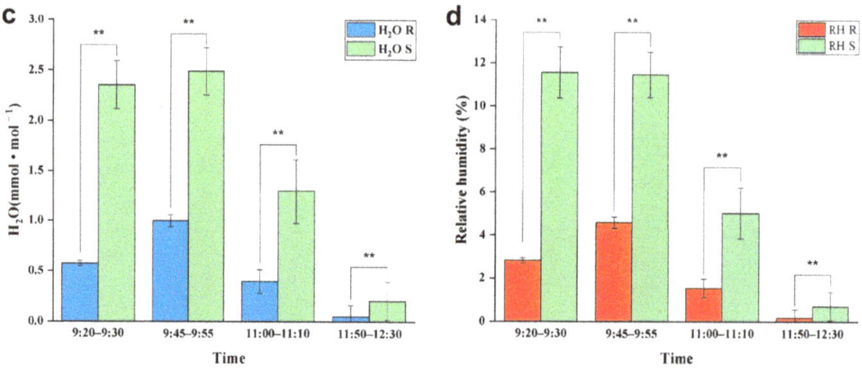

Figure 10. Response of *E. angustifolia* leaves to environment. (**a**) Change in temperature on the leaves with time (Tleaf is the temperature on the leaves; Tair is the temperature in the environment); (**b**) change in CO_2 on the leaves with time (CO_2 S is the CO_2 on the leaves; CO_2 R is the CO_2 in the environment); (c) change in H_2O on the leaves with time (H_2O S is the CO_2 on the leaves; H_2O R is the H_2O in the environment); (**d**) change in relative humidity on the leaves with time (RH S is the relative humidity on the leaves; RH R is the relative humidity in the environment). *p* values were determined by Student's *t*-test (** $p < 0.01$).

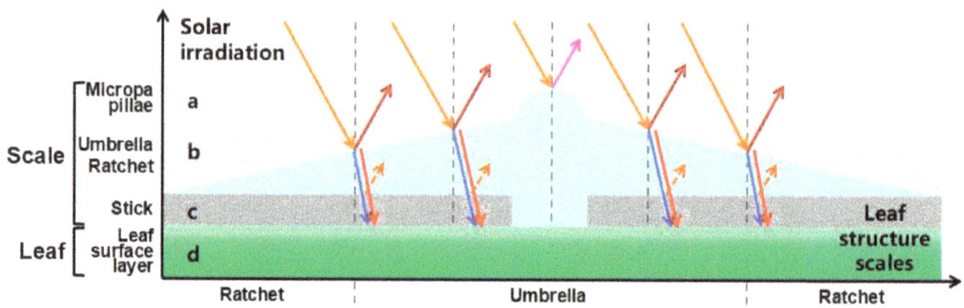

Figure 11. Screening of umbrella ratchet trichomes on leaf surface and mechanism of sunlight. The different shaded areas show the selection range a–c of different parts on the umbrella ratchet trichome for sunlight. All the arrows show the sun's rays on the trichome. a. The microemulsion on the top of the umbrella ratchet trichome reflects the ultraviolet rays (solid purple arrow). b. The umbrella surface part of the umbrella like ratchet trichome reflects the near-infrared ray (solid dark red arrow), and the blue light (blue dotted arrow) and red light (red dotted arrow) reach the blade surface through the umbrella surface part. c. Gap between umbrella-shaped ratchet trichomes and the leaf surface. d. Leaf surface. The gray vertical dotted line is the normal line.

4. Discussion

4.1. Occult Precipitation Is an Important Source of Water in Arid Regions

In arid regions with little precipitation and strong sunlight radiation, the relative air humidity fluctuates between 30% and 90% with the changing temperature [29]. The natural precipitation and occult precipitation (e.g., nucleation droplets and fog droplets) are two different sources for plant water collection [30], of which natural precipitation is the main source [31]. Although the water amount of natural precipitation is large, the duration is often short, with a long dry interval between two precipitation peaks. Occult precipitation accounts for 13% of the total surface water sources as an important source of water for plants to survive during the intermittent period between natural precipitation [32]. Different from natural precipitation, the water amount of occult precipitation, in which water is adsorbed on the surface of plants, is small [33]. However, its duration is longer and more stable than that of natural precipitation, which is of great significance to the survival of plants and small animals in arid regions [34].

Natural *E. angustifolia* forests are widely distributed in arid and semi-arid regions, and the abaxial surface of *E. angustifolia* leaves is fully covered with umbrella-shaped ratchet trichomes. In this study, observing *E. angustifolia* leaves on cloudy and windless mornings, we found that the abaxial surface of the leaves was moist. This indicates that the abaxial surface of *E. angustifolia* leaves can intercept the occult precipitation from the air, which is an important mechanism for *E. angustifolia* to adapt to the arid environment.

4.2. Umbrella-Shaped Ratchet Trichomes on the E. angustifolia Leaf Surface Can Capture and Directionally Transport Water Droplets

The wetting process in solid materials exists widely in nature; the homogeneous rough surface is hydrophilic [35], while the non-homogeneous rough surface is hydrophobic [36]. Materials with anisotropic wetting characteristics can control the movement behavior of droplets in a specific direction, enabling directional droplet transport [37]. A droplet can be driven by factors related to asymmetric gradients [38,39], thermodynamics [40,41], and chemistry [42,43]. For example, droplets on the surface of asymmetric objects can be driven to move toward regions with a larger radius due to the Laplace pressure generated by the asymmetric gradients.

The phenomenon that water accumulates on the leaf surface and enters the leaf interior is called foliar water uptake (FWU), which can be facilitated with the assistance of stomata, cuticles [44], trichomes [45], or hydathodes [46]. Fog droplets usually harbor a smaller diameter (1–15 μm) compared to rain droplets (0.5–5 mm) [47]. On a nanostructure interface, a single microdroplet with a diameter of 2–4 μm gradually forms an approximately spherical droplet with a diameter of 4–6 μm [48,49]. A single umbrella-shaped ratchet trichome on the *E. angustifolia* leaf surface exhibited a 3D interface composed of anisotropic 1D and 2D structures (Figure 2a). The 1D ratchet structure was composed of a plurality of radial cones with an acute apex angle ($\theta_1 = 7.46 \pm 1.72°$) and a micro-/nanostructure surface (Figure 2b,f). The conical structure [8,38,50] produced the Laplace pressure difference between two asymmetric gradients of the droplet. The capillary force (Figure 2b,f) generated by the angle between two adjacent conical spines ($\theta_2 = 15.48 \pm 2.64°$) could further drive the droplets to the base of the spine. On the surface of 2D materials, the directional transport of liquids can be achieved through surface-energy gradients and Laplace pressure gradients. It has been reported that the water-collection ability is enhanced with the size of the patterned surface with a star-shape decrease from 1000 to 250 μm [51]; the patterned surface with a wedge-shaped array structure can also efficiently collect water [52]. The 2D umbrella surface structure of the ratchet trichomes is a patterned surface composed of multiple wedge-shaped groove structures (Figure 2a), which can also enhance the water-collection ability. In addition, since CO_2 diffuses 10,000 times slower in water than in air [53], there is a strong selective pressure on land plant leaves to enhance water repellency. Trichomes on the leaf surface have a great influence on water repellency of the leaf (i.e., the extent to which droplets form on the leaf surface), as well as on the retention of droplets on the leaf [54]. The micropapillary structure on the apex of the trichomes (Figure 2b) increased the heterogeneous roughness of the leaf, making the leaf superhydrophobic (Figure 3).

The water-collection process of the 3D interface structure of the trichomes on the *E. angustifolia* leaf surface exhibited a hierarchical asymmetric effect. During the nucleation process, the droplets first randomly condensed on the spine surface (Figure 4b,c and Video S1) and then converged toward the base of the spine, where the droplets aggregated to form larger droplets and moved from the semi-conical structure to the patterned surface with wedge-shaped grooves. Then, the larger droplets moved along the patterned surface (Figure 5b–f and Video S2), enabling the directional transport of droplets from micro-trichome to macro-trichome (Figures 6b–f and 7, Video S3). However, the water-collection efficiency of trichomes, either a single trichome or multiple trichomes, has not been thoroughly investigated in this study, which needs to be further examined in the future. Compared with other water-collection systems, such as 1D fibers [8,9], highly irregular 2D surfaces [55], and 3D leaf–trichome hierarchical structures [27], each umbrella-shaped ratchet trichome on the *E. angustifolia* leaf surface is an independent microsystem with a 3D interface structure composed of 1D and 2D structures that can collect water hierarchically (Figure 8), enabling the capture of water droplets and the directional transport of water. In the natural environment, the apex of the trichomes on the abaxial surface of *E. angustifolia* leaves is facing the ground, like many umbrella-shaped containers hanging upside down. The water droplets in the air aggregate into larger droplets via the spines on the periphery of the umbrella-shaped structure and are transported to the umbrella-shaped container for temporary storage through the umbrella surface, which strengthens the collection and retention of water droplets and prevents the collected large droplets from falling on the ground and being lost.

4.3. *Leaves Are the Main Plant Organ Responding to Solar Radiation*

The spectrum of sunlight has a wide range, including UV (100–400 nm, 5%), visible (400–700 nm, 46%), and infrared rays (>700 nm, 49%), among which NIR lights (700–1100 nm) are the major source of heat on the Earth's surface [56,57]. Plants mainly absorb blue (400–520 nm) and red lights (610–720 nm) during photosynthesis and are most sensitive to NIR radiation [58]. The heat generated after the absorption of NIR radiation can drive the leaves to transpire [59,60]. In high-temperature regions with strong sunlight radiation, the cooling mechanism of plants is usually at the expense of losing transpiration-related water on the leaf surface. In other plants, the pubescent trichomes on the leaf surface can increase reflectivity, preventing damage to plants from radiative overheating [10,61]. Therefore, plant cooling in high-temperature regions is more challenging than heating.

The surfaces of plant leaves are rarely completely flat, and their epidermal cells show a convex curvature (Figure 2c), resulting in slightly rounded, rounded, papillary, or even sharply conical shapes. It has been reported that the light-collection performance of plants depends on the morphology of the microstructure array on the surface of plant tissues [62]. Specifically, the leaves or petals of plants can combine the micro-/nanostructures of epidermal cells to reduce the reflection loss and redirect incident photons, eventually inducing the light-focusing effect. For example, the most common conical cell structure can reduce the specular reflection on the petal surface, which is attributed to the multiple reflections between the micropapillae on the apex of conical cells. The authors of designed a "silicon pyramid" array with a height of 20 μm and a vertex spacing of 4 μm on the surface of a solar panel [63]. Compared with the solar panel without the array, the temperature of the solar panel with the silicon pyramid array can be reduced by 17.6 K, and it also shows superhydrophobicity and self-cleaning properties. The leaf epidermis acts as the first layer of the optical boundary and plays a key role in controlling the entry of light into the leaf. Some studies have pointed out that leaf surface trichomes can reduce the damage to the proteins of mesophyll cells caused by incident UV radiation [64]. Other studies have pointed out that the density of trichomes on the *E. angustifolia* leaf surface has an impact on the reflectance index of the visible lights and infrared lights. Upon dehairing (trichomes), considerable decreases in the visible reflectance (400–700 nm) were accompanied by slight decreases in the infrared reflectance (700–1000 nm) [26]. In this study, the *E. angustifolia* leaves showed a strong reflection in the UV and NIR regions (Figure 9c). Studies have shown that the height (h), diameter (w), and radius (r) of the micropapillae on the apex of the conical cells on the petal surface can all affect the absorption of visible and UV [65]. The greater the micropapillary height (20.00 ± 4.00 to 79.90 ± 8.70 μm), the stronger the UV-light absorption; the larger the micropapillary apex angle (20° to 118°), the stronger the UV-light reflection [66]. There are micropapillae on the apex of the umbrella-shaped ratchet trichomes (Figure 1d,e), with a height of 30.00 ± 9.10 μm, a basal diameter of 58.20 ± 9.30 μm, an aspect ratio of 0.3–0.7, and an apex angle of 115.30 ± 9.30° (Figure 2d). Therefore, the micropapillae on the apex of the umbrella-shaped trichomes can reflect UV, avoiding damage to proteins of mesophyll cells from UV and assisting the visits of pollinators [67]. In contrast to reflectance at shorter wavelengths, the MIR reflectance spectra cuticular development is the main driving factor [68]. The multiple spines on the periphery of the ratchet trichome may also diffract lights, while the trichome can transmit red and blue lights and reflect NIR (Figure 11). It is possible that the air gap of the hollow structure of the trichomes can improve the directional reflectivity of NIR lights since the hollow structure provides a second planar interface, from which light can be specularly reflected. Moreover, the wedge-shaped structure patterning on the surface of the umbrella-shaped trichomes produces a combined optical response that is highly directional, which is also the key to light reflection and needs to be further validated in the future. In summary, the umbrella-shaped ratchet trichomes on the *E. angustifolia* leaf surface have the ability to filter sunlight.

5. Conclusions

In this study, we first reported that the trichomes on the leaf surface of *E. angustifolia*, a sandy sub-tree species that lives in arid and semi-arid regions, have excellent structural characteristics absorbing occult precipitation with a special umbrella-shaped ratchet structure, and the peripheral conical structure can effectively capture water droplets in the air. The radial cones can capture water droplets, which are transported directionally through the wedge-shaped groove structure of the umbrella surface to the central point of the umbrella handle and aggregate into larger droplets. These larger droplets eventually reach the base of the umbrella handle and enter the leaf interior through the stomata. In addition, trichomes also harbor a special structure that can reflect sunlight. At noon on a cloudless day, the leaf curls or turns over; thus, the umbrella-shaped ratchet trichomes on the abaxial surface of the leaf are exposed to the sunlight. The trichomes can selectively transmit blue and red lights and reflect the UV and NIR lights, reducing leaf temperature. This study is of significance for understanding how plants with medium-sized leaves respond to environmental stresses in arid regions, providing a valuable reference for bionic development and research in atmospheric water harvesting, seawater desalination, energy management, microfluidic control, and daytime radiative cooling.

Supplementary Materials: The following supporting information can be downloaded at: https://www.mdpi.com/article/10.3390/biology12071024/s1, Video S1. Water-collection video of a single trichome on the *Elaeagnus angustifolia* leaf surface; Video S2. Water-collection video of a single trichome on the *Elaeagnus angustifolia* leaf surface with local magnification; Video S3. Water-collection video of multiple trichomes on the *Elaeagnus angustifolia* leaf surface.

Author Contributions: Z.B.: Writing—original draft, Data curation, Formal analysis, Conceptualization, Methodology, Investigation. X.Z.: Methodology, Investigation. X.T.: Writing—review and editing, Methodology, Funding acquisition, Project administration, Supervision. All authors have read and agreed to the published version of the manuscript.

Funding: This research was funded by the Water Conservancy Science and Technology Project of Jiangsu Province (No. ZQ2018107), the National Natural Science Foundation of China (No. 32260312, No. 31870598). the Natural Science Foundation of Ningxia China (No. 2021AAC03190, No. 2023AAC03294) and the Baijitan Scientific Research Project of the North Minzu University (No. SKBJT202204).

Institutional Review Board Statement: Not applicable.

Informed Consent Statement: Not applicable.

Data Availability Statement: This study did not generate a datasets.

Conflicts of Interest: The authors declare no conflict of interest.

References

1. Kasten, F.; Czeplak, G. Solar and terrestrial radiation dependent on the amount and type of cloud. *Sol. Energy* **1980**, *24*, 177–189. [CrossRef]

2. Prăvălie, R. Drylands extent and environmental issues. A global approach. *Earth-Sci. Rev.* **2016**, *161*, 259–278. [CrossRef]
3. Dregne, H.E. Land degradation in the drylands. *Arid. Land Res. Manag.* **2002**, *16*, 99–132. [CrossRef]
4. Vicente-Serrano, S.M.; Gouveia, C.; Camarero, J.J.; Beguería, S.; Trigo, R.; López-Moreno, J.I.; Azorín-Molina, C.; Pasho, E.; Lorenzo-Lacruz, J.; Revuelto, J. Response of vegetation to drought time-scales across global land biomes. *Proc. Natl. Acad. Sci. USA* **2013**, *110*, 52–57. [CrossRef] [PubMed]
5. Breshears, D.D.; Cobb, N.S.; Rich, P.M.; Price, K.P.; Allen, C.D.; Balice, R.G.; Romme, W.H.; Kastens, J.H.; Floyd, M.L.; Belnap, J. Regional vegetation die-off in response to global-change-type drought. *Proc. Natl. Acad. Sci. USA* **2005**, *102*, 15144–15148. [CrossRef]
6. Miriti, M.N.; Rodríguez-Buriticá, S.; Wright, S.J.; Howe, H.F. Episodic death across species of desert shrubs. *Ecology* **2007**, *88*, 32–36. [CrossRef]
7. Pivovaroff, A.L.; Pasquini, S.C.; De Guzman, M.E.; Alstad, K.P.; Stemke, J.S.; Santiago, L.S. Multiple strategies for drought survival among woody plant species. *Funct. Ecol.* **2016**, *30*, 517–526. [CrossRef]
8. Ju, J.; Bai, H.; Zheng, Y.; Zhao, T.; Fang, R.; Jiang, L. A multi-structural and multi-functional integrated fog collection system in cactus. *Nat. Commun.* **2012**, *3*, 1247. [CrossRef]
9. Pan, Z.; Pitt, W.G.; Zhang, Y.; Wu, N.; Tao, Y.; Truscott, T.T. The upside-down water collection system of *Syntrichia caninervis*. *Nat. Plants* **2016**, *2*, 16076. [CrossRef]
10. Ye, C.; Li, M.; Hu, J.; Cheng, Q.; Jiang, L.; Song, Y. Highly reflective superhydrophobic white coating inspired by poplar leaf hairs toward an effective "cool roof". *Energy Environ. Sci.* **2011**, *4*, 3364–3367. [CrossRef]
11. Weyl, P.; Asadi, G.A.; Cristofaro, M.; Vidovic, B.; Petanovic, R.; Marini, F.; Schaffner, U. The host range and impact of *Aceria angustifoliae* (Eriophyidae), a potential biological control agent against Russian olive, *Elaeagnus angustifoliae* (Elaeagnaceae) in North America. *Biocontrol Sci. Technol.* **2020**, *30*, 85–92. [CrossRef]
12. Sun, M.; Lin, Q. A revision of *Elaeagnus* L. (Elaeagnaceae) in mainland China. *J. Syst. Evol.* **2010**, *48*, 356–390. [CrossRef]
13. Bailey, L.H. *The Standard Cyclopedia of Horticulture*; Macmillan: London, UK, 1914; Volume 2.
14. Collette, L.K.; Pither, J. Russian-olive (*Elaeagnus angustifolia*) biology and ecology and its potential to invade northern North American riparian ecosystems. *Invasive Plant Sci. Manag.* **2015**, *8*, 1–14. [CrossRef]
15. Safdari, L.; Khadivi, A. Identification of the promising oleaster (*Elaeagnus angustifolia* L.) genotypes based on fruit quality-related characters. *Food Sci. Nutr.* **2021**, *9*, 5712–5721. [CrossRef] [PubMed]
16. Zhang, X.; Li, G.; Du, S. Simulating the potential distribution of *Elaeagnus angustifolia* L. based on climatic constraints in China. *Ecol. Eng.* **2018**, *113*, 27–34. [CrossRef]
17. Liu, B.; Zhao, W.; Jin, B. The response of sap flow in desert shrubs to environmental variables in an arid region of China. *Ecohydrology* **2011**, *4*, 448–457. [CrossRef]
18. Dong, S.; Wan, S.; Kang, Y.; Li, X. Establishing an ecological forest system of salt-tolerant plants in heavily saline wasteland using the drip-irrigation reclamation method. *Agric. Water Manag.* **2021**, *245*, 106587. [CrossRef]
19. Sudnik, B.; Wójcikowska, I.M.; Slim, P.A.; Moraczewski, I.R. Impact of the Invasive Species *Elaeagnus angustifolia* L. on Vegetation in Pontic Desert Steppe Zone (Southern Ukraine). *Pol. J. Ecol.* **2009**, *57*, 269–281.
20. Abramova, L.; Mustafina, A.; Golovanov, Y.; Zhigunov, O.Y.; Anishchenko, I.; Shigapov, Z.K. Features of the Biology and Ecology of *Elaeagnus angustifolia* L. in Southern Ural. *Contemp. Probl. Ecol.* **2021**, *14*, 446–455. [CrossRef]
21. Liu, Z.; Zhu, J.; Yang, X.; Wu, H.; Wei, Q.; Wei, H.; Zhang, H. Growth performance, organ-level ionic relations and organic osmoregulation of *Elaeagnus angustifolia* in response to salt stress. *PLoS ONE* **2018**, *13*, e0191552. [CrossRef]
22. Wang, Q.; Ruan, X.; Huang, J.; Xu, N.; Yan, Q. Intra-specific genetic relationship analyses of *Elaeagnus angustifolia* based on RP-HPLC biochemical markers. *J. Zhejiang Univ. Sci. B* **2006**, *7*, 272–278. [CrossRef]
23. Weryszko-Chmielewska, E.; Chernetskyy, M. Structure of trichomes from the surface of leaves of some species of Kalanchoë Adans. *Acta Biol. Cracoviensia Ser. Bot.* **2005**, *47*, 15–22.
24. Bacelar, E.A.; Correia, C.M.; Moutinho-Pereira, J.M.; Gonçalves, B.C.; Lopes, J.I.; Torres-Pereira, J.M. Sclerophylly and leaf anatomical traits of five field-grown olive cultivars growing under drought conditions. *Tree Physiol.* **2004**, *24*, 233–239. [CrossRef] [PubMed]
25. Klich, M.a.G. Leaf variations in *Elaeagnus angustifolia* related to environmental heterogeneity. *Environ. Exp. Bot.* **2000**, *44*, 171–183. [CrossRef] [PubMed]
26. Levizou, E.; Drilias, P.; Psaras, G.K.; Manetas, Y. Nondestructive assessment of leaf chemistry and physiology through spectral reflectance measurements may be misleading when changes in trichome density co-occur. *New Phytol.* **2005**, *165*, 463–472. [CrossRef]
27. Andrews, H.; Eccles, E.; Schofield, W.; Badyal, J. Three-dimensional hierarchical structures for fog harvesting. *Langmuir* **2011**, *27*, 3798–3802. [CrossRef]
28. Lawrence, E.H.; Stinziano, J.R.; Hanson, D.T. *Using the Rapid A-Ci Response (RACiR) in the Li-Cor 6400 to Measure Developmental Gradients of Photosynthetic Capacity in Poplar*; 0140-7791; Wiley Online Library: Hoboken, NJ, USA, 2019.
29. Lee, A.; Moon, M.-W.; Lim, H.; Kim, W.-D.; Kim, H.-Y. Water harvest via dewing. *Langmuir* **2012**, *28*, 10183–10191. [CrossRef]
30. Zangvil, A. Six years of dew observations in the Negev Desert, Israel. *J. Arid. Environ.* **1996**, *32*, 361–371. [CrossRef]
31. Kidron, G.J. Dew moisture regime of endolithic and epilithic lichens inhabiting limestone cobbles and rock outcrops, Negev Highlands, Israel. *Flora* **2000**, *195*, 146–153. [CrossRef]

32. Kidron, G.J.; Herrnstadt, I.; Barzilay, E. The role of dew as a moisture source for sand microbiotic crusts in the Negev Desert, Israel. *J. Arid. Environ.* **2002**, *52*, 517–533. [CrossRef]
33. Zhang, J.; Zhang, Y.; Downing, A.; Cheng, J.; Zhou, X.; Zhang, B. The influence of biological soil crusts on dew deposition in Gurbantunggut Desert, Northwestern China. *J. Hydrol.* **2009**, *379*, 220–228. [CrossRef]
34. Moffett, M.W. An Indian ant's novel method for obtaining water. *Natl. Geogr. Res.* **1985**, *1*, 146–149.
35. Robert, N.W. Resistance of solid surfaces to wetting by water. *Ind. Eng. Chem.* **1936**, *28*, 988–994.
36. Cassie, A.; Baxter, S. Wettability of porous surfaces. *Trans. Faraday Soc.* **1944**, *40*, 546–551. [CrossRef]
37. Bliznyuk, O.; Jansen, H.P.; Kooij, E.S.; Zandvliet, H.J.; Poelsema, B. Smart design of stripe-patterned gradient surfaces to control droplet motion. *Langmuir* **2011**, *27*, 11238–11245. [CrossRef]
38. Lorenceau, E.; Quéré, D. Drops on a conical wire. *J. Fluid Mech.* **2004**, *510*, 29–45. [CrossRef]
39. Zhang, J.; Han, Y. Shape-gradient composite surfaces: Water droplets move uphill. *Langmuir* **2007**, *23*, 6136–6141. [CrossRef]
40. Yarin, A.L.; Liu, W.; Reneker, D.H. Motion of droplets along thin fibers with temperature gradient. *J. Appl. Phys.* **2002**, *91*, 4751–4760. [CrossRef]
41. Mettu, S.; Chaudhury, M.K. Motion of drops on a surface induced by thermal gradient and vibration. *Langmuir* **2008**, *24*, 10833–10837. [CrossRef]
42. Chaudhury, M.K.; Whitesides, G.M. How to make water run uphill. *Science* **1992**, *256*, 1539–1541. [CrossRef]
43. Daniel, S.; Chaudhury, M.K.; Chen, J.C. Fast drop movements resulting from the phase change on a gradient surface. *Science* **2001**, *291*, 633–636. [CrossRef] [PubMed]
44. Guzmán-Delgado, P.; Laca, E.; Zwieniecki, M.A. Unravelling foliar water uptake pathways: The contribution of stomata and the cuticle. *Plant Cell Environ.* **2021**, *44*, 1728–1740. [CrossRef]
45. Emery, N.C. Foliar uptake of fog in coastal California shrub species. *Oecologia* **2016**, *182*, 731–742. [CrossRef] [PubMed]
46. Martin, C.; von Willert, D.J. Leaf epidermal hydathodes and the ecophysiological consequences of foliar water uptake in species of Crassula from the Namib Desert in southern Africa. *Plant Biol.* **2000**, *2*, 229–242. [CrossRef]
47. Garland, J. Some fog droplet size distributions obtained by an impaction method. *Q. J. R. Meteorol. Soc.* **1971**, *97*, 483–494. [CrossRef]
48. Rykaczewski, K.; Scott, J.; Fedorov, A.G. Electron beam heating effects during environmental scanning electron microscopy imaging of water condensation on superhydrophobic surfaces. *Appl. Phys. Lett.* **2011**, *98*, 093106. [CrossRef]
49. Rykaczewski, K.; Scott, J.H.J.; Rajauria, S.; Chinn, J.; Chinn, A.M.; Jones, W. Three dimensional aspects of droplet coalescence during dropwise condensation on superhydrophobic surfaces. *Soft Matter* **2011**, *7*, 8749–8752. [CrossRef]
50. Zheng, Y.; Bai, H.; Huang, Z.; Tian, X.; Nie, F.-Q.; Zhao, Y.; Zhai, J.; Jiang, L. Directional water collection on wetted spider silk. *Nature* **2010**, *463*, 640–643. [CrossRef] [PubMed]
51. Bai, H.; Wang, L.; Ju, J.; Sun, R.; Zheng, Y.; Jiang, L. Efficient water collection on integrative bioinspired surfaces with star-shaped wettability patterns. *Adv. Mater.* **2014**, *26*, 5025–5030. [CrossRef]
52. Wang, X.; Zeng, J.; Li, J.; Yu, X.; Wang, Z.; Zhang, Y. Beetle and cactus-inspired surface endows continuous and directional droplet jumping for efficient water harvesting. *J. Mater. Chem. A* **2021**, *9*, 1507–1516. [CrossRef]
53. Kempe, S. Carbon in the freshwater cycle. *Glob. Carbon Cycle* **1979**, *13*, 317–342.
54. Brewer, C.; Smith, W.; Vogelmann, T. Functional interaction between leaf trichomes, leaf wettability and the optical properties of water droplets. *Plant Cell Environ.* **1991**, *14*, 955–962. [CrossRef]
55. Roth-Nebelsick, A.; Ebner, M.; Miranda, T.; Gottschalk, V.; Voigt, D.; Gorb, S.; Stegmaier, T.; Sarsour, J.; Linke, M.; Konrad, W. Leaf surface structures enable the endemic Namib desert grass *Stipagrostis sabulicola* to irrigate itself with fog water. *J. R. Soc. Interface* **2012**, *9*, 1965–1974. [CrossRef] [PubMed]
56. Zong, J.-Y.; Zhou, X.-J.; Hu, Y.-F.; Yang, T.-B.; Yan, D.-X.; Lin, H.; Lei, J.; Li, Z.-M. A wearable multifunctional fabric with excellent electromagnetic interference shielding and passive radiation heating performance. *Compos. Part B Eng.* **2021**, *225*, 109299. [CrossRef]
57. Pakdel, E.; Xie, W.; Wang, J.; Kashi, S.; Sharp, J.; Zhang, Q.; Varley, R.J.; Sun, L.; Wang, X. Superhydrophobic natural melanin-coated cotton with excellent UV protection and personal thermal management functionality. *Chem. Eng. J.* **2022**, *433*, 133688. [CrossRef]
58. Pakdel, E.; Naebe, M.; Sun, L.; Wang, X. Advanced functional fibrous materials for enhanced thermoregulating performance. *ACS Appl. Mater. Interfaces* **2019**, *11*, 13039–13057. [CrossRef]
59. Bilger, H.-W.; Schreiber, U.; Lange, O. Determination of leaf heat resistance: Comparative investigation of chlorophyll fluorescence changes and tissue necrosis methods. *Oecologia* **1984**, *63*, 256–262. [CrossRef] [PubMed]
60. Henrion, W.; Tributsch, H. Optical solar energy adaptations and radiative temperature control of green leaves and tree barks. *Sol. Energy Mater. Sol. Cells* **2009**, *93*, 98–107. [CrossRef]
61. Sandquist, D.R.; Ehleringer, J.R. Intraspecific variation of leaf pubescence and drought response in *Encelia farinosa* associated with contrasting desert environments. *New Phytol.* **1997**, *135*, 635–644. [CrossRef]
62. Wang, H.; Liang, Y.; Cheng, S.; Li, B.; Li, A.; Du, G.; Hu, W. Bio-inspired nanostructures for enhanced light management. *J. Vac. Sci. Technol. B* **2017**, *35*, 06GJ02. [CrossRef]
63. Zhu, L.; Raman, A.; Wang, K.X.; Abou Anoma, M.; Fan, S. Radiative cooling of solar cells. *Optica* **2014**, *1*, 32–38. [CrossRef]

64. Karabourniotis, G.; Bornman, J.F. Penetration of UV-A, UV-B and blue light through the leaf trichome layers of two xeromorphic plants, olive and oak, measured by optical fibre microprobes. *Physiol. Plant.* **1999**, *105*, 655–661. [CrossRef]
65. Hünig, R.; Mertens, A.; Stephan, M.; Schulz, A.; Richter, B.; Hetterich, M.; Powalla, M.; Lemmer, U.; Colsmann, A.; Gomard, G. Flower power: Exploiting plants' epidermal structures for enhanced light harvesting in thin-film solar cells. *Adv. Opt. Mater.* **2016**, *4*, 1487–1493. [CrossRef]
66. Schulte, A.J.; Mail, M.; Hahn, L.A.; Barthlott, W. Ultraviolet patterns of flowers revealed in polymer replica–caused by surface architecture. *Beilstein J. Nanotechnol.* **2019**, *10*, 459–466. [CrossRef] [PubMed]
67. Klomberg, Y.; Dywou Kouede, R.; Bartoš, M.; Mertens, J.E.; Tropek, R.; Fokam, E.B.; Janeček, Š. The role of ultraviolet reflectance and pattern in the pollination system of *Hypoxis camerooniana* (Hypoxidaceae). *AoB Plants* **2019**, *11*, plz057. [CrossRef]
68. Richardson, A.D.; Aubrecht, D.M.; Basler, D.; Hufkens, K.; Muir, C.D.; Hanssen, L. Developmental changes in the reflectance spectra of temperate deciduous tree leaves and implications for thermal emissivity and leaf temperature. *New Phytol.* **2021**, *229*, 791–804. [CrossRef] [PubMed]

biology

MDPI

Review

Current Progress and Future Trends in Carbon Sources and Sinks in Farmland Ecosystems: A Bibliometric Analysis (2002–2023)

Yugong Pang [1], Menghao Zhang [1], Hesen Zhong [1], Tibihenda Cevin [1,2], Chuanzhun Sun [3], Shoutao Zhang [1], Xinyu Li [1], Jun Dai [1], Chengshuai Liu [1] and Chi Zhang [1,*]

1 College of Natural Resources and Environment, South China Agricultural University, Guangzhou 510642, China; pangyugong1123@163.com (Y.P.); 15737316382@163.com (M.Z.); 15775071468@163.com (H.Z.); kevootibe@gmail.com (T.C.); 19953345795@163.com (S.Z.); lxinyu224gd@163.com (X.L.); jundai@scau.edu.cn (J.D.); csliu@scau.edu.cn (C.L.)
2 Tanzania Agricultural Research Institute, Dodoma 1571, Tanzania
3 College of Public Management, South China Agricultural University, Guangzhou 510642, China; suncz@scau.edu.cn
* Correspondence: zhangchi2012@scau.edu.cn

Simple Summary: Farmland ecosystems play a vital role in global carbon cycling and climate change, yet understanding how these systems balance carbon absorption and emissions remains challenging. This study reviewed 1411 research articles (2002–2023) to identify key findings and challenges in this field. Research has grown steadily over the past two decades, with China, the U.S., and Germany leading global efforts. Recent advances show that soil microbes are central to locking carbon into soils. In contrast, science-based practices like conservation tillage can significantly boost carbon storage. Scientists have also focused on reducing greenhouse gases from farming and using satellite technology to track carbon changes. However, regional differences in carbon processes and inconsistent methods to estimate carbon sources versus sinks create uncertainties. For example, some regions naturally store more carbon, but existing frameworks struggle to explain these variations. Therefore, future research should deeply explore the complex interactions within soil carbon sequestration mechanisms in farmland ecosystems and optimize carbon accounting methods. By clarifying these mechanisms, our findings provide crucial theoretical support for global strategies to achieve agricultural "carbon neutrality", ensuring farmland can mitigate climate change while sustainably producing food.

Abstract: Farmland ecosystems, as the most active carbon pool, are integral to global climate change and carbon cycling. Therefore, systematically studying the roles of carbon sources and sinks in farmland ecosystems is essential to deepening our understanding of the carbon cycle and meeting the goals of "peak carbon emissions" and "carbon neutrality" in agriculture. Using the Web of Science database, this study reviewed 1411 articles (2002–2023) via bibliometric analysis to identify key research themes, trends, future priorities and address suggestions for future directions in farmland ecosystem carbon sources and sinks. The main results include the following: (1) Over the past 22 years, global research in this field has shown a consistent growth trend, with remarkably rapid expansion in the past three years. China, the United States, and Germany are the most influential countries. As the research scope has expanded, the field has evolved into an interdisciplinary domain. (2) The diversity of this research area has become enriched, and the research content is becoming more refined and systematic. The main research topics focus on carbon sequestration, soil organic carbon (SOC), farmland management, greenhouse gas (GHG) emissions, carbon stocks, ecosystem services, land use changes,

 https://doi.org/10.3390/biology14040365

climate change, and spatiotemporal heterogeneity. (3) Current research hotspots primarily focus on studying soil microbial carbon sequestration mechanisms, the application of remote sensing technologies, and reducing GHG emissions to achieve "carbon neutrality". While existing studies have systematically elucidated carbon sequestration mechanisms mediated by soil aggregates, microorganisms, and minerals, critical knowledge gaps persist. Regional disparities in the relative contributions of these mechanisms remain unresolved, compounded by methodological inconsistencies in carbon assessment that introduce substantial uncertainties. Although farmland management practices are identified as pivotal drivers of carbon flux variation, the interactive effects of anthropogenic interventions and natural factors on ecosystem-scale carbon balance require further mechanistic exploration. This review provides a comprehensive reference for further study on carbon sources and sinks of farmland ecosystems and devising effective emission reduction strategies.

Keywords: farmland ecosystem; carbon sources; carbon sink; carbon sequestration; soil organic carbon (SOC); bibliometric analysis

1. Introduction

With the global economy shifting towards achieving net-zero greenhouse gas (GHG) emissions, reducing GHG emissions has emerged as one of the most critical challenges confronting the agricultural sector [1,2]. Farmland ecosystems account for approximately 12% of the global land area and play a dual role as atmospheric carbon sources and sinks [3]. On one hand, farmland-related agricultural activities contribute approximately 23% to 30% of global GHG emissions [4]. Moreover, global farmland carbon stocks range from 128 to 165 Pg C, representing 8% to 10% of the global carbon stock [5,6]. Regardless of their role as a source or sink of GHG, farmland ecosystems significantly impact atmospheric CO_2 concentrations [7,8]. Controlling GHG emissions from farmland and enhancing carbon sequestration in these ecosystems are effective strategies for mitigating climate change and achieving the goals of "peak carbon" and "carbon neutrality".

Scholars have contributed significantly to advancing our understanding of carbon sources and sinks in farmland ecosystems. In examining the carbon sequestration effects and potential of farmland management practices, researchers have explored the impacts of various tillage practices, fertilization methods, and water management patterns on carbon sequestration and emission reduction in farmland soils. West et al. employed a full-carbon-cycle analysis methodology to compare conventional versus no-tillage practices, showing that no-tillage practices offer more significant advantages in enhancing carbon sequestration and reducing emissions [9]. Similarly, Jiang et al. evaluated the impacts of varying nitrogen fertilizer application rates on carbon sequestration and emission reduction in single-season rice production through field experiments conducted in Zhejiang, China. They found that applying 225 $kg \cdot ha^{-1}$ of nitrogen fertilizer in this region could simultaneously enhance grain yield, reduce GHG emissions, and increase system carbon sequestration [10]. Moreover, various methods have been developed to estimate carbon sources and sinks, including the emission factor method, life cycle method, and modeling techniques. Advancements in geographical information system (GIS) and remote sensing technology have facilitated the study of spatiotemporal variation of carbon sources and sinks within farmland ecosystems. For instance, Hei et al. used multi-source remote sensing data to reveal the spatiotemporal variation trends of net primary productivity (NPP) in farmland ecosystems [11]. In addition, some scholars have conducted reviews on carbon source and sink research in farmland ecosystems, but these are often limited to specific research directions or the progress within a specific country. For example, Li et al. conducted a comprehensive review of the uncertainties and discrepancies associated with estimating agricultural GHG emissions in China [12]. Similarly, Liu et al. reviewed GHG measurement methods, carbon source and sink assessment models for farmland ecosystems, as well as the key factors influencing the dynamics of carbon sources and sinks [13]. Nevertheless, their analysis of assessment methods was relatively limited, focusing solely on the emission factor method without providing comparative analyses of other approaches. Furthermore, Li et al. systematically reviewed accounting methods for carbon sources and sinks in farmland ecosystems [14]. Research on carbon sources and sinks in farmland ecosystems represents a global challenge that intersects with multiple disciplines. Focusing solely on specific aspects or studies from a single country makes it difficult to achieve a comprehensive understanding of existing research characteristics and emerging trends. Therefore, a thorough, detailed, and objective summary of existing research in this field is crucial for its future development.

Bibliometric analysis is a well-established method for quantitatively evaluating research progress and identifying developmental trends within specific fields of study [15,16]. This study applies bibliometric analysis to systematically review the literature on carbon sources and sinks in the farmland ecosystem, aiming to identify the existing knowledge domain, recognize emerging trends, and guide future research. This paper mainly addresses four primary objectives: (1) It aims to provide an overview of studies on carbon sources and sinks in farmland ecosystems published from 2002 to 2023. (2) It summarizes the main research topics and principal theories in this field. (3) It analyzes the current research hotspots and future development trends in this field. (4) It identifies existing research gaps and proposes recommendations for future studies.

2. Materials and Methods

In this study, we employed CiteSpace.6.2.R7 and VOSviewer.1.6.20 for bibliometric analysis. CiteSpace is a Java-based tool that supports in-depth data mining, knowledge mapping networks, and visualization of research evolution, including the exploration of emerging research hotspots and frontiers [15]. VOSviewer is a bibliometric tool for knowledge units from the literature with unique advantages in clustering techniques and network mapping visualization, useful for identifying central themes in scientific research in the following areas [17]. The data for this study were sourced from the Web of Science (WOS) database, restricted to the Web of Science Core Collection (WOSCC). Our search parameters in Web of Science were "Topic: ("carbon emission" OR "carbon absorption" OR "carbon sink" OR "carbon source" OR "carbon sequestration") AND ("cropland" OR "farmland" OR "cropland ecosystem")", refined by the following: Document type: (Article or Review); Timespan: 2002–2023; Web of Science Index: Social Sciences Citation Index (SSCI) and Science Citation Index Expanded (SCI-EXPANDED). The search resulted in 1513 articles (Figure 1). All these articles were downloaded on 19 February 2024. To address the issue of irrelevant records from the database, a manual selection process was conducted in two rounds by one Master's student and two PhD students (Figure 1). In the first round, the Master's student eliminated articles irrelevant to our topic by reading the titles, abstracts, and keywords of 1513 articles. In the second round, the two PhD students read the full texts of the remaining articles and determined whether they were relevant to our topic before making further exclusions, resulting in 1411 valid documents as data for analysis (article: 1346; review: 65). The data were then imported into CiteSpace and VOSviewer software for the analysis and visualization of various indicators, while OriginPro 2024b and Microsoft Excel 2016 were used to assist with the graphing (Figure 1).

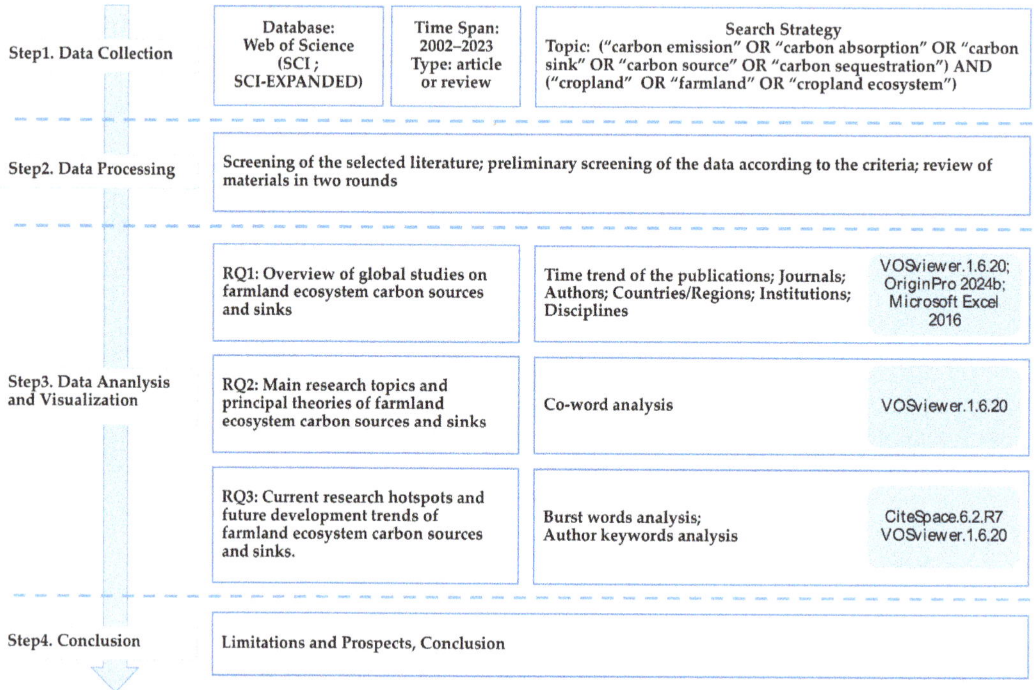

Figure 1. Research framework of this study.

3. Results and Discussion

3.1. Characteristics of the Publications

Figure 2 shows the annual publication volume and publication trends in research on carbon sources and sinks in farmland ecosystems from 2002 to 2023. Only 99 studies (99; 7%) were published before 2008. Afterward, the number of publications increased, potentially related to the start of the Bali Road Map negotiations in 2007. The negotiations of the "Bali Road Map" incorporated all developing countries into the emission reduction framework and proposed a mechanism of "measurable, reportable, and verifiable" (MRV) actions to drive carbon emission reductions. This directly heightened global attention to the dynamics of carbon sources and sinks, promoting research in this field. Thereby, the adoption of the Paris Agreement in 2015, which clarified countries' emission reduction commitments, provided a strong push for research in this field. By 2019, more than half of the publications (730; 51.74%) had been published. The period from 2021 to 2023 is the most productive (carbon neutrality has been proposed globally), with over 40% of the studies (568; 40.2%) published in this period (Figure 2).

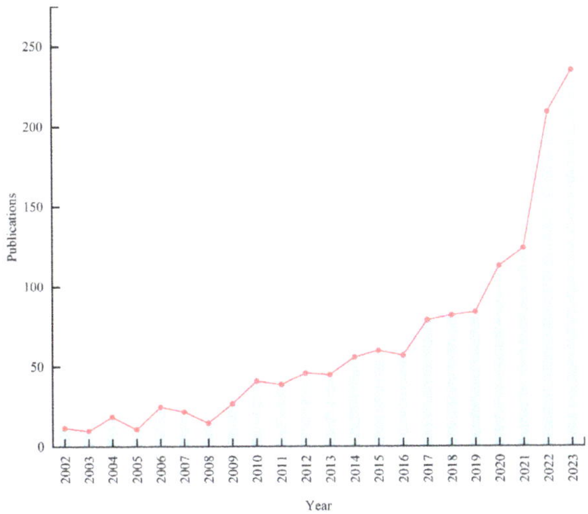

Figure 2. Distribution of publications on carbon sources and sinks of farmland ecosystems published from 2002 to 2023.

3.2. Publication Output Analysis

From 2002 to 2023, 309 journals published research articles on carbon sources and sinks in farmland ecosystems (data from the WOS platform). *Science of the Total Environment* (93 publications; 2800 citations), *Agriculture Ecosystems & Environment* (74 publications; 4903 citations) and *Global Change Biology* (93 publications; 8270 citations) were the journals with the highest numbers of publications and citation rates (Table 1). *Global Change Biology* had 1.69 times more citations than *Agriculture Ecosystems & Environment* and 2.9 times more citations than *Science of the Total Environment*. This highlights these journals' strong international academic influence in the field of carbon sources and sinks in farmland ecosystems, particularly that of *Global Change Biology*, which stands out as an important platform for academic exchange. Additionally, the studies in these journals primarily focus on the environmental sciences, ecosystems, and climate change, reflecting the increased attention to carbon sources and sinks in farmland ecosystems in terms of their ecological and environmental impacts in the context of global climate change.

Table 1. Top 10 journals with published articles in the field of carbon sources and sinks in farmland ecosystems.

Ranking	Journal	Number of Publications	Proportion of Publications (%)	Total Citations
1	*Science of the Total Environment*	93	6.59	2800
2	*Agriculture Ecosystems & Environment*	74	5.24	4903
3	*Global Change Biology*	56	3.97	8270
4	*Catena*	43	3.05	1203
5	*Journal of Cleaner Production*	40	2.83	939
6	*Ecological Indicators*	39	2.76	1375
7	*Land*	35	2.48	168
8	*Sustainability*	34	2.41	360
9	*Geoderma*	31	2.20	1434
10	*Land Degradation & Development*	30	2.13	1197

3.3. Analysis by Author, Institution, Country (Region), and Subject Area

In this study, we employed VOSviewer software for author cluster analysis to identify the most representative scholars in the field of carbon sources and sinks in farmland ecosystems (Table 2) and their collaborations (Figure 3). We set a minimum of five publications per author, with 73 out of 6549 authors meeting the criteria. For each of the 73 authors, the total link strength of co-authorship with others was calculated, and those with the highest link strength were selected. The authors were clustered into eight categories, each marked with a different color in Figure 2. Although Professor Pete Smith (20 publications) and Professor Fu Bojie (11 publications) are the top contributors in terms of the number of publications, they have fewer collaborations (Figure 3). On the other hand, Tian Hanqin, Ingrid Kögel-Knabner, and Martin Wiesmeier both have a notable number of publications and have collaborated extensively with other researchers.

Table 2. Top 10 authors in the field of carbon sources and sinks of farmland ecosystems.

Author	Number of Publications	Total Citations in the Literature	Average Year of Publication
Smith, Pete	20	2824	2016
Tian, hanqin	12	870	2014
Fu, bojie	11	1473	2013
Koegel-knabner, Ingrid	11	821	2016
Wiesmeier, Martin	11	828	2017
Ciais, Philippe	10	582	2020
Deng, Lei	10	1054	2016
Don, Axel	10	916	2018
Kuemmerle, Tobias	10	1054	2015
Prishchepov, alexander V.	10	643	2017

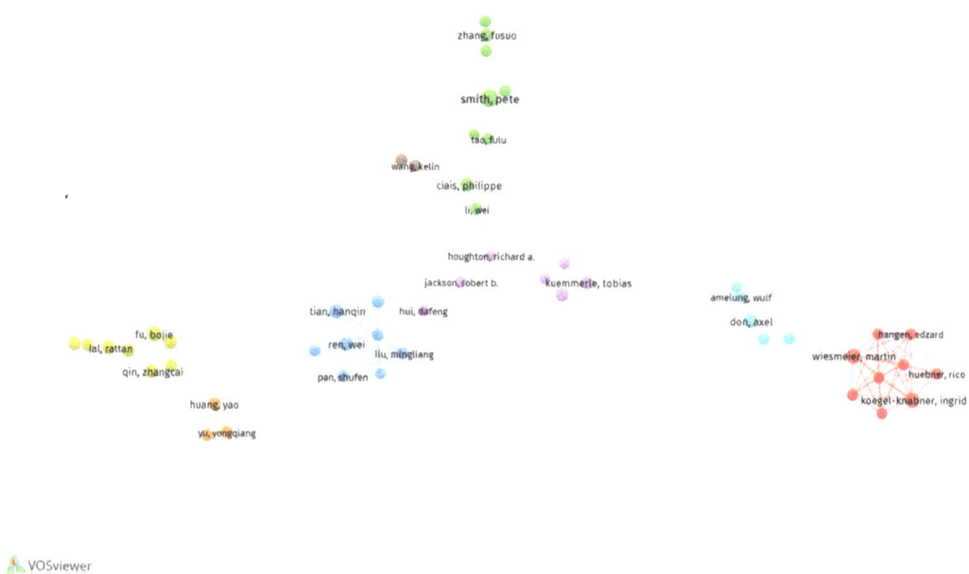

Figure 3. Authors' co-authorship network analysis. Each cluster represents the network of Authors working together.

Publishing organizations are the driving force behind academic research and can provide insight into institutions' importance in a particular field. A total of 1673 organizations worldwide have published studies on carbon sources and sinks in farmland ecosystems, reflecting the global interest in research on the carbon cycle in farmland ecosystems. Over the last two decades, 27 organizations have published more than 15 publications each (Figure 4), with the Chinese Academy of Science (402 publications; 28.49%) having the highest output, followed by the Northwest A&F University (83 publications; 6%), Beijing Normal University (44 publications; 3%), China Agricultural University (43 publications; 3%), and the Chinese Academy of Agricultural Sciences (43 publications; 3%). It is noteworthy that 15 of the 27 institutions are from China, reflecting the nation's commitment to reducing carbon and emissions in farmland ecosystems and indicating the vital role played by Chinese research institutions in this field.

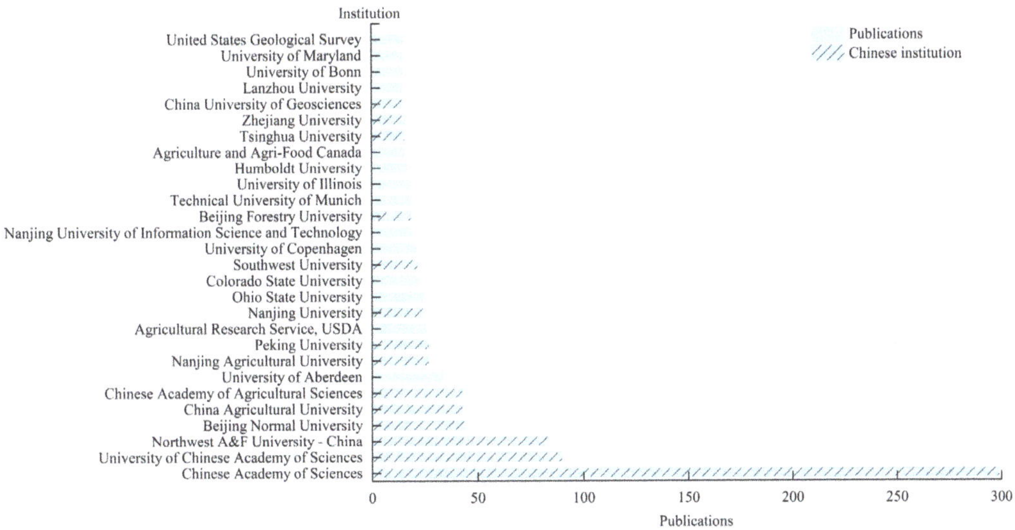

Figure 4. Research institutions or organizations with 15 or more publications.

Collaboration in scientific research enhances the capacity for disciplinary integration, increases the impact of research, and promotes deeper theoretical mechanisms [12]. Publications co-authored by Chinese scholars in collaboration with researchers from other countries represent 39.08% of China's total publication output, underscoring China's extensive engagement in international cooperative efforts in this field (Figure 5). In addition, the United States, Germany, Canada, and Australia are also countries that cooperate closely with other countries in scientific research (Figure 5). Most of the countries with a high number of publications featuring international collaborations are major agricultural countries with advanced scientific and technological capabilities, and all of them are parties to the Paris Agreement, which indicates that they are committed to reducing carbon emissions [18]. This suggests that the international environment, politics, and the strength of scientific research have shaped the evolution of this research arena.

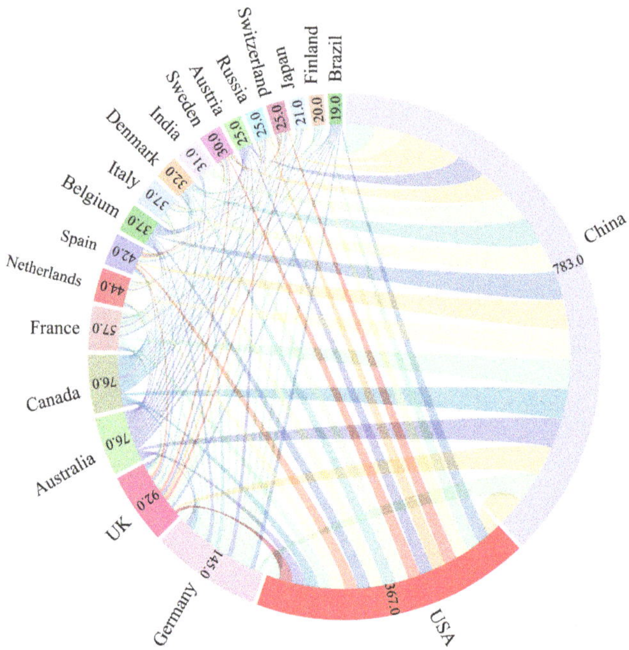

Figure 5. Top 20 countries by volume of collaborative research. The lines of different colours represent cooperation between different countries.

Interdisciplinarity has emerged as a significant approach in modern scientific research [19]. Engaging in interdisciplinary exchanges and collaborations not only advances systematic theoretical research but also propels the rapid development of the field [20]. In the subject matter of farmland ecosystems, the study of carbon sources and sinks is primarily dominated by the disciplines of environmental science, ecology, and soil science, with environmental science being markedly predominant (Figure 6). This dominance reflects the role of farmland ecosystems in responding to global climate change. Ecology and soil science have always been essential disciplines, underscoring the ongoing attention to ecosystem services, biodiversity conservation, and soil carbon sequestration. Over time, the field of water resources has highlighted the study of the interplay between climate change and agricultural soil hydrological processes, attracting growing attention. Moreover, the integration of biotechnology, microbiology, botany, and remote sensing has provided new research perspectives and technical approaches for understanding the carbon sequestration mechanisms within farmland ecosystems.

3.4. Research Theme Analysis

Keywords are a brief summary of the core content of studies, capable of reflecting the main research content in the field of study [21]. This study employed the co-word analysis functionality of VOSviewer to identify the primary research directions within the field. With a minimum of 10 set as thresholds, 228 out of 5752 keywords were selected (Figure 7). In this analysis, the circle size represents keyword frequency, and the line thickness indicates the strength of keyword associations. Keyword clustering separates the current research topic areas of carbon sources and sinks in farmland ecosystems into four categories, and four overarching themes, marked in different colors, become apparent (Figure 7).

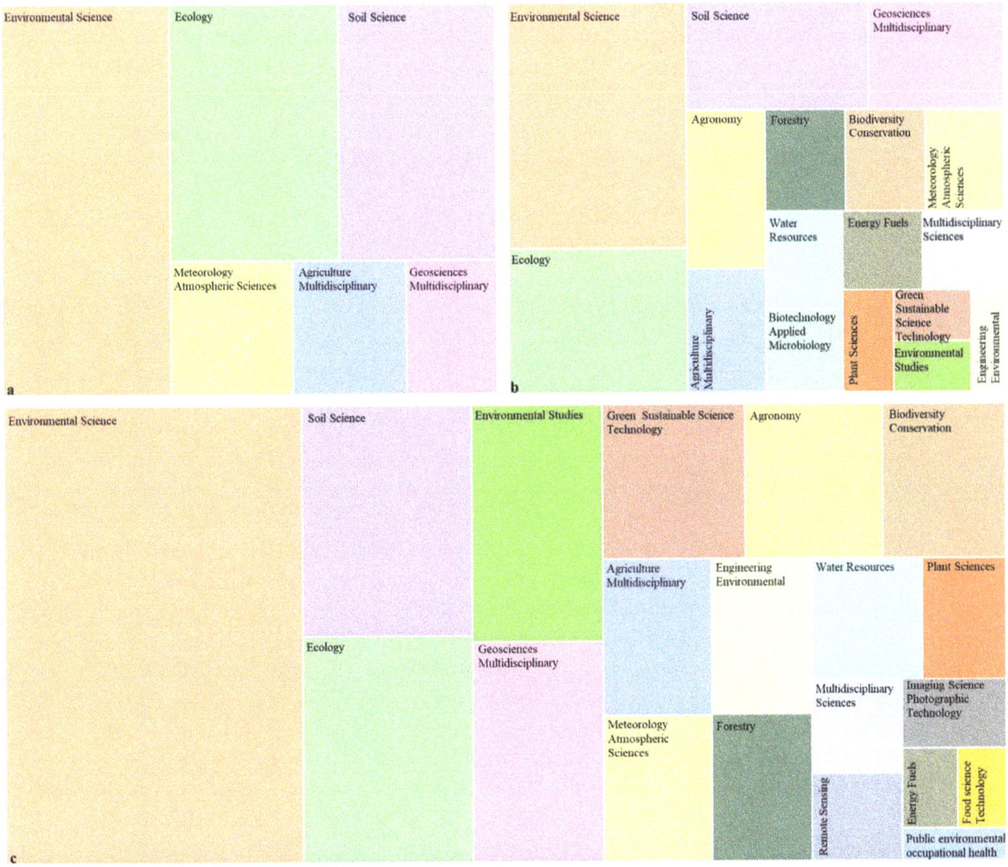

Figure 6. Disciplines with over ten publications on carbon sources and sinks of global farmland ecosystems: (**a**) 2002–2009; (**b**) 2010–2016; (**c**) 2017–2023. Different colours represent different disciplines.

3.4.1. Study on Soil Organic Carbon Sequestration Mechanisms

Theme 1 focuses on carbon sequestration, SOC, carbon storage, soil organic matter (SOM), and their dynamics (Figure 7). Based on the total link strength of node connections, this research cluster is primarily focused on SOC dynamics, the mechanisms of SOC sequestration, and the stocks of SOC. Increasing the SOC content in agricultural fields has a dual effect: it enhances soil productivity and mitigates the rise in atmospheric CO_2 concentration [5]. The dynamic processes of SOC input, decomposition, migration, transformation, and accumulation in soils are influenced by various factors, including natural conditions, human activities, and time. For example, earthworm bioturbation affects soil carbon storage, particularly near earthworm burrows, although their activity does not significantly increase subsoil carbon [22]. However, it facilitates the transport of fresh organic matter to deeper soil layers [22]. When natural ecosystems are converted into farmland, the rate of SOC loss varies depending on the type of cropping system established, with perennial woody plants resulting in lower organic carbon loss compared to annual crops [23]. The commonly used methods for studying SOC dynamics include isotope labeling and modeling approaches [24,25]. Isotope labeling provides direct measurement and tracking of organic carbon dynamics, but it is costly [26]. Modeling integrates a large number of climate and soil data to simulate crop growth and the turnover of SOC, enabling large-scale predictions and analyses. Nevertheless, the accuracy of this approach is influenced by the parameters and quantity of input data [27]. Over the years, scholars have extensively advanced in understanding the mechanisms of SOC sequestration. Six et al. systematically summarized the main factors influencing the dynamics of soil aggregates and organic matter [28]. They proposed a conceptual model of the "life cycle" of a macroaggregate, highlighting the formation of new microaggregates and the interplay between organic carbon accumulation and mineralization. This process involves the cyclical process of aggregate formation, turnover, and disruption. In the formation phase, fresh residues enter the soil, and intra-aggregate particulate organic matter (iPOM) is incorporated into macroaggregates. During turnover, iPOM decomposed into finer particles, which are then bound by clay and microbial products, forming more stable microaggregates. Disruption occurs when the degradation of microbial products and other binding substances leads to a reduction in the stability of macroaggregates, causing their breakdown, releasing microaggregates, and initiating a new cycle of macroaggregate formation [29]. This cycle shows that the turnover of aggregates is closely related to the dynamic changes in particulate organic matter [28,29]. Christensen challenged the traditional size-based classification of aggregates, proposing a theory of structural division with three levels: Level 1 structure—soil organic–mineral complexes; Level 2 structure—secondary complexes formed by the reaggregation of primary organomineral complexes; and Level 3 structure—intact soil in situ [30]. This theory provides a multi-scale perspective for a deeper understanding of organic matter turnover and storage in soil. Previous studies have suggested that plant-derived carbon is the primary initial source for the formation of SOM, which undergoes microbial decomposition reactions and is subsequently transformed into stable SOC [31]. Advances in high-throughput sequencing and omics technologies have revealed a crucial role of soil microorganisms in the formation and stabilization of SOC. Liang et al. proposed the "microbial carbon pump" (MCP) theory, suggesting that microbial-derived carbon is an essential source of SOC pools alongside plant-derived residues. The theory primarily consists of three aspects: First, soil microorganisms decompose or transform macromolecular plant organic residues through the ex vivo modification pathway and transport them into the soil. Second, soil microorganisms convert small molecular substrates into their biomass through in vivo turnover, with the process of microbial growth, reproduction, and death cycling and assimilation, thereby delivering microbial-derived organic carbon into the soil. Third, microbial communities continuously produce microbial-derived carbon through the iterative process of growth and death, which is driven by the "entombing effect" [32]. The proposal of this theory clarifies the regulatory mechanisms by which soil microorganisms influence soil carbon sequestration, thereby providing a new perspective for research on carbon cycling in farmland ecosystems [32]. Xiao et al. proposed the theory of the "soil mineral carbon pump" (MnCP), which suggests that soil minerals can interact with SOM through various abiotic processes such as adsorption, occlusion, aggregation, redox reactions, and polymerization. Adsorption refers to SOM adhering to mineral surfaces, while occlusion utilizes SOM being trapped within mineral crystals. In addition, aggregation occurs when minerals and organic matter bind to form aggregates. Redox reactions involve SOM reacting with minerals via direct or indirect electron transfer, resulting forming radicalized organic compounds that promote polymerization. Finally, polymerization is catalyzed by metal oxides or clay minerals, converting the reducing sugars or free amino acids into complex aromatic hydrocarbon macromolecules [33]. These interactions form organo-mineral complexes that lower the bioavailability and degradation rate of SOM, thus converting labile SOC derived from plants or microorganisms into more stable forms. This theory enhances our understanding of the abiotic mechanisms of soil carbon sequestration [33]. The MCP and the MnCP exhibit a "sequential" temporal relationship, where the microbial process of plant-derived carbon transformation and organic carbon production is considered the "upstream" processes of SOC formation. Soil minerals play a stabilizing role in the SOC formed through MCP, representing the "downstream" process. However, soil minerals can also stabilize SOC that has not undergone MCP transformation. Therefore, both the MCP and MnCP are interconnected, and each has unique characteristics. Regardless of how the "dual pumps" interact, each contribution to SOC sequestration will be influenced by both natural factors and human activities [34,35]. Future research on carbon cycling in farmland ecosystems should prioritize examining the synergistic and interactive effects of the "dual pumps" in the transformation and sequestration of SOC. Furthermore, assessing SOC stock change across global, national, and regional scales is vital for evaluating the potential of farmland ecosystems in mitigating climate change [34]. Scholars have estimated SOC stocks at various scales, including profile, plot, regional, and global scales, through long-term field experiments, literature reviews, and modeling approaches (both empirical and mechanistic). These efforts have helped quantify the global and regional SOC stocks and their carbon sequestration capacity [6].

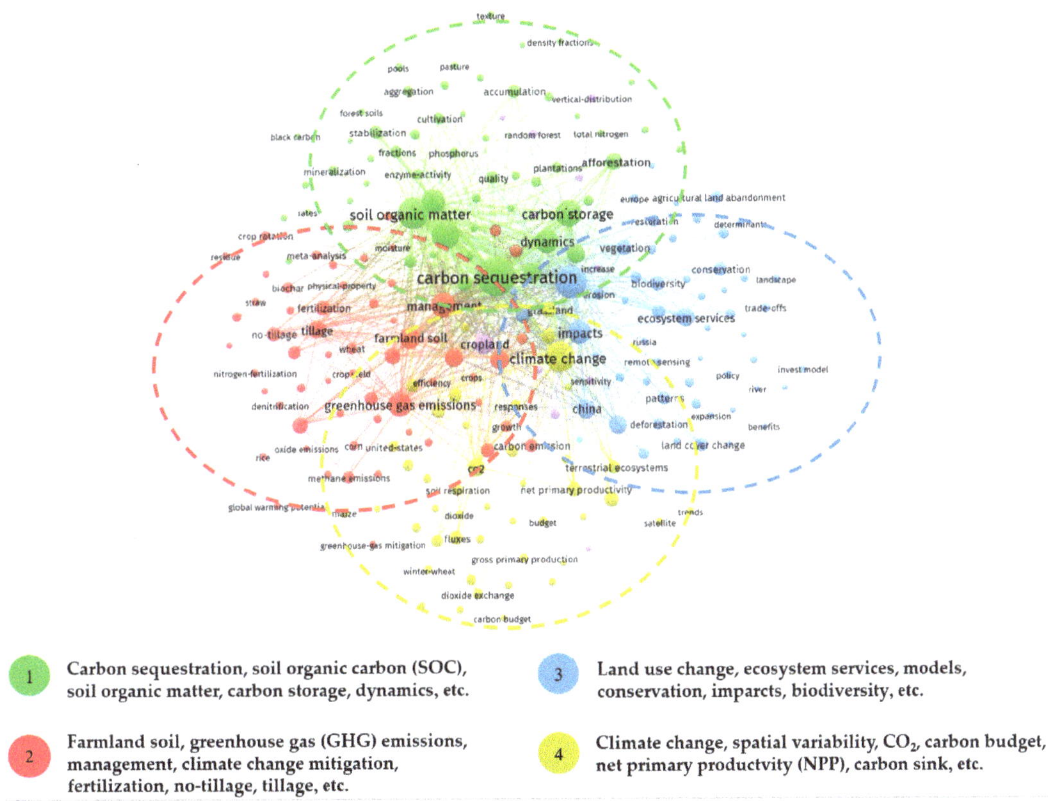

Figure 7. Co-occurrence map of keyword clusters. Four major themes in the literature are based on research topics defined by all the keywords. The circle size represents the occurrence of keywords, and high occurrences usually have high link strengths.

3.4.2. Research on Greenhouse Gas Monitoring and Estimation

Theme 2 focuses on farmland management, GHG emissions, ecosystems, and climate change mitigation (Figure 7). The total link strength of node connections indicates that the majority of studies revolve around the GHG emissions and carbon sequestration effects of different agricultural management practices. The Kyoto Protocol recognized agricultural land management as one of the key measures for reducing GHG emissions during the first commitment period, advocating the adoption of sustainable farming practices to enhance soil carbon sequestration [35,36]. This has garnered widespread attention, with scholars conducting extensive research to verify that agricultural management practices enhance soil carbon sequestration capacity and effectively reduce GHG emissions. Among these practices are fertilization methods (organic fertilizer applications, precision fertilization) [37], tillage practices (no-tillage, conservation tillage) [38], optimizing crop rotation schemes [39], returning crop residues to the soil, and so on [40,41]. By 2011, conservation tillage practices had been implemented on 1.25 billion hectares of farmland worldwide, excluding Antarctica, with widespread adoption across all continents [42–45]. Countries such as Brazil, Argentina, and Paraguay in South America have actively promoted conservation tillage through policy initiatives [46]. By the end of the 20th century, the adoption of conservation tillage practices and the Conservation Reserve Program (CRP) in the United States transformed agricultural land from a carbon source into a carbon sink [47,48]. In a long-term field experiment, Tian et al. found that conservation tillage does not increase GHG emissions in the context of global warming. Instead, it enhances microbial growth efficiency, promotes the accumulation of fungal necromass carbon, and facilitates the formation and sequestration of SOC [49]. This confirms that conservation tillage is a widely recognized and effective measure for carbon sequestration and GHG reduction globally. However, compared to tillage practices, the promotion costs and technical challenges associated with agricultural management measures such as fertilization management and cropping systems are relatively high. Consequently, there is still no consensus on the effectiveness of these measures in reducing GHG emissions and enhancing carbon sequestration, necessitating further research and exploration. Scholars have developed measurement methods that can be broadly categorized into two types to evaluate GHG emissions under different agricultural management practices: direct and indirect methods. Direct methods primarily include the chamber method and gas chromatography (GC) [50], while indirect methods mainly consist of the micrometeorological method and the mass balance method [51]. The chamber method is widely used for measuring soil CO_2, CH_4, and N_2O emissions [14]. For example, Chi et al. used this method to examine the effects of no-till and conventional tillage cropping systems on carbon and water fluxes in the inland Pacific Northwest of the United States [52]. The findings revealed that no-till farming could reduce CO_2 emissions and contribute to the sustainable development of the farmland ecosystem. Compared to the chamber method, the micrometeorological method is an open-ended and effective method for studying the material transport and energy exchange between vegetation and the atmosphere [53]. Additionally, this method plays a crucial role in analyzing ecological processes and determining parameters for remote sensing models [54]. The most commonly used micrometeorological technique is the vorticity correlation method, which calculates gas exchange flux within the system by measuring turbulence parameters [14]. For example, Xiao et al. utilized eddy covariance flux and meteorological data to study annual carbon fluxes in the Heihe River Basin of China [55]. The results demonstrated that annual carbon fluxes in forest and farmland plots were higher than those in grassland plots, while fluxes in coastal wetland plots were comparable to or slightly exceeded those in forest plots. Furthermore, assessing GHG emissions across farmland management practices and regions using different carbon emission accounting methods has become a crucial approach for quantifying GHG emissions. The primary GHG accounting methods currently include direct measurement, emission factor, mass balance, life cycle, and modeling approaches (e.g., farmland ecosystem carbon cycle model). Each method has distinct advantages and limitations (Table 3), and the selection of an appropriate method should align with the research objectives.

Table 3. Different GHG emission accounting methods [12–14,51,52,55].

Method	Advantages	Shortcomings	Scale	Applications
Measurement method	① Easy to operate; ② results are precise and reliable.	① Data acquisition challenges; ② expensive human and material resources; ③ vulnerability.	Microscopic	Microscopic or simple ecosystem.
Emission factor method	① Easily operated and understood; ② established accounting formulas, activity data, and emission factor databases are available; ③ straightforward data collection.	Lack of capacity for emission system changes.	Macroscopic; mesoscopic; microscopic	Socioeconomic emission sources; typically complex or simplified natural emission sources or carbon sinks.
Mass balance method	Clearly distinguishes between emission sources.	① Cumbersome intermediate procedures for emissions to be taken into account; ② prone to systematic errors; ③ obtaining data is difficult.	Macroscopic; mesoscopic	Emission equipment is frequently updated and highly precise; natural emission sources are complex.
Life cycle method	Assesses the total GHG emissions arising from all activities and inputs across the full life cycle of a production or consumption process.	Cumbersome process.	Mesoscopic; microscopic	Features a relatively complete and systematic production process.
Modeling method	① Quantitatively distinguishes the contribution of different factors to changes in carbon sources and sinks; ② predicts future changes in carbon sources and sinks; ③ reflects the material cycling processes of ecosystems.	① Complex model structure with difficult parameter adjustments; ② limited or simplified consideration of ecosystem management's impact on carbon cycling processes; ③ many models neglect lateral carbon transfer processes, such as watershed transport.	Macroscopic; mesoscopic; microscopic	Large regions, various administrative divisions, field scales, and multi-scale ecosystems.

3.4.3. Research on Ecosystem Services

Theme 3 concentrates on land use change, ecosystem services, conservation, and models (Figure 7). The total link strength of node connections indicates that most studies focus on the loss of ecosystem services caused by land use changes and the internal relationships within farmland ecosystem services. Land use and cover changes can significantly alter the energy balance and material cycling processes of ecosystems, leading to changes in ecosystem functions and affecting the supply of ecosystem services [56]. Urban expansion has encroached upon a large amount of farmland, resulting in a series of ecological issues such as reduced food production, carbon sequestration, and biodiversity loss [57]. The scarcity of farmland ecosystem services has become increasingly prominent, and their multifunctionality has garnered attention, leading to widespread interest in the study of these services and their valuation [58]. Farmland ecosystems provide various services, but people often aim to maximize the utilization of a single ecosystem service [56]. However, when multiple ecosystem services are utilized competitively, it typically leads to trade-offs within the ecosystem services themselves [59]. For example, crop yield increases often come at the cost of other ecosystem services [60]. However, sustainable human development requires the synergistic advancement of multiple ecosystem services [61]. As a result, achieving coordinated development across these services and developing scientific trade-off decision-making methods have become central research directions. Moreover, systematic evaluation methods have been developed to quantify and assess the dynamics of farmland ecosystem services under different management practices. These methods mainly include participatory approaches, sample plot observation, empirical models, mechanistic models, and value assessment methods. Each method focuses on the evaluation of different farmland ecosystem service functions based on their distinct principles or approaches [60]. In addition, scholars have proposed a range of assessment methods to quantify the dynamics of these services and evaluate their change under different management and agricultural practices. The sample plot observation method is commonly used to evaluate farmland ecosystem services such as crop pollination and pest control [62]. The Cycle model focuses on the water, carbon, and nitrogen balance of soil-crop systems under climate and farmland management influences. It is primarily used to evaluate farmland ecosystem service functions related to food production (provisioning services) and climate change mitigation (regulating services) [63].

3.4.4. Spatiotemporal Evolution and Driving Factors of Carbon Sources and Sinks in Farmland Ecosystems

Theme 4 focuses on research related to climate change, carbon sinks, primary productivity, CO_2, and spatial variability (Figure 7). According to the total link strength of node connections and clustering results, this theme centers on exploring the spatiotemporal heterogeneity of carbon stocks and carbon emissions in farmland ecosystems, alongside their drivers in the context of climate change. The carbon pools in farmland ecosystems mainly consist of the soil carbon pool (measured by SOC) and the crop carbon pool (assessed by NPP and crop biomass) [64]. The primary GHG emission sources in these ecosystems include soil respiration, methane (CH_4) emissions from paddy fields, nitrous oxide (N_2O) emissions from agricultural soils, and GHG emissions resulting from agricultural inputs and farm practices [65–68]. The carbon cycle in farmland ecosystems is a complex process influenced by factors such as climate, topography, crop type, soil properties, and anthropogenic activities [5,69]. Variations in these factors can modify carbon exchanges among different carbon pools within the ecosystem, leading to "carbon source" or "carbon sink" effects [70]. Moreover, numerous studies have demonstrated that the factors controlling carbon sources and sinks in farmland ecosystems exhibit varying influences across spatiotemporal scales, highlighting the scale effect in these control factors [71–73]. At the microscale, soil properties and farming management practices are key drivers of GHG emissions and SOC sequestration [74], as shown by numerous laboratory and field experiments [2,73]. At the mesoscale, such as in basins and geographical regions, climate, topography, and anthropogenic activities primarily influence GHG emissions and SOC sequestration [73]. For example, Smith et al. demonstrated that changes in farmland management were the primary drivers of declining SOC levels in farmland across England and Wales during the late 20th century [75]. The impacts of climate change and anthropogenic activities are more pronounced at the macroscale, such as global and continental [73,76]. The research conducted by Zhou et al. indicated that temperature is leading factor driving the spatiotemporal variability in future topsoil SOC storage [77]. Furthermore, high-latitude regions, which are more warming rapidly experience the most pronounced effects on topsoil SOC loss [77–79]. According to Wang et al., the efficiency of nitrogen fertilizer use is the primary contributor to global N_2O emissions [80]. However, climate change, particularly rising temperatures, accelerates soil microbial metabolic activity, enhancing soil organic nitrogen decomposition and increasing N_2O emissions [81]. Through a meta-analysis, Yu et al. found that non-continuous flooding significantly reduces GHG emissions from paddy fields worldwide [81]. Research on carbon sequestration in farmland crops remains underdeveloped due to ongoing controversy surrounding the carbon pools in these systems [11,82]. There is a need for further investigation into the spatiotemporal variation patterns and driving factors of crop carbon pools. Additionally, there is currently no consensus on how different factors influence carbon sources and sinks at various scales within farmland ecosystems. Many studies tend to focus on the impact of specific categories or individual factors rather than the broader spatiotemporal heterogeneity of these processes.

3.5. Research Hotspot Evolution and Trend Analysis

3.5.1. Research Hotspot Evolution Analysis

Burst words refer to terms that suddenly become hotspots or are more frequent during a specific period, signifying the research hotspots at different stages within a particular field [83,84]. Using the CiteSpaces burst analysis function, the top 25 burst keywords in the study of carbon sources and sinks in farmland ecosystems were identified (Figure 8). Key terms mainly include carbon sequestration, SOM, models, cropland, microbial biomass, productivity, and cover, each with a burst strength exceeding 6.0, highlighting focal areas in this research field. Carbon sequestration and SOM exhibited the most extended and intense burst durations, indicating a significant focus on SOC sequestration in farmland soils, which aligns with the findings of the thematic analysis. Similarly, models showed high burst intensity during the same period, reflecting the prominence of modeling research as a mainstream methodological approach in this field. Additionally, tillage practices, including conservation and conventional tillage, demonstrated strong burst strength, highlighting increasing attention to farmland management practices and their potential for carbon reduction and sequestration. Since 2018, microbial biomass, ecosystem services, and carbon emissions have emerged as prominent frontier research areas. After the proposed concept of "carbon neutrality" was introduced globally, China became a hotspot for research, indicating the high importance the Chinese government and scholars place on the carbon neutrality goal. Furthermore, by reviewing the literature, we found that the number of publications from China surged after 2018. Between 2018 and 2023, Chinese scholars contributed 25.02% of the papers in this field (353 articles). The frequent occurrence of "China" as a research label for case study regions also substantiates China's strong commitment to carbon sequestration and emission reduction. In addition, remote sensing technology emerged in this field in 2021 and quickly gained attention, enabling the acquisition of multi-scale data on carbon fluxes. This advancement has promoted research on the spatiotemporal variations of carbon sequestration in farmland ecosystems.

3.5.2. Research Fronts and Research Trends

Author keywords, specified by authors during article publication, directly convey their understanding and definition of the research topic, effectively highlighting the key focus areas and emerging trends within a specific research field. Therefore, this paper explores the current research frontiers and trends through the temporal map of author keyword co-occurrence using VOSviewer. Setting the minimum occurrence frequency of keywords to 8 and the minimum link strength to 8, of all the 3633 author keywords, 95 met the threshold. The top 10 author keywords were carbon sequestration (N = 340), SOC (N = 217), land use change (N = 205), climate change (N = 100), cropland (N = 80),

GHG emissions (N = 61), climate change mitigation (N = 52), ecosystem services (N = 69), and agriculture (N = 37) (Figure 9). The total link strengths of "carbon sequestration" and "SOC" were 536 and 397, respectively, far surpassing other keywords and indicating they are the core keywords, highly correlated with other keywords. The temporal evolution of keywords showed that research on carbon sources and sinks in farmland ecosystems has changed over time. Initially, studies primarily focused on the carbon sequestration potential and effects of farmland soils, while later research shifted attention to soil organic carbon loss resulting from land use changes. Recent studies mainly emphasize reducing carbon emissions and carbon footprints to achieve agricultural carbon neutrality. Biochar has gained significant attention as an essential carbon sequestration measure for its potential in carbon sequestration due to its ability to store carbon and reduce emissions. Farmland ecosystem services, closely linked to sustainable human development, are a critical area of research. The Invest model is an essential tool for assessing the value of ecosystem services.

Figure 8. Evolution of the top 25 keywords during 2002–2023. Strength is a measure of the degree of a burst event. The larger the value, the more significant the keyword is in the research field. The red line represents the year with active burst words, and the cyan-blue line represents the year with inactive burst words.

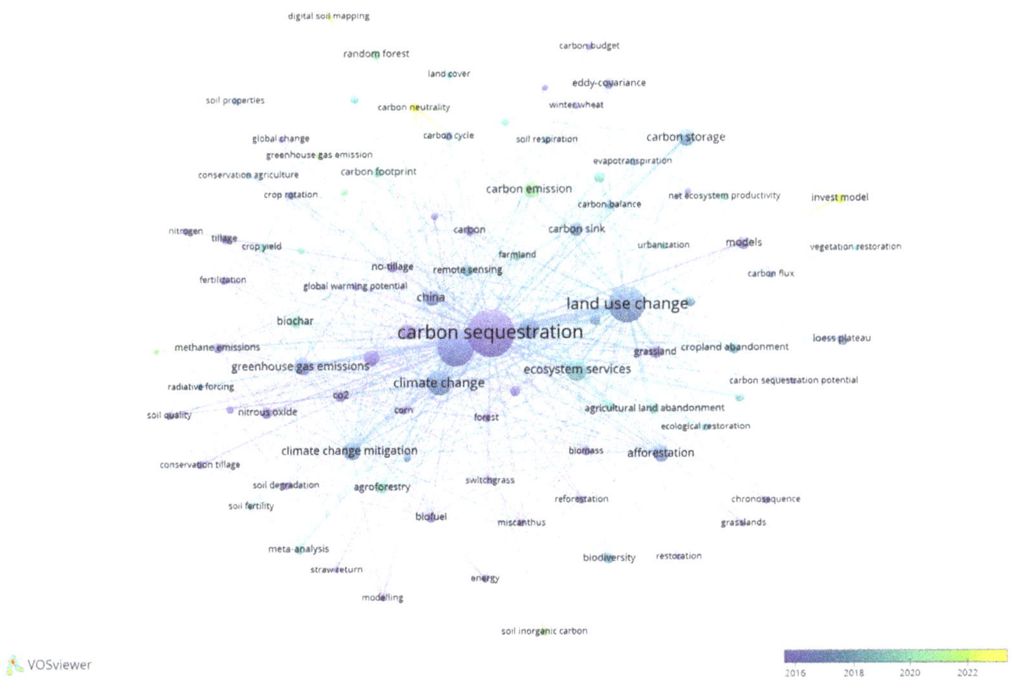

Figure 9. Temporal evolution trend of author keywords in carbon sources and sinks research of farmland ecosystem. The size of the circles represents the frequency of keyword occurrences, different colors represent different years, and lines indicate the link strength between keywords.

4. Future Research Directions

Farmland ecosystems are crucial in climate change mitigation, enhancing carbon sequestration, and reducing GHG emissions to achieve carbon neutrality. Based on the current challenge in carbon sources and sinks of farmland ecosystem research, we propose the following focuses for future research.

Further in-depth exploration of soil carbon sequestration mechanisms in farmland ecosystems is needed. Various factors, including crops, microorganisms, soil fauna, agricultural management practices, and parent material, all influence soil carbon sequestration in the farmland ecosystem. While the role of soil minerals and microorganisms in carbon sequestration is understood, research on the dominant roles and interactions of the MCP and MnCP in SOC storage under different climate types and agricultural practices in the context of global warming is still lacking. Furthermore, although it is well established that microorganisms regulate the chemical composition of SOC through processes such as "ex vivo modification" and "in vivo turnover" and control the dynamic storage of soil-stable organic carbon via "priming effects" and "entombment effects", soil fauna also influence the formation and accumulation of microbially derived organic carbon. However, there is currently a lack of understanding and assessment regarding the impact of soil fauna on the entire process chain of microbially derived carbon sequestration.

The focus should be on studying the interactive effects of drivers on the spatiotemporal changes in carbon sources and sinks within farmland ecosystems. Due to variations in management practices and natural geographic environments, the regulatory mechanisms and cycling patterns of carbon sources and sinks differ significantly across regions. Most studies on the spatiotemporal variation in farmland SOC pools focus on specific factors influencing carbon sources and sinks. However, the carbon cycle in farmland ecosystems involves multiple factors, such as temperature, precipitation, topography, soil respiration, and farming management practices. These factors interact in intricate ways, collectively influencing carbon sequestration and emissions in farmland ecosystems. There is a notable lack of research on the spatiotemporal variation in crop carbon pools in farmland ecosystems. Some scholars argue that crops have minimal carbon sink effect due to the short growing period of crops and the release of CO_2 back into the atmosphere after harvest [85,86]. Consequently, crops are often assumed to contribute zero carbon sequestration when estimating the carbon stock of farmland ecosystems [6,87]. However, some scholars have proposed that crops play an important role in carbon sequestration within the system before harvest. Furthermore, after crops are harvested, agricultural activities continue, and the corresponding carbon is transferred from the atmosphere to the soil and other components of the system [11,68]. It is worth noting that recent research has shown that the pathways and forms of crop carbon sequestration have shifted in the context of global change. Firstly, the construction of high-standard farmland has improved the productivity of farmland in China [88]. Secondly, organic carbon from crop residues and straw decomposition is accelerated into the soil, increasing SOC accumulation and forming a relatively stable carbon sink [89]. Thirdly, the widespread use of agricultural fertilizers has significantly increased crop growth rates, boosted food production, and accelerated the sequestration of CO_2 from the atmosphere into the farm ecosystem [90]. In conclusion, the farmland crop carbon pool can be considered a tangible carbon reservoir with significant carbon absorption functions, like the vegetation carbon pools in other ecosystems. Therefore, future research should focus on the combined effects of multiple factors, such as temperature rise, changes in precipitation, shifts in agricultural management practices, and topography, at global and regional scales to comprehensively understand the mechanisms behind the spatiotemporal changes and heterogeneity of carbon sources and sinks in farmland ecosystems under climate change.

Improving carbon accounting methods for farmland ecosystems. Different methods used for carbon accounting at the regional scale lead to significant uncertainties, making it challenging to compare results across various approaches. It is recommended to assess the weakness and strength of different methods. A "multi-data, multi-process, multi-scale, multi-method" integration should be adopted to establish a "sky-space-ground" system for measuring carbon sources and sinks in farmland ecosystems to improve estimation approaches. This includes strengthening remote sensing observations (vegetation data), developing a comprehensive network to monitor atmospheric CO_2 concentrations, and enhancing soil carbon pool monitoring and inventory, thus providing a solid data foundation for farmland ecosystems.

Emphasize research on the internal relationships and value assessment of farmland ecosystem services. In the context of global warming and the escalating food security crisis, adjusting farmland land use intensity, optimizing spatial land layout, and promoting the coordinated development of provisioning, regulating, and supporting services in the farmland ecosystem are vital for mitigating climate change, ensuring food security, preserving biodiversity, and supporting regional sustainable development.

5. Conclusions

In the context of climate change, research on carbon sources and sinks in farmland ecosystems has become one of the most pressing global environmental issues. This study employs bibliometric methods using CiteSpace and VOSviewer, based on the Web of Science database, to conduct a bibliometric analysis and systematic review of farmland ecosystem carbon source and sink research. This study also offers a reproducible methodological framework for constructing knowledge maps of complex ecological issues by providing a data-driven foundation for defining research directions and development pathways. The main research findings are presented below.

Over the past 22 years, research on farmland ecosystem carbon sources and sinks has shown continuous growth, with a notable increase in the last three years. China, the United States, and Germany have made significant contributions and frequently engaged in international collaborations with representative scholars, including Ingrid Kögel-Knabner, Martin Wiesmeier, and Hanqin Tian. This field has developed into a multidisciplinary research domain, with *Global Change Biology* and *Agriculture, Ecosystems & Environment* emerging as key academic exchange platforms.

The research primarily focuses on soil organic carbon (SOC) sequestration mechanisms and GHG measurement methodologies, leading to establishing differentiated assessment systems that confirm the significant carbon sequestration and emission reduction benefits of conservation tillage. However, climate change, land use transformation, and variations in farmland management practices result in pronounced spatiotemporal heterogeneity in carbon source–sink effects. Among these factors, climate change is the primary determinant of large-scale carbon balance, while farmland management practices play a crucial regulatory role in multi-scale effects.

Nevertheless, current research still has limitations. The dominant roles and interactive mechanisms of soil aggregates, mineral-associated carbon, and microbial metabolic products under different regional conditions remain unclear. Disparities in models and methodologies contribute to high uncertainty in assessments, and the interactive mechanisms among driving factors are yet to be fully elucidated. Future research should deepen investigations into soil organic carbon stabilization mechanisms, refine carbon source–sink accounting methodologies, and enhance the analysis of interactions among driving factors. These efforts will provide theoretical support for sustainable agricultural development and achieving "dual carbon" goals.

Author Contributions: Conceptualization, Y.P. and C.Z.; methodology, Y.P., M.Z. and C.Z.; data curation, Y.P., S.Z., M.Z. and H.Z.; software, Y.P.; validation, Y.P.; formal analysis, Y.P. and M.Z.; writing—original draft preparation, Y.P.; writing—review and editing, Y.P., M.Z., H.Z., T.C., C.S., X.L., J.D., C.L. and C.Z.; supervision, C.Z.; funding acquisition, C.Z. and C.S. All authors have read and agreed to the published version of the manuscript.

Funding: We gratefully acknowledge the funding provided by the National Natural Science Foundation of China (Grant No. 42471288), the Agricultural Science and Technology Development and Resource-Environment Protection Management Project of Guangdong Province (2022KJ161), and the Scientific Research Project of the Guangdong Provincial Department of Natural Resources: Key Technology Research for Construction in Centralized Cultivated Land Remediation Areas in Guangdong Province (GDZRZYKJ2024009).

Data Availability Statement: Publications on carbon sources and sinks in farmland ecosystems referenced in this work were sourced from Web of Science (https://webofscience.clarivate.cn/wos/woscc/advanced-search), accessed on 19 February 2024.

Acknowledgments: We cordially thank the anonymous referees and the editors for their helpful suggestions on the earlier draft of our paper.

Conflicts of Interest: The authors declare no conflicts of interest.

Abbreviations

The following abbreviations are used in this manuscript:

SOC	Soil organic carbon
GHG	Greenhouse gas
GIS	Geographical Information System
NPP	Net primary productivity
WOS	Web of Science
WOSCC	Web of Science Core Collection
SSCI	Social Sciences Citation Index
SCI-EXPANDED	Science Citation Index Expanded
SOM	Soil organic matter
iPOM	Intra-aggregate particulate organic matter
MCP	Microbial carbon pump
MnCP	Soil mineral carbon pump
CRP	Conservation Reserve Program
GC	Gas chromatography

References

1. Fang, J.; Yu, G.; Liu, L.; Hu, S.; Chapin, F.S. Climate Change, Human Impacts, and Carbon Sequestration in China. *Proc. Natl. Acad. Sci. USA* **2018**, *115*, 4015–4020. [CrossRef]
2. Della Chiesa, T.; Northrup, D.; Miguez, F.E.; Archontoulis, S.V.; Baum, M.E.; Venterea, R.T.; Emmett, B.D.; Malone, R.W.; Iqbal, J.; Necpalova, M.; et al. Reducing Greenhouse Gas Emissions from North American Soybean Production. *Nat. Sustain.* **2024**, *7*, 1608–1615. [CrossRef]
3. Jobbágy, E.G.; Jackson, R.B. The Vertical Distribution of Soil Organic Carbon and Its Relation to Climate and Vegetation. *Ecol. Appl.* **2000**, *10*, 423–436. [CrossRef]
4. Intergovernmental Panel on Climate Change (IPCC) (Ed.) Land–Climate Interactions. In *Climate Change and Land: IPCC Special Report on Climate Change, Desertification, Land Degradation, Sustainable Land Management, Food Security, and Greenhouse Gas Fluxes in Terrestrial Ecosystems*; Cambridge University Press: Cambridge, UK, 2022; pp. 131–248.
5. Lal, R. Soil Carbon Sequestration Impacts on Global Climate Change and Food Security. *Science* **2004**, *304*, 1623–1627. [CrossRef] [PubMed]
6. Yang, Y.H.; Shi, Y.; Sun, W.J.; Chang, J.F.; Zhu, J.X.; Chen, L.Y.; Wang, X.; Guo, Y.P.; Zhang, H.T.; Yu, L.F.; et al. Trestrial carbon sinks in China and around the world and their contribution to carbon neutrality. *Sci. Sin.* **2022**, *52*, 534–574. [CrossRef]
7. Clune, S.; Crossin, E.; Verghese, K. Systematic Review of Greenhouse Gas Emissions for Different Fresh Food Categories. *J. Clean. Prod.* **2017**, *140*, 766–783. [CrossRef]
8. Lal, R. Global Potential of Soil Carbon Sequestration to Mitigate the Greenhouse Effect. *Crit. Rev. Plant Sci.* **2003**, *22*, 151. [CrossRef]
9. West, T.O.; Marland, G. Net Carbon Flux from Agricultural Ecosystems: Methodology for Full Carbon Cycle Analyses. *Environ. Pollut.* **2002**, *116*, 439–444. [CrossRef]
10. Jiang, Z.; Zhong, Y.; Yang, J.; Wu, Y.; Li, H.; Zheng, L. Effect of Nitrogen Fertilizer Rates on Carbon Footprint and Ecosystem Service of Carbon Sequestration in Rice Production. *Sci. Total Environ.* **2019**, *670*, 210–217. [CrossRef]
11. Wei, H.; Wu, L.; Chen, D.; Yang, D.; Du, J.; Xu, Y.; Jia, J. Rapid Climate Changes Responsible for Increased Net Global Cropland Carbon Sink during the Last 40 Years. *Ecol. Indic.* **2024**, *166*, 112465. [CrossRef]
12. Li, H.; Jin, X.; Zhao, R.; Han, B.; Zhou, Y.; Tittonell, P. Assessing Uncertainties and Discrepancies in Agricultural Greenhouse Gas Emissions Estimation in China: A Comprehensive Review. *Environ. Impact Assess. Rev.* **2024**, *106*, 107498. [CrossRef]
13. Liu, X.; Wang, S.; Zhuang, Q.; Jin, X.; Bian, Z.; Zhou, M.; Meng, Z.; Han, C.; Guo, X.; Jin, W.; et al. A Review on Carbon Source and Sink in Arable Land Ecosystems. *Land* **2022**, *11*, 580. [CrossRef]
14. Li, M.; Peng, J.; Lu, Z.; Zhu, P. Research Progress on Carbon Sources and Sinks of Farmland Ecosystems. *Resour. Environ. Sustain.* **2023**, *11*, 100099. [CrossRef]
15. Chen, C.; Song, M. Visualizing a Field of Research: A Methodology of Systematic Scientometric Reviews. *PLoS ONE* **2019**, *14*, e0223994. [CrossRef]
16. Chen, C. CiteSpace II: Detecting and Visualizing Emerging Trends and Transient Patterns in Scientific Literature. *J. Am. Soc. Inf. Sci. Technol.* **2006**, *57*, 359–377. [CrossRef]
17. Van Eck, N.J.; Waltman, L. Software Survey: VOSviewer, a Computer Program for Bibliometric Mapping. *Scientometrics* **2010**, *84*, 523–538. [CrossRef] [PubMed]

18. Zheng, X.; Lu, Y.; Yuan, J.; Baninla, Y.; Zhang, S.; Stenseth, N.C.; Hessen, D.O.; Tian, H.; Obersteiner, M.; Chen, D. Drivers of Change in China's Energy-Related CO_2 Emissions. *Proc. Natl. Acad. Sci. USA* **2020**, *117*, 29–36. [CrossRef]
19. Ledford, H. How to Solve the World's Biggest Problems. *Nature* **2015**, *525*, 308–311. [CrossRef]
20. Okamura, K. Interdisciplinarity Revisited: Evidence for Research Impact and Dynamism. *Palgrave Commun.* **2019**, *5*, 141. [CrossRef]
21. Xie, H.; Wen, Y.; Choi, Y.; Zhang, X. Global Trends on Food Security Research: A Bibliometric Analysis. *Land* **2021**, *10*, 119. [CrossRef]
22. Don, A.; Steinberg, B.; Schöning, I.; Pritsch, K.; Joschko, M.; Gleixner, G.; Schulze, E.-D. Organic Carbon Sequestration in Earthworm Burrows. *Soil Biol. Biochem.* **2008**, *40*, 1803–1812. [CrossRef]
23. Beillouin, D.; Corbeels, M.; Demenois, J.; Berre, D.; Boyer, A.; Fallot, A.; Feder, F.; Cardinael, R. A Global Meta-Analysis of Soil Organic Carbon in the Anthropocene. *Nat. Commun.* **2023**, *14*, 3700. [CrossRef] [PubMed]
24. Rumpel, C.; Kögel-Knabner, I. Deep Soil Organic Matter—A Key but Poorly Understood Component of Terrestrial C Cycle. *Plant Soil* **2011**, *338*, 143–158. [CrossRef]
25. Poeplau, C.; Don, A.; Vesterdal, L.; Leifeld, J.; Van Wesemael, B.; Schumacher, J.; Gensior, A. Temporal Dynamics of Soil Organic Carbon after Land-Use Change in the Temperate Zone–Carbon Response Functions as a Model Approach. *Glob. Change Biol.* **2011**, *17*, 2415–2427. [CrossRef]
26. Feng, W.; Shi, Z.; Jiang, J.; Xia, J.; Liang, J.; Zhou, J.; Luo, Y. Methodological Uncertainty in Estimating Carbon Turnover Times of Soil Fractions. *Soil Biol. Biochem.* **2016**, *100*, 118–124. [CrossRef]
27. Huang, Y.; Lu, X.; Shi, Z.; Lawrence, D.; Koven, C.D.; Xia, J.; Du, Z.; Kluzek, E.; Luo, Y. Matrix Approach to Land Carbon Cycle Modeling: A Case Study with the Community Land Model. *Glob. Change Biol.* **2018**, *24*, 1394–1404. [CrossRef]
28. Six, J.; Bossuyt, H.; Degryze, S.; Denef, K. A History of Research on the Link between (Micro)Aggregates, Soil Biota, and Soil Organic Matter Dynamics. *Soil Tillage Res.* **2004**, *79*, 7–31. [CrossRef]
29. Six, J.; Elliott, E.T.; Paustian, K. Soil Macroaggregate Turnover and Microaggregate Formation: A Mechanism for C Sequestration under No-Tillage Agriculture. *Soil Biol. Biochem.* **2000**, *32*, 2099–2103. [CrossRef]
30. Christensen, B.T. Physical Fractionation of Soil and Structural and Functional Complexity in Organic Matter Turnover. *Eur. J. Soil Sci.* **2001**, *52*, 345–353. [CrossRef]
31. Oglesby, R.T.; Christman, R.F.; Driver, C.H. The Biotransformation of Lignin to Humus—Facts and Postulates. *Adv. Appl. Microbiol.* **1968**, *9*, 171–184. [CrossRef]
32. Liang, C.; Schimel, J.P.; Jastrow, J.D. The Importance of Anabolism in Microbial Control over Soil Carbon Storage. *Nat. Microbiol.* **2017**, *2*, 17105. [CrossRef]
33. Xiao, K.Q.; Zhao, Y.; Liang, C.; Zhao, M.; Moore, O.W.; Otero-Fariña, A.; Zhu, Y.G.; Johnson, K.; Peacock, C.L. Introducing the Soil Mineral Carbon Pump. *Nat. Rev. Earth Environ.* **2023**, *4*, 135–136. [CrossRef]
34. Liao, Q.; Zhang, X.; Li, Z.; Pan, G.; Smith, P.; Jin, Y.; Wu, X. Increase in Soil Organic Carbon Stock over the Last Two Decades in China's Jiangsu Province. *Glob. Change Biol.* **2009**, *15*, 861–875. [CrossRef]
35. Triberti, L.; Nastri, A.; Giordani, G.; Comellini, F.; Baldoni, G.; Toderi, G. Can Mineral and Organic Fertilization Help Sequestrate Carbon Dioxide in Cropland? *Eur. J. Agron.* **2008**, *29*, 13–20. [CrossRef]
36. Greene, L.A. United Nations Framework Convention on Climate Change. *Environ. Health Perspect.* **2000**, *108*, A353. [CrossRef]
37. Forte, A.; Fagnano, M.; Fierro, A. Potential Role of Compost and Green Manure Amendment to Mitigate Soil GHGs Emissions in Mediterranean Drip Irrigated Maize Production Systems. *J. Environ. Manag.* **2017**, *192*, 68–78. [CrossRef]
38. Roger, C.; Bowman, M.; McFadden, J.; Smith, D.; Wallander, S. *Tillage Intensity and Conservation Cropping in the United States*; U.S. Department of Agriculture, Economic Research Service: Washington, DC, USA, 2018; Volume 197, pp. 1–8. Available online: https://www.ers.usda.gov/publications/pub-details/?pubid=90200 (accessed on 19 October 2024).
39. Sun, X.; Qian, L.; Cao, Y.; Wang, M.; Li, N.; Pang, R.; Si, T.; Yu, X.; Zhang, X.; Zuza, E.J.; et al. Exploration of the Optimal Low-Carbon Peanut Rotation System in South China. *Agric. Syst.* **2024**, *221*, 104145. [CrossRef]
40. Bhattacharyya, P.; Roy, K.S.; Neogi, S.; Adhya, T.K.; Rao, K.S.; Manna, M.C. Effects of Rice Straw and Nitrogen Fertilization on Greenhouse Gas Emissions and Carbon Storage in Tropical Flooded Soil Planted with Rice. *Soil Tillage Res.* **2012**, *124*, 119–130. [CrossRef]
41. Lugato, E.; Bampa, F.; Panagos, P.; Montanarella, L.; Jones, A. Potential Carbon Sequestration of European Arable Soils Estimated by Modelling a Comprehensive Set of Management Practices. *Glob. Change Biol.* **2014**, *20*, 3557–3567. [CrossRef]
42. Benites, J.R.; Derpsch, R.; McGarry, D. The current Status and Future Growth Potential of Conservation Agriculture in the World Context. In Proceedings of the "Soil Management for Sustainability", 16th ISTRO Conference, Brisbane, Australia, 13–19 July 2003; pp. 120–129.
43. Corbeels, M.; Cardinael, R.; Naudin, K.; Guibert, H.; Torquebiau, E. The 4 per 1000 Goal and Soil Carbon Storage under Agroforestry and Conservation Agriculture Systems in Sub-Saharan Africa. *Soil Tillage Res.* **2019**, *188*, 16–26. [CrossRef]
44. Kassam, A.; Friedrich, T.; Derpsch, R. Global Spread of Conservation Agriculture. *Int. J. Environ. Stud.* **2019**, *76*, 29–51. [CrossRef]

45. Friedrich, T.; Derpsch, R.; Kassam, A. Overview of the Global Spread of Conservation Agriculture. *Field Actions Sci. Rep.* **2012**, *76*, 29–51.
46. Ryan, J.; Rashid, A.; Torrent, J.; Yau, S.K.; Ibrikci, H.; Sommer, R.; Erenoglu, E.B. Micronutrient Constraints to Crop Production in the Middle East-West Asia Region: Significance, Research, and Management. *Adv. Agron.* **2013**, *122*, 1–84. [CrossRef]
47. Allmaras, R.R.; Schomberg, H.H. Conservation Tillage Unforeseen Advantage. *Resour. Eng. Technol. Sustain. World* **1999**, *6*, 7–8.
48. Follett, R.F. Soil Management Concepts and Carbon Sequestration in Cropland Soils. *Soil Tillage Res.* **2001**, *61*, 77–92. [CrossRef]
49. Tian, J.; Dungait, J.A.J.; Hou, R.; Deng, Y.; Hartley, I.P.; Yang, Y.; Kuzyakov, Y.; Zhang, F.; Cotrufo, M.F.; Zhou, J. Microbially Mediated Mechanisms Underlie Soil Carbon Accrual by Conservation Agriculture under Decade-Long Warming. *Nat. Commun.* **2024**, *15*, 377. [CrossRef]
50. Pumpanen, J.; Kolari, P.; Ilvesniemi, H.; Minkkinen, K.; Vesala, T.; Niinistö, S.; Lohila, A.; Larmola, T.; Morero, M.; Pihlatie, M.; et al. Comparison of Different Chamber Techniques for Measuring Soil CO_2 Efflux. *Agric. For. Meteorol.* **2004**, *123*, 159–176. [CrossRef]
51. Li, H.; Zhang, H.B.; Li, G.C. *Evaluation Method of Farmland Carbon Sequestration and Emission Reduction Technology*, 1st ed.; China Agricultural Science and Technology Press: Beijing, China, 2020; p. 67.
52. Chi, J.; Waldo, S.; Pressley, S.; O'Keeffe, P.; Huggins, D.; Stöckle, C.; Pan, W.L.; Brooks, E.; Lamb, B. Assessing Carbon and Water Dynamics of No-till and Conventional Tillage Cropping Systems in the Inland Pacific Northwest US Using the Eddy Covariance Method. *Agric. For. Meteorol.* **2016**, *218–219*, 37–49. [CrossRef]
53. Yuan, Y.; Dai, Q.D.; Wang, M.W. Greenhouse Gas Fluxes Dataset Effected by Land-Use Conversion from Double Rice Cropping to Vegetables in Southern China. *J. Glob. Change Data Discov.* **2018**, *2*, 99–106. [CrossRef]
54. Baldocchi, D.; Falge, E.; Gu, L.; Olson, R.; Hollinger, D.; Running, S.; Anthoni, P.; Bernhofer, C.; Davis, K.; Evans, R.; et al. FLUXNET: A New Tool to Study the Temporal and Spatial Variability of Ecosystem–Scale Carbon Dioxide, Water Vapor, and Energy Flux Densities. *Bull. Am. Meteorol. Soc.* **2001**, *82*, 2415–2434. [CrossRef]
55. Xiao, J.; Sun, G.; Chen, J.; Chen, H.; Chen, S.; Dong, G.; Gao, S.; Guo, H.; Guo, J.; Han, S.; et al. Carbon Fluxes, Evapotranspiration, and Water Use Efficiency of Terrestrial Ecosystems in China. *Agric. For. Meteorol.* **2013**, *182–183*, 76–90. [CrossRef]
56. Wu, X.; Wang, S.; Fu, B.; Liu, Y.; Zhu, Y. Land Use Optimization Based on Ecosystem Service Assessment: A Case Study in the Yanhe Watershed. *Land Use Policy* **2018**, *72*, 303–312. [CrossRef]
57. Zhang, Z.; Shen, Z.; Liu, L.; Zhang, Y.; Yu, C.; Cui, L.; Gao, Y. Integrating Ecosystem Services Conservation into the Optimization of Urban Planning Policies in Eco-Fragile Areas: A Scenario-Based Case Study. *Cities* **2023**, *134*, 104200. [CrossRef]
58. Xie, G.D.; Xiao, Y. Review of agro-ecosystem services and their values. *Chin. J. Eco-Agric.* **2013**, *21*, 645–651. [CrossRef]
59. Rodríguez, J.P.; Beard, T.D.; Bennett, E.M.; Cumming, G.S.; Cork, S.J.; Agard, J.; Dobson, A.P.; Peterson, G.D. Trade-Offs across Space, Time, and Ecosystem Services. *Ecol. Soc.* **2006**, *11*, 28. [CrossRef]
60. Liu, Q.; Sun, X.; Wu, W.; Liu, Z.; Fang, G.; Yang, P. Agroecosystem Services: A Review of Concepts, Indicators, Assessment Methods and Future Research Perspectives. *Ecol. Indic.* **2022**, *142*, 109218. [CrossRef]
61. Bennett, E.M.; Peterson, G.D.; Gordon, L.J. Understanding Relationships among Multiple Ecosystem Services. *Ecol. Lett.* **2009**, *12*, 1394–1404. [CrossRef]
62. Kremen, C.; Williams, N.M.; Bugg, R.L.; Fay, J.P.; Thorp, R.W. The Area Requirements of an Ecosystem Service: Crop Pollination by Native Bee Communities in California. *Ecol. Lett.* **2004**, *7*, 1109–1119. [CrossRef]
63. Schipanski, M.E.; Barbercheck, M.; Douglas, M.R.; Finney, D.M.; Haider, K.; Kaye, J.P.; Kemanian, A.R.; Mortensen, D.A.; Ryan, M.R.; Tooker, J.; et al. A Framework for Evaluating Ecosystem Services Provided by Cover Crops in Agroecosystems. *Agric. Syst.* **2014**, *125*, 12–22. [CrossRef]
64. Zhao, M.Y.; Liu, Y.X.; Zhang, X.Y. A Review of Research Advances on Carbon Sinks in Farmland Ecosystems. *Acta Ecol. Sin.* **2022**, *42*, 9405–9416. [CrossRef]
65. Huang, Y.; Tang, Y. An estimate of greenhouse gas (N_2O and CO_2) mitigation potential under various scenarios of nitrogen use efficiency in Chinese Croplands. *Glob. Change Biol.* **2010**, *16*, 2958–2970. [CrossRef]
66. Qian, H.; Zhu, X.; Huang, S.; Linquist, B.; Kuzyakov, Y.; Wassmann, R.; Minamikawa, K.; Martinez-Eixarch, M.; Yan, X.; Zhou, F.; et al. Greenhouse Gas Emissions and Mitigation in Rice Agriculture. *Nat. Rev. Earth Environ.* **2023**, *4*, 716–732. [CrossRef]
67. Tangen, B.A.; Finocchiaro, R.G.; Gleason, R.A. Effects of Land Use on Greenhouse Gas Fluxes and Soil Properties of Wetland Catchments in the Prairie Pothole Region of North America. *Sci. Total Environ.* **2015**, *533*, 391–409. [CrossRef] [PubMed]
68. Chen, X.; Ma, C.; Zhou, H.; Liu, Y.; Huang, X.; Wang, M.; Cai, Y.; Su, D.; Muneer, M.A.; Guo, M.; et al. Identifying the Main Crops and Key Factors Determining the Carbon Footprint of Crop Production in China, 2001–2018. *Resour. Conserv. Recycl.* **2021**, *172*, 105661. [CrossRef]
69. Mummey, D.L.; Smith, J.L.; Bluhm, G. Assessment of Alternative Soil Management Practices on N_2O Emissions from US Agriculture. *Agric. Ecosyst. Environ.* **1998**, *70*, 79–87. [CrossRef]
70. Chen, B.; Lu, Q.; Wei, L.; Fu, W.; Wei, Z.; Tian, S. Global Predictions of Topsoil Organic Carbon Stocks under Changing Climate in the 21st Century. *Sci. Total Environ.* **2024**, *908*, 168448. [CrossRef] [PubMed]

71. Mishra, U.; Hugelius, G.; Shelef, E.; Yang, Y.; Strauss, J.; Lupachev, A.; Harden, J.W.; Jastrow, J.D.; Ping, C.L.; Riley, W.J.; et al. Spatial Heterogeneity and Environmental Predictors of Permafrost Region Soil Organic Carbon Stocks. *Sci. Adv.* **2021**, *7*, eaaz5236. [CrossRef]
72. Li, J.; Pei, J.; Pendall, E.; Fang, C.; Nie, M. Spatial Heterogeneity of Temperature Sensitivity of Soil Respiration: A Global Analysis of Field Observations. *Soil Biol. Biochem.* **2020**, *141*, 107675. [CrossRef]
73. Wiesmeier, M.; Urbanski, L.; Hobley, E.; Lang, B.; von Lützow, M.; Marin-Spiotta, E.; van Wesemael, B.; Rabot, E.; Ließ, M.; Garcia-Franco, N.; et al. Soil Organic Carbon Storage as a Key Function of Soils-A Review of Drivers and Indicators at Various Scales. *Geoderma* **2019**, *333*, 149–162. [CrossRef]
74. Conant, R.T.; Ryan, M.G.; Ågren, G.I.; Birge, H.E.; Davidson, E.A.; Eliasson, P.E.; Evans, S.E.; Frey, S.D.; Giardina, C.P.; Hopkins, F.M.; et al. Temperature and Soil Organic Matter Decomposition Rates–Synthesis of Current Knowledge and a Way Forward. *Glob. Change Biol.* **2011**, *17*, 3392–3404. [CrossRef]
75. Smith, P.; Chapman, S.J.; Scott, W.A.; Black, H.I.J.; Wattenbach, M.; Milne, R.; Campbell, C.D.; Lilly, A.; Ostle, N.; Levy, P.E.; et al. Climate Change Cannot Be Entirely Responsible for Soil Carbon Loss Observed in England and Wales, 1978–2003. *Glob. Change Biol.* **2007**, *13*, 2605–2609. [CrossRef]
76. Amundson, R.; Berhe, A.A.; Hopmans, J.W.; Olson, C.; Sztein, A.E.; Sparks, D.L. Soil and Human Security in the 21st Century. *Science* **2015**, *348*, 1261071. [CrossRef]
77. Zhou, J.; Wang, Y.; Tong, Y.; Sun, H.; Zhao, Y.; Zhang, P. Regional Spatial Variability of Soil Organic Carbon in 0-5 m Depth and Its Dominant Factors. *CATENA* **2023**, *231*, 107326. [CrossRef]
78. Crowther, T.W.; Todd-Brown, K.E.O.; Rowe, C.W.; Wieder, W.R.; Carey, J.C.; Machmuller, M.B.; Snoek, B.L.; Fang, S.; Zhou, G.; Allison, S.D.; et al. Quantifying Global Soil Carbon Losses in Response to Warming. *Nature* **2016**, *540*, 104–108. [CrossRef]
79. Huo, Y.; Mi, G.; Zhu, M.; Chen, S.; Li, J.; Hao, Z.; Cai, D.; Zhang, F. Carbon Footprint of Farming Practices in Farmland Ecosystems on the North and Northeast China Plains. *J. Environ. Manag.* **2024**, *354*, 120378. [CrossRef]
80. Wang, Q.; Zhou, F.; Shang, Z.; Ciais, P.; Winiwarter, W.; Jackson, R.B.; Tubiello, F.N.; Janssens-Maenhout, G.; Tian, H.; Cui, X.; et al. Data-Driven Estimates of Global Nitrous Oxide Emissions from Croplands. *Natl. Sci. Rev.* **2020**, *7*, 441–452. [CrossRef] [PubMed]
81. Jiang, Y.; Carrijo, D.; Huang, S.; Chen, J.; Balaine, N.; Zhang, W.; van Groenigen, K.J.; Linquist, B. Water Management to Mitigate the Global Warming Potential of Rice Systems: A Global Meta-Analysis. *Field Crops Res.* **2019**, *234*, 47–54. [CrossRef]
82. Kremen, C.; Miles, A. Ecosystem Services in Biologically Diversified versus Conventional Farming Systems: Benefits, Externalities, and Trade-Offs. *Ecol. Soc.* **2012**, *17*, 40. [CrossRef]
83. Zhang, Y.; Huang, K.; Yu, Y.; Yang, B. Mapping of Water Footprint Research: A Bibliometric Analysis during 2006–2015. *J. Clean. Prod.* **2017**, *149*, 70–79. [CrossRef]
84. Chen, C. Searching for Intellectual Turning Points: Progressive Knowledge Domain Visualization. *Proc. Natl. Acad. Sci. USA* **2004**, *101* (Suppl. S1), 5303–5310. [CrossRef]
85. Fang, J.Y.; Guo, Z.D.; Piao, S.L.; Chen, A.P. Terrestrial vegetation carbon sinks in China, 1981-2000. *Sci. China Ser. D Earth Sci.* **2007**, *6*, 804–812. [CrossRef]
86. Pacala, S.W.; Hurtt, G.C.; Baker, D.; Peylin, P.; Houghton, R.A.; Birdsey, R.A.; Heath, L.; Sundquist, E.T.; Stallard, R.F.; Ciais, P.; et al. Consistent Land and Atmosphere-Based U.S. Carbon Sink Estimates. *Science* **2001**, *292*, 2316–2320. [CrossRef] [PubMed]
87. Yu, G.R.; He, N.P.; Wang, Q.F. *Theoretical Basis and Comprehensive Assessment: Theoretical Basis and Comprehensive Assessment*, 1st ed.; Science Press: Beijing, China, 2013; pp. 128–133.
88. Ye, F.; Wang, L.; Razzaq, A.; Tong, T.; Zhang, Q.; Abbas, A. Policy Impacts of High-Standard Farmland Construction on Agricultural Sustainability: Total Factor Productivity-Based Analysis. *Land* **2023**, *12*, 283. [CrossRef]
89. Zhang, W.; Furtado, K.; Wu, P.; Zhou, T.; Chadwick, R.; Marzin, C.; Rostron, J.; Sexton, D. Increasing Precipitation Variability on Daily-to-Multiyear Time Scales in a Warmer World. *Sci. Adv.* **2021**, *7*, eabf8021. [CrossRef]
90. Velimirovic, A.; Jovovic, Z.; Przulj, N. From Neolithic to Late Modern Period: Brief History of Wheat. *Genetika* **2021**, *53*, 407–417. [CrossRef]

MDPI AG
Grosspeteranlage 5
4052 Basel
Switzerland
Tel.: +41 61 683 77 34

Biology Editorial Office
E-mail: biology@mdpi.com
www.mdpi.com/journal/biology

www.ingramcontent.com/pod-product-compliance
Lightning Source LLC
Chambersburg PA
CBHW041949260926
51850CB00019B/2269
* 9 7 8 3 7 2 5 8 5 9 6 1 0 *